LEGAL
INDEP
CONTR

Third Edition

2003 Cumulative Supplement

This supplement supersedes all previous supplements.

LEGAL GUIDE TO INDEPENDENT CONTRACTOR STATUS
Third Edition

2003 Cumulative Supplement

Robert W. Wood

Robert W. Wood Professional Corporation
San Francisco

A PANEL PUBLICATION
ASPEN PUBLISHERS, INC.

A Division of Aspen Publishers, Inc.
A Wolters Kluwer Company
www.aspenpublishers.com

Permissions
Panel Publishers
1185 Avenue of the Americas
New York, NY 10036

Printed in the United States of America

ISBN 0-7355-3145-5

2 3 4 5 6 7 8 9 0

HIGHLIGHTS

Independent contractor status is becoming more complex today because of the number of significant developments affecting the use of independent contractors, and associated liability. On the administrative and judicial fronts, there have been many new cases and rulings. On the legislative front, there have been diverse legislative adjustments. *Legal Guide to Independent Contractor Status* is an invaluable tool that guides the employer in grappling with the tough and unforgiving rules involving questions of tax liability and employee benefits.

2003 Cumulative Supplement

The 2003 Cumulative Supplement includes:

- Analysis of a Ninth Circuit decision, *Deja Vu-Lynnwood Inc. v. United States*, awarding litigation costs to a nightclub that treated its exotic dancers as nonemployees for tax purposes since the government's reasons for pursuing the assessments were not substantially justified.
- Discussion of two Tax Court cases, *Veterinary Surgical Consultants, P.C. v. Commissioner* and *Yeagle Drywall Co. v. Commissioner* in which workers performed substantial services for an S corporation and received remuneration for those services and were held to be employees for federal employment tax puposes.
- Discussion of the consequences concerning the employment status of a construction worker and physician as determined by two district courts, and of an insurance agent as determined by the IRS in a technical assistance memorandum.
- Explanation of the Tax Court decision in *Ewens and Miller Inc. v. Commissioner*, on an issue of first impression, holding that the court had jurisdiction over additions to tax relating to employment taxes, that a bakery's workers were employees, and that the bakery was not entitled to relief under section 530 of the Revenue Act of 1978.
- Most recent versions of the IRS Employer's Tax Guide and Employer's Supplement Tax Guide including the 2003 Wage Withholding and Advance Earned Income Credit Payment tables.

For questions concerning this shipment, billing, or other customer service matters, call our Customer Service Department at 1-800-234-1660. For toll-free ordering, please call 1-800-638-8437.

10/02

About Panel Publishers

Panel Publishers—comprising the former Little, Brown and Company's Professional Division, Wiley Law Publications, the Bureau of Business Practice, Summers Press, Harcourt Professional Publishing, and Loislaw—is a leading publisher of authoritative and timely treatises, practice manuals, information services, and journals written by specialists to assist attorneys, financial and tax advisors, and other business professionals. Our mission is to provide practical solution-based how-to information keyed to the latest legislative, judicial, and regulatory developments.

We offer publications in the areas of compensation and benefits, pensions, payroll, employment, civil rights, taxation, estate planning, financial planning, and elder law.

Other Panel products treating employment law issues include:

AIDS and the Law
Americans with Disabilities Act Handbook
Civil Rights in the Workplace
Employee Dismissal Law and Practice
Employment Discrimination: Law and Practice
Employment Law Answer Book
Employment Law Update
Employment Relationships: Law and Practice
Equal Employment Opportunity Compliance Guide
Handbook of Section 1983 Litigation
Mandated Benefits Compliance Guide
Pregnancy Discrimination and Parental Leave Handbook
Public Employee Discharge and Discipline
Representing Plaintiffs in Title VII Actions
Section 1983 Litigation: Claims and Defenses
Section 1983 Litigation: Federal Evidence
Section 1983 Litigation: Forms
Section 1983 Litigation: Jury Instructions
Section 1983 Litigation: Statutory Attorney's Fees
Sexual Harassment in the Workplace
State by State Guide to Employee Leave and Disability
State by State Guide to Human Resource Law

PANEL PUBLISHERS
A Division of Aspen Publishers, Inc.
Practical Solutions for Legal and Business Professionals
www.aspenpublishers.com

SUBSCRIPTION NOTICE

This Panel product is updated on a periodic basis with supplements to reflect important changes in the subject matter. If you purchased this product directly from Panel, we have already recorded your subscription for the update service.

If, however, you purchased this product from a bookstore and wish to receive future updates and revised or related volumes billed separately with a 30-day examination review, please contact our Customer Service Department at 1-800-234-1660, or send your name, company name (if applicable), address, and the title of the product to:

PANEL PUBLISHERS
A Division of Aspen Publishers, Inc.
7201 McKinney Circle
Frederick, MD 21704

ABOUT THE AUTHOR

Robert W. Wood practices law with Robert W. Wood, P.C., in San Francisco (information at <www.robertwwood.com>). Admitted to the bars of California, New York, Arizona, Wyoming, Montana, and the District of Columbia and qualified as a solicitor in England and Wales, he is a Certified Specialist in Taxation and is the author of 28 books. His most recent book is *Taxation of Damage Awards and Settlement Payments* (2d ed. © 1998), published by Tax Institute (e-mail: *info@taxinstitute.com*). This publication is designed to provide accurate and authoritative information in regard to the subject matter covered and is sold with the understanding that neither the publisher nor the author is engaged in rendering legal, accounting, or other professional services.

ABOUT THE CONTRIBUTOR

Barbara S. de Oddone is the managing shareholder of the Santa Rosa, California, office of Littler Mendelson, P.C., a nationwide law firm that devotes its practice exclusively to representing employers in employment and labor relations matters. She is a graduate of Wellesley College and the University of California School of Law at Davis. During her 23 years with Littler, she has concentrated on representing employers in wrongful discharge and employment discrimination litigation and preventive measures to avoid labor and employment problems. She also represents employers in litigation under the Labor Management Relations Act and the Railway Labor Act, as well as unfair labor practice proceedings and collective bargaining. She routinely counsels employers on measures to avoid litigation and assists with employment agreements, personnel policies and procedures, wage and hour matters, and independent contractor issues. She is an author and a lecturer on discrimination, wrongful discharge, and independent contractor issues. Ms. de Oddone serves on the Board of Directors of Sonoma County Legal Aid, as well as the Board of Trustees of VNA & Hospice Foundation of Northern California. She was appointed to the Sonoma County Workplace Investment Board in January 2000. She is a past member of the Board of Directors of the Legal Aid Society of San Francisco (1980–1997) and served as an arbitrator for the Committee on Arbitration of Fee Disputes of the Bar Association of San Francisco for over a decade.

TABLE OF CONTENTS

CHAPTER 2

TAX CONSIDERATIONS FOR EMPLOYERS

§ 2.04 FAILURE TO WITHHOLD OR PAY OVER TAX—COLLECTION OF PENALTIES

[B] 100 Percent Penalty—IRC § 6672

[2] Definition of Responsible Party

[a] In General

Page 39, add at end of section:

The Trust Fund Recovery Assessment (as it is now euphemistically called) imposed under § 6672 of the Code continues to be asserted with considerable vigor by the IRS. In *Rocha v. United States*,[56.1] a U.S. magistrate judge found that an individual was a "responsible person" who had willfully failed to pay over employee withholding taxes from several intertwined companies.

Steven Rocha operated Rocha Enterprises, and Rocha provided bookkeeping and payroll services for Portland Mechanical Corporation (PMC). Rocha and PMC President Rick Mercer agreed to a plan under which Rocha Enterprises provided payroll services to PMC and hired PMC's employees so PMC could take advantage of Rocha's better worker's compensation rates. Rocha created Portland Mechanical Systems, which employed the former PMC employees, even though PMC continued to bid and contract for work for the employees in PMC's name. Using Systems' bank account, Rocha prepared and signed paychecks and payroll withholding and tax deposit checks for the former PMC workers.

Rocha did not have signature authority over the PMC bank account that was used to fund Systems' payroll. Mercer failed to cover the tax payments but did cover employee wages. Rocha, working with the IRS, paid Systems' quarterly taxes, but Mercer failed to fund the payroll. Rocha asked the IRS to levy the amount due to Systems from PMC's assets and to apply the levy to the unpaid taxes. Mercer liquidated PMC and sold its assets to a newly formed company, Aloha Sheet Metal Inc., jointly owned by Mercer and Rocha. PMC then assigned the installment payments to the government to satisfy its tax debt. When the installment payments stopped, the IRS placed a levy on PMC for money owed. The IRS also assessed a § 6672 penalty against Rocha Enterprises, Rocha personally, and Aloha. Rocha paid the penalty and filed a refund suit.

The court held that Rocha was a responsible person within the meaning of § 6672 and that he had willfully failed to pay over employee withholding taxes to the government. The court rejected Rocha's argument that because he was able to pay the taxes only when Mercer properly funded Systems' account, he was not a responsible party. The court held that Rocha had enough control over the funds to be liable for the taxes. Rocha had the authority to decide whether he should use the funds Mercer provided for the next month's wages to pay the taxes or the wages. Because Rocha chose to pay the wages, he was liable for the penalty. The court also found that Rocha's actions were willful because he continued to pay employee wages when he knew taxes were owed.

However, there are still some disputes about the legal standards that should be applied. For example, in *United States v. Chapman*,[56.2] the Ninth Circuit, in an unpublished *per curiam* memorandum opinion, reversed a district court holding that the lower court applied an incorrect standard to determine that an individual was not a responsible person under § 6672. The IRS brought suit under § 6672 against Richard Sibbrel and David Chapman, as responsible persons, for failure to collect, account for, and pay over employee taxes. A U.S. district court held for Sibbrel and the government appealed.

The Ninth Circuit reversed, holding that the district court erred in determining that Sibbrel was not a responsible person under § 6672 and in applying improper legal standards. The circuit court noted that the district court had looked to whether Sibbrel knew that the taxes were unpaid and so found in error, citing *Davis v. United States*,[56.3] in which the court held that responsibility is a matter of status, duty, and authority, not knowledge. The Ninth Circuit also held that the district court erred in basing its conclusion in part on the fact that paying taxes was not part of Sibbrel's "functional" responsibility, because Sibbrel had the authority to exercise significant control over the corporation's financial affairs, even if he did not do so.

Next, the Ninth Circuit held that the district court had erred in believing that Sibbrel was precluded from being held responsible because David Chapman held *greater* control in the corporation. The Ninth Circuit noted that § 6672 does not confine liability for unpaid taxes to a single officer with the greatest or the closest control or authority over corporate affairs.

Finally, the Ninth Circuit noted that it was unable to find a basis for the district court's premise that a person cannot be held responsible for the nonpayment of taxes once the IRS takes action to recover the money by way of a lien. The court cited the factors in *United States v. Jones*,[56.4] as factors that the district court should consider on remand.

A U.S. district court held that a corporate officer was a responsible person who willfully failed to pay over employee withholding taxes and that the underlying assessment against her was valid, even though it listed an incorrect Social Security number.[56.5]

Beverly Frey, an officer with Pinpoint Communications, was responsible for the company's day-to-day accounting functions, had access to financial records, generated financial reports for the board of directors, and had check-signing authority. The company failed to meet its payroll tax obligations and filed for Chapter 11 bankruptcy protection. The IRS assessed Frey $285,000 for the overdue payroll taxes, seeking to hold her liable as a responsible person. Frey made a payment toward the assessment and sued to recover the remainder. The government counterclaimed to recover the full assessment and also filed a third-party complaint against Patrick Bromley, Pinpoint's chief executive officer. Bromley settled the claim, and the government's action against him was dismissed. Frey contended that the original assessment against her was invalid because it listed an incorrect Social Security number. She argued that Bromley had authority to pay the taxes and that his control of thc operations at Pinpoint prevented her from being a responsible person.

U.S. District Judge Sidney A. Fitzwater rejected Frey's arguments, first finding that an assessment notice that contains a technical error, such as an incorrect Social Security number, is valid if the individual was not misled by the error. The court found that Frey had offered no evidence that she was misled. The court next held that Frey was a responsible person because of the nature of her Pinpoint responsibilities. The court explained that the focus of § 6672 was not on whether Bromley exercised more authority than Frey but on whether Frey's status, duty, and authority gave her power to pay Pinpoint's delinquent payroll taxes. The court found that she had had that power. Finally, the court held that Frey's failure to pay over the taxes was willful because she knew the taxes were due but paid other creditors ahead of the government.

[56.1] No. 00-72-HU (D. Or., Mar. 26, 2001), Tax Analysts Doc. No. 2001-10783, 2001 TNT 73-16.

[56.2] No. 99-56470, No. 99-56862 (9th Cir., Apr. 17, 2001), Tax Analysts Doc. No. 2001-12375, 2001 TNT 85-9.

[56.3] 961 F.2d 867 (9th Cir. 1992).

[56.4] 33 F.3d 1137 (9th Cir 1994).

[56.5] Beverly Frey v. United States, 2001 U.S. Dist. (N.D.TX) LEXIS 5786.

[c] Application of Penalty Assessment Policies

Page 42, add at end of section:

In *Van Camp & Bennion, P.S. v. United States*,[65.1] a law firm was found liable for penalties for failure to pay withholding taxes. A U.S. magistrate judge, on remand, held that a law firm was liable for late payment and filing penalties because it failed to show reasonable cause for failing to pay its withholding taxes. Walter Van Camp and Irving Bennion formed a law firm, Van Camp and Bennion P.S. (VC&B). The firm failed to pay withholding taxes for the last three quarters of 1989 and the last two quarters of 1991. The IRS made assessments, and the corporation paid the taxes and sought refunds.

VC&B argued that Van Camp and Bennion were independent contractors for whom no withholding was required. The IRS found that Van Camp and Bennion were employees and were liable for federal employment taxes paid to them. Based on this characterization, the firm argued that it wasn't liable for negligence penalties or failure to timely deposit and pay the withholding taxes.

The district court concluded that Van Camp exercised sole authority to make major corporate decisions and was liable for the penalties. The Ninth Circuit affirmed in part but remanded on the question of whether his personal problems prevented the corporation from paying over the employment taxes. Also at issue was whether Van Camp reasonably relied on the firm's accountant or if it delegated the duty to pay taxes to an agent.

The court granted summary judgment to the government, holding that VC&B failed to establish reasonable cause to avoid the imposition of the penalties. The court concluded that although Van Camp's illness caused him fatigue and fevers, there was always someone at the firm with the authority to sign checks and pay taxes. The court found that although the firm had financial troubles caused by income fluctuations, there was enough money in the bank to make deposits for the employment taxes. The court concluded that had VC&B exercised ordinary business care, the taxes could have been paid. Finally, the court held that the firm's reliance on its accountant to pay the taxes wasn't reasonable because the firm never delegated the authority to the accountant nor was the firm prevented from making the payments because of the accountant's services. The court dismissed VC&B's complaint with prejudice.

[65.1] No. CS-94-409-CI (E.D. Wash. Jan. 28, 2002).

§ 2.05 DEFINITION OF WAGES

[C] Different Definitions of Wages

Page 49, add at end of section:

In *United States v. Cleveland Indians Baseball Co.*,[91.1] the Supreme Court voted 9–0 to assess Major League baseball clubs with much higher employment tax bills on back pay payments associated with a settlement. The dispute arose between the Cleveland Indians baseball team and the IRS over assessments on back wages paid to baseball players under an arbitration agreement. The Supreme Court concluded that such payments are subject to FICA and FUTA taxes by reference to the year in which the wages are in fact paid (not by reference to years to which the back pay award applied).

In 1990 the Major League Baseball Players Association claimed that Major League baseball clubs had breached the collective bargaining agreement regarding the free agency rights of baseball players in 1986 through 1988. After an arbitration panel issued a series of rulings against the clubs, the parties settled in 1990, with the clubs agreeing to pay $280 million to a custodial account for distribution to players.

Under the agreement, the Cleveland Indians owed eight players a total of $610,000 in salary damages for 1986, and it owed 14 players a total of $1,457,848 in salary damages for 1987. The Indians received and distributed the funds in 1994. None of the players affected by the settlement were employed by the Cleveland Indians in the year of distribution.

The baseball club paid FICA and FUTA taxes on the distributed funds as if they were wages for services rendered in 1994. The FICA payments totaled $99,382, and the FUTA payments were $1,008. The club sought a refund on the theory that the payments were wages when the club breached the agreement. The claim was denied and the baseball club filed suit in district court, relying largely on *Bowman v. United States*.[91.2] The club's arguments were also buttressed by the Fifth Circuit's opinion in *Johnston v. Harris County Flood Control District*.[91.3] In both cases, the appellate courts had held that back wages are subject to FICA taxes at the rate in effect when the wages are earned. In *Bowman* the Sixth Circuit said that "[a] settlement for back wages should not be allocated to the period when the employer finally pays but should be allocated to the periods when the regular wages were not paid as usual," quoting *Social Security Board v. Nierotko*.[91.4]

After the baseball club filed suit for a refund, a district court in Ohio entered judgment for the Indians and ordered the government to refund the FICA and FUTA taxes paid on the settlement disbursements designated as back wages, including interest from when the payments were made. In *Cleveland Indians Baseball Co. v. United States*,[91.5] the district court relied on *Bowman* as circuit precedent and, in doing so, rejected the government's contention that *Bowman* had been wrongly decided.

On appeal, the government pointed out that two appellate courts have held that wages are to be taxed for FICA purposes in the year they are actually received, citing *Walker v. United States*,[91.6] and *Hemelt v. United States*.[91.7] In those two cases, the appellate courts disagreed with the reasoning in the Sixth Circuit's decision in *Bowman*.

The government petitioned for *en banc* reversal of the decision in *Bowman*. The petition was denied, and the court referred the case to a three-judge panel, which issued a *per curiam* opinion in *Cleveland Indians Baseball Co. v. United States*.[91.8] In that opinion, the Sixth Circuit noted that circuit precedent in *Bowman* indicated that settlements for back wages were to be allocated to the period they were earned or should have been paid, not to the period in which they were ultimately disbursed.

In light of the appellate conflicts, the Supreme Court granted certiorari. In mid-April, in *United States v. Cleveland Indians Baseball Co.*,[91.9] in an opinion written by Justice Ruth Bader Ginsburg, the Supreme Court held that the award of back pay under the settlement agreement was subject to FICA and FUTA taxes in the year the settlement award was paid, not in the year the wages should have been paid.

The Court observed in its opinion that both FICA and FUTA tax rates and the wage base rates have increased over time. That meant, the Court said, that allocating the settlement payments to 1986 and 1987 would generate no additional FICA or FUTA tax liability for the baseball club or its former employees. In contrast, it noted, treating the back wages as taxable in the year paid, 1994, would subject both the baseball club and the former employees to significant tax liability. The Court also noted that both sides had "plausible" arguments.

The government contended that, based on the text of the FICA and FUTA tax provisions, wages are taxed according to the calendar year in which they are actually paid, regardless of when they should have been paid. Citing §§ 3111(a), 3121(a), and 3306(b)(1), the government directed the Court's attention to "the statute's constant reference to wages paid during a calendar year as the touchstone for determining the applicable tax rate and wage base."[91.10]

Despite the plain-language argument advocated by the government, the Court agreed with the baseball club that the language in *Social Security Board v. Nierotko*,[91.11] undermined the government's position. The court explained that in *Nierotko* it had concluded that when determining a wrongfully discharged worker's eligibility for Social Security benefits, a back pay award had to be allocated as wages to the calendar quarters of the year "when the regular wages were not paid as usual."[91.12] The Court said that the *Nierotko* court had found no conflict between an allocation-back rule for back pay and language tying benefits eligibility to the number of calendar quarters "in which" a minimum amount of "wages" had been paid.[91.13]

The Supreme Court reasoned that the *Nierotko* court's allocation implicitly construed "wages paid" as incorporating regular wages that should have been paid but were not paid as usual. Consequently, the Court believed the FICA and FUTA tax provisions prescribing tax rates based on wages paid during a calendar year did not have a plain meaning that precluded the allocation of back pay to the year it should have been paid. However, the Court then diverged from the baseball club's stance. Simply because the 1939 "wages paid" language for benefits eligibility rules accommodated an allocation-back rule for back pay wages, the Court said, did not mean that the identical 1939 "wages paid" language for tax purposes must be construed in the same manner. "*Nierotko* dealt specifically and only with Social Security benefits eligibility, not with taxation," the Court said.[91.14]

The Court suggested that the earlier holding in *Nierotko* reflected concern that the benefits scheme created in 1939 would be disserved by allowing an employer's wrongdoing to reduce the quarters of coverage an employee would otherwise be entitled to claim toward eligibility. The Court pointed to the government's brief, which stated—and to which the baseball club had agreed—that Social Security contributions, unlike private pension contributions, do not create in the contributor a property right to benefits against the government. Further, the Court stressed that wages, rather than tax contributions, are the statutory basis for calculating an individual's benefits. As such, the Court reasoned, *Nierotko* "does not compel" symmetrical construction of the "wages paid" language in the respective taxation and benefits eligibility contexts. Also, the Court was unpersuaded that *Nierotko*'s treatment of back pay had been incorporated into the tax provisions when the Social Security Act was amended after the *Nierotko* decision.

The baseball club had argued that the government's refusal to allocate back wages to the year they should have been paid created an inequity in taxation and "incentives for strategic behavior." The Court conceded that the dual arguments were "not without force." However, the Supreme Court observed that there was no structural unfairness in taxation, compared to the structural inequity outlined in *Nierotko*.[91.15]

In this instance, the Supreme Court believed the government's rule was not clearly incompatible with the statutory scheme. The Court acknowledged that sometimes the government's rule disadvantages the taxpayer, and at other times it works to the disadvantage of the public fisc. The anomalous rules, the Court said, must be considered in light of Congress's evident interest in reducing complexity and minimizing administrative confusion within the FICA and FUTA tax schemes. It commented, "The most we can say is that Congress intended the tax provisions to be both efficiently administrable and fair, and that this case reveals the tension that sometimes exists when Congress seeks to meet those twin aims."[91.16]

Although the regulations do not specifically address back pay, the IRS "has consistently interpreted them to require taxation of back wages according to the year the wages are actually paid, regardless of when those wages were earned or should have been paid," the Supreme Court explained.[91.17] Consequently, the regulations deserved "substantial judicial deference." As a result, the Supreme Court reversed the Sixth Circuit on the issue, holding that, for FICA and FUTA tax purposes, back wages should be attributable to the year in which they were actually paid.

In a concurring opinion, Justice Antonin Scalia explained that the issue was whether damage awards compensating an employee for lost wages should be regarded for tax purposes as wages paid when the award is received, or rather as wages paid when they would have been paid *but for* the employer's unlawful actions. In this instance, the text of the FICA and FUTA provisions did not apply, Justice Scalia said. In *Nierotko*, "we provided one rule for purposes of the benefits provisions. The Internal Revenue Service has since provided another rule for purposes of the tax provisions," he said.[91.18] Because the Service's interpretation was reasonable, he concurred in the Court's decision.

The United States District Court, on remand from the Sixth Circuit, entered final judgment for the government in *Cleveland Indians Baseball Co. v. United States.*[91.19] The Sixth Circuit vacated its previous judgment and remanded the case to the district court to enter final judgment in the government's favor. The United States District Court, complying with the Sixth Circuit's order, entered final judgment in the government's favor.

Conclusion. Although the amounts may be minimal, this government victory in *Cleveland Indians* will probably affect many individuals and companies involved in numerous labor, employment, and other discrimination claims under federal law. The Supreme Court's decision also confirms the Service's interpretation of the proper tax treatment of back pay awards and highlights the Court's reluctance to disturb the Service's reasonable interpretation of the relevant tax statutes. This gives greater impetus to notions of judicial deference to agency interpretation, and buttresses long-standing agency rules.

[91.1] Tax Analysts Doc. No. 2001-11045, 2001 TNT 75-7.

[91.2] 824 F.2d 528 (6th Cir. 1987), 87 TNT 160-7.

[91.3] 869 F.2d 1565 (5th Cir. 1989), *cert. denied*, 493 U.S. 1019 (1990).

[91.4] 327 U.S. 358, 370 (1946).

[91.5] 83 A.F.T.R.2d ¶ 99-398 (D. Ohio 19__), 1999 TNT 25-8.

[91.6] 202 F.3d 1290 (10th Cir. 2000), 2000 TNT 26-23.

[91.7] 122 F.3d 204 (4th Cir. 1997), 97 TNT 159-30.

[91.8] 85 A.F.T.R.2d ¶ 2000-663 (6th Cir. 2000), 2000 TNT 95-19.

[91.9] 87 A.F.T.R.2d ¶ 2001-798 (2001), 2001 TNT 75-7.

[91.10] Beverly Frey v. United States, 87 A.F.T.R.2d ¶ 2309 (N.D. Tex. 2001).

[91.11] 327 U.S. 358 (1946).

[91.12] Beverly Frey v. United States, 87 A.F.T.R.2d ¶ 2309 (N.D. Tex. 2001).

[91.13] *Id.*

[91.14] *Id.*

[91.15] *Id.*

[91.16] *Id.*

[91.17] *Id.*

[91.18] *Id.*

[91.19] No. 1:96CV2240 (N.D. Ohio July 20, 2001).

Page 61, add after § 2.07:

§ 2.08 PARTNERS AS EMPLOYEES

There is nothing extraordinary about the proposition that a worker may be an employee of a partnership. However, it *would* seem extraordinary if a person could be a member of a partnership and also be an employee. This dichotomy was recently brought into focus by a legal memorandum released by the IRS. That IRS Internal Legal Memorandum,[153] deals with questions raised by a number of Forms SS-8 (the forms that typically are sent in by workers or companies to request employee vs. independent contractor status rulings). These particular Forms SS-8 involved several different types of claims. The first was a claim by a worker that for federal tax purposes, although there was a valid partnership, the worker was an employee of the partnership and not a *bona fide* member of the partnership. The second Form SS-8 considered in the legal memorandum was a claim by a worker that a partnership of which he was a purported partner was not a *bona fide* partnership for federal tax purposes, but instead that one of the partners (typically a general partner) was actually the employer of all the other purported partners of the partnership.

The question of whether a person is a partner in a partnership (for federal tax purposes) is a question of federal tax law. Although a person may be recognized as a partner under state law, this is not necessarily determinative of whether he or she will be treated as a partner for federal tax purposes.

The IRS position is that *bona fide* partners of a partnership cannot also be employees of that partnership for purposes of the employment tax provisions of the Code. A Form SS-8 in this particular instance was intended to elicit the information needed to determine the proper classification of the worker under the common law tests (set out in Revenue Ruling 87-41, 1987-1 C.B. 296) used for determining a worker's status as contractor or employee.

Although this legal memorandum bears a January 26, 2001 release date, and can be found as an IRS Internal Legal Memorandum,[154] the document is not generally publicly available. The memorandum discusses published revenue rulings, including the long-standing position of the IRS announced in Revenue Ruling 69-184, 1969-1 C.B. 256. That letter ruling states, in pertinent part:

> *Bona fide* members of a partnership are not employees of the partnership within the meaning of the Federal Insurance Contributions Act (the Federal Unemployment Tax Act), and the collection of income tax at source on wages. . . . [S]uch a partner who devotes his time and energies in the conduct of the trade or business of the

> partnership, or in providing services to the partnership as an independent contractor, is, in either event, a self-employed individual rather than an individual who, under the usual common law rules applicable in determining the employer-employee relationship, has the status of an employee.[155]

ILM 200117003 concludes that to qualify as an employee for purposes of employment tax provisions (such as FICA, FUTA, and income tax withholding), the worker must be classified both as not a *bona fide* partner *and* as an employee under the common law control test (as delineated in Revenue Ruling 87-41).

[153] ILM 200117003, Tax Analysts Doc. No. 2001-12043, 2001 TNT 83-23, released Jan. 26, 2001.

[154] *Id.*

[155] *See* Revenue Ruling 69-184, 1969-1 C.B. 256.

§ 2.09 TREATMENT OF STOCK OPTIONS

Whether one is dealing with independent contractors or employees, it is increasingly common for stock options to be some of the compensation awarded. Generally speaking, this type of compensation will be available to employees more than to independent contractors. Furthermore, as is discussed in more detail in this section, certain types of stock options can be provided *only* to employees, independent contractors being ineligible under the federal income tax law for these "qualified" options. Employees may be provided either qualified or nonqualified options. Even independent contractors, however, can be provided nonqualified options. The following discussion briefly gives the basic rules for providing stock options.

[A] Nonqualified Stock Options

There are two types of stock options, nonqualified options and incentive stock options (ISOs). Whether one believes the tax benefits of ISOs are as great as they are cracked up to be, it is certainly indisputable that there are a number of qualification restrictions on ISOs that make them unattractive in many circumstances. ISOs, for example, are subject to many approval requirements, timing and duration requirements, exercise rules, percentage tests, and a whole host of other limitations (the major categories of which are listed later in this section). For these and other reasons, nonqualified options constitute the bulk of the options floating around corporate America.

Nonqualified options (NSOs) are best defined by exclusion. They encompass all options that do not meet the special requirements for ISOs. NSOs may be granted to both employees and nonemployees alike in exchange for their services (so independent contractors or consultants are acceptable recipients). There are no restrictions on the options, making them infinitely flexible.

The tax rules are straightforward. NSOs granted in connection with the performance of services are not taxable when granted unless they carry a readily ascertainable fair market value. NSOs with a readily ascertainable fair market value are generally only those traded on an established securities market.

When an NSO is exercised and stock is received, the holder is taxed on the difference between the price paid for the stock (the option exercise price) and the market value. This treatment applies whether or not the optionholder hangs onto the stock after the exercise or immediately sells it.[156] The income from the exercise of the options is not only income, but also constitutes compensation for services.[157] Thus, employment taxes (and withholding rules) apply.

A good deal of NSO planning involves trying to avoid this ordinary income/compensation rule. An exception specifies that income will not be recognized on the date the NSO is exercised if the stock received is both subject to a substantial risk of forfeiture and nontransferable. In this event, the recipient of the stock will not be taxed until either of these two conditions lapses.[158]

An exception to this rule for NSOs applies when the employee *elects* to include the value of the option in income at the date of grant, even though it is subject to a substantial risk of forfeiture. The employee makes a § 83(b) election, a one-page form that essentially says "I want to be taxed now." Predictably, these forms are typically filed only when the value of the option (valued without regard to the restrictions on the option) is quite low, or in some cases zero.

It has long been true that a traditional goal of a § 83(b) election is aggressive: to take as little as possible into income as ordinary income. Then, by virtue of the § 83(b) election, the balance (that will eventually be realized when the option is exercised and the stock is later sold) will all be capital gain. Assuming the stock is held for the requisite holding period, it will be long-term capital gain.

Moreover, by virtue of the § 83(b) election having been made, the timing of the taxation will differ. Some small amount of tax (or in some cases even zero) may be payable at the time the § 83(b) election is made. If the election is made, the exercise of the options will not be a taxable event. Instead, the exercise will simply be a purchase (more like the ISO rules discussed in subsection [B]), but the spread between the option exercise price and the then value of the stock will not constitute income. All of this makes the § 83(b) election a fairly nifty device.

Significantly, NSOs that do not have a readily ascertainable fair market value are not considered property for purposes of § 83. Thus, a § 83(b) election would not be available for such NSOs. The term "readily ascertainable fair market value" is defined in Reg. § 1.83-7. Nevertheless, there has been litigation over the question whether NSOs can qualify for the § 83(b) election. NSOs that do not have a readily ascertainable fair market value are simply not considered property for purposes of § 83—hence, a § 83(b) election is not available. It is pretty clear that the § 83(b) election simply does not work in such a case.

Consider the case of *Cramer v. Commissioner.*[159] The *Cramer* case squarely considers this issue—which may mean that at least someone thought the issue was worth arguing about and trying in Tax Court—and flatly refuses to give the taxpayer a break. It is an interesting case, and there were several quite prominent counsel at the taxpayer's counsel table. They argued, quite persuasively, that the options in *Cramer* did have a fair market value and that the § 83(b) election was available. It is hard to disagree with the statute, though. Section 83(e)(3) flatly states that § 83 "shall not apply to . . . the transfer of an option without a readily ascertainable fair market value." Although the legislative history does not say much, if anything, about this, Judge Cohen considered this provision, as well as the regulations and case law, in concluding that there was no value in the case at bar and that the § 83(b) election was unavailable. Interestingly, the Tax Court's opinion in *Cramer* cites Brookes Billman, who wrote "Nonstatutory Stock Options," 383 *Tax Management* A-8 (1984). Billman and others are cited for the proposition that commentators believe the 1976 Act's legislative history indicates Congress's disapproval of the regulations that specify when the value of an option is "readily ascertainable." Judge Cohen evidently does not put much stock in these commentators.

Indeed, Judge Cohen also flatly rejected the taxpayer's arguments in *Cramer* that the fact that a § 83(b) election was made bootstraps the option into *definitely* having a readily ascertainable fair market value. The taxpayers cited *Walt Disney Productions v. United States*,[160] for the proposition that the 1976 Conference Report (legislative history subsequent to the § 83 regulations) should be given considerable weight. According to Judge Cohen, no matter how much weight is given to all of this, the argument just does not fly.

It is quite interesting that the taxpayers in *Cramer* also argued that their advisors relied on *Alves v. Commissioner*,[161] in determining the tax treatment of the options. The taxpayers in *Cramer* cited the dissent in *Alves* for the proposition that the court agreed that zero income could be reported at the time of the grant of an option, and that capital gains could then be reported on its sale.

Alves, clearly an important (and even bellwether) case, involved the question of whether § 83 applied when the taxpayer had paid fair market value for the stock when he received it, and therefore received no income when the stock was transferred to him. Judge Cohen in *Cramer* flatly said that the *Alves* holding bears no relationship to the *Cramer* facts; in the latter taxpayers attempted to report a fair market value of zero on receipt of their *options*, not stock. The issue, Judge Cohen stated, is simply whether the options had a readily ascertainable fair market value when they were transferred. The *Cramer* opinion is quite long and quite detailed, and is worth reading.

Another point about § 83(b) elections deserves mention, simply because so many mistakes are made here—including those by professionals. A § 83(b) election must be made within 30 days of the grant of any restricted property. The election must be filed within this 30-day period, and a copy of the election must accompany the taxpayer's return for the year in which the options were granted.

For tax purposes, when an NSO is issued, the company has not yet "paid" anything until the time it is taxable to the employee. There is a predictable reciprocity here. Assuming that the NSO is subject to restrictions (as most are), there is no income to the employee, and no deduction to the company, until these restrictions lapse. Or, as in the *Cramer* case, if the NSO is exercised and the option has a spread between exercise price and fair market value, that amount of spread must be taken into income as wages by the employee/optionholder. Of course, this generates a corresponding deduction for the spread to the company.

Financial Statement Treatment of NSOs. The tax treatment of NSOs is pretty straightforward. Fortunately, that is also the case with their financial statement treatment. A company is generally not required to take a charge against earnings at the time the NSOs are granted. It is only when they are exercised (and compensation is payable) that a charge to earnings is required.[162]

If a company follows APB 25 (and has not adopted FAS 123), when a nonqualified option is exercised, there is no charge to earnings, as long as the NSO was a "market value" option at the date of grant. Thus, if the grant price equaled the fair market value at the date of grant, there is no charge to earnings. The only book entry made is that paid-in capital is credited, and federal income tax receivable is debited to reflect the permanent tax benefit to the company (which does not affect earnings).

[B] Tax and Accounting Treatment of ISOs

The ISO rules are quite complex. There are a few fundamentals about ISOs that are worth noting, particularly inasmuch as they stand in rather stark contrast to the NSO rules. NSOs, as discussed in subsection [A], come with almost a complete lack of restrictions. ISOs must meet all of the following requirements:

- ISOs must be granted under a plan, which must set forth the aggregate number of shares that may be issued as option shares (specified as either a specific number or a formula).
- The plan must specify the employees (or class of employees) eligible to receive them. (*Note:* they must be employees, not independent contractors.)
- The plan must be approved by shareholders within 12 months before (or 12 months after) plan adoption. The shareholders must also approve amendments to the plan increasing the aggregate number of shares that can be issued, or changing eligibility requirements. These shareholder approvals must comply with the corporate charter, bylaws, and state laws as applicable.
- The ISOs must be granted within 10 years from the *earlier* of the date the plan is adopted by the board or the date the plan is approved by the shareholders.
- The options must be exercised within 10 years of grant. For 10 percent shareholders, the option must be exercised within five years from grant.

- The option price must be no less than the fair market value of the stock when the option is granted (but see later in this section for the 110 percent rule for 10 percent shareholders/employees).
- The employee cannot own 10 percent or more of the company's voting stock at the time the option is granted. This rule does not apply if the option price is at least 110 percent of the fair market value of the shares subject to the option, and if the option is exercisable within five years from the date of grant.
- The option must not be transferable by the employee except on death, and will be exercisable during the lifetime of the employee only by the employee.
- The aggregate fair market value of the stock (as of the time the option is granted), for which ISOs are exercisable for the first time by the employee during any calendar year, cannot exceed $100,000.

Obviously, this is a fairly onerous and extensive list of requirements, particularly when one compares it to the almost nonexistent requirements for NSOs. Among other reasons, this lengthy list of potential pitfalls is why ISOs can be a real pain in the neck. But, for those who are offered ISOs, the difference in tax treatment can be significant.

[1] Basic Tax Treatment of ISO

There is no tax when the company grants the ISO. There is also no tax (no regular income tax, that is) when the ISO is exercised. This "no tax on grant, no tax on exercise" mantra has gotten a fair number of ISO participants in trouble. The not-so-hidden but sometimes surprisingly painful reason is the alternative minimum tax (AMT). There can be AMT when ISOs are exercised. The only *regular* tax possible with an ISO is when the underlying shares acquired pursuant to the option are sold. When these shares are sold, they can qualify for capital gain treatment (and assuming the long-term holding period is met, long-term capital gain). There is even a trick about this long-term capital gain (see subsection [2]).

The AMT is quite difficult to compute. Suffice to say that the entire spread between the ISO exercise price and the fair market value of the underlying stock as of the date of exercise is considered a tax preference (translation: it goes into the AMT computation). Unless the employee exercises the ISOs and immediately sells the stock, the employee will be faced with what may be a significant AMT liability.

[2] Sale of ISO Shares

As if the ISO rules were not complicated enough, it is important to clarify the preceding statement about long-term gain on sale of shares. You might think that long-term capital gain treatment would be assured if you exercise an ISO, wait a year and a day, and then sell your shares. As with just about everything else concerning ISOs, it is not quite that simple. There are two fundamental types of sales of shares acquired as ISOs.

The first is a *qualifying sale*. As its name suggests, this is obviously the "good" kind of sale. The second is a *disqualifying sale*, which carries an equally obvious moniker. The employee will recognize *capital* gain or loss if the sale is made in a qualifying sale, meaning that two requirements must be met. The sale must be made at least two years after the date of grant, *and* one year after the shares are transferred to the employee (*i.e.*, one year after the ISOs are exercised).

The sale is a *disqualifying* one if it is made within two years from the date of grant *or* within one year after the shares are transferred to the employee. Many an employee has received a grant of ISOs, exercised the options, held the stock for just over a year, and then blithely sold it, assuming that it would receive long-term capital gain treatment. No go, if the sale of shares does not occur at least two years from the date the ISOs were granted. In this disqualifying sale, the spread upon exercise is treated as compensation income. The balance of the gain is long- or short-term capital gain, depending on the holding period of the shares acquired.

Note especially the and/or conjunctions in the last few paragraphs. It is truly surprising how many ISO holders exercise and then unwittingly get these disposition rules wrong (ouch!).

[3] Treatment by Employer

It should come as no surprise that, for tax purposes, the employer is not entitled to an income tax deduction when the ISO is granted, or even when it is exercised. Indeed, the employer is not even entitled to a deduction when the employee sells the shares acquired through the exercise of an ISO in a qualifying disposition. A qualifying disposition is a good thing for an employee (because it imports capital gain treatment), but it is a bad thing for the employer (no tax deduction). From the employer's perspective, a disqualifying disposition (because it has compensation income, at least to a certain extent) will result in the employer becoming entitled to a tax deduction (for the amount of that compensation).

[4] Accounting Treatment of ISOs

The accounting treatment of ISOs has been controversial. The basic accounting treatment of ISOs is one reason companies do not like ISOs (especially when they are trying to make their earnings for financial statement purposes look good—and what company doesn't?). The basic value of ISOs is a charge to earnings at the time the ISOs are granted, even though no tax deduction is available (and no compensation is actually treated as paid to the employee/ISO recipient) until the ISOs are exercised.

[C] To Deduct or Capitalize Option Deal Costs

In this post-*INDOPCO* era, it hardly takes a microscope to spot deduction-vs.-capitalization issues in any acquisition. One fear is always that a variety of types of costs will be required to be capitalized rather than deducted. In the context of stock options that are canceled as part of the deal (either ISOs or NSOs), one question is whether the payment attributable to such a cancellation is deductible.

The amount at stake would be the amount of cash or property used to cancel the options. Assuming that the company issues a W-2 form (treating the cancellation payment as wages) to each pertinent employee, there is little question that the payment is properly deductible. Happily, the gremlins of *INDOPCO* have apparently not reached this particular area. Indeed, in a surprising show of largesse, the IRS even has ruled in letter rulings that the following items can also be deducted notwithstanding *INDOPCO*:

- a bonus paid to optionholders to compensate them for the loss of ISO status as a result of the deal
- the cancellation of unvested options
- the portions of any cancellation payment attributable to a premium paid for stock of the target.

All these result in deductible expenditures (not capital expenditures). Of course, one cannot get carried away with this logic. If the options were issued as a result of the sale or acquisition of the business, or in connection with some other transaction that is by its nature capital, then the payments may not be currently deductible. This is our old friend the origin-of-the-claims doctrine (which applies in litigation recoveries, among other areas) raising its head again.[163]

How are NSOs and ISOs treated in transactions? If one sets aside as a subset the golden parachute rules, there is still plenty to know and do when dealing with outstanding ISOs and/or NSOs held either by the acquiring or the target company.

In many transactions, the buyer and target will agree that the target's obligations under its options plans will be assumed by the buyer. Often, substitute options to purchase the buyer's stock will be swapped for the outstanding options to purchase the target stock. Generally, the buyer will be able to make this substitution so that the employee/optionholders are not taxable on this substitution itself. In such a substitution, the target's optionholders will generally be able to preserve the gain inherent in their old target options, while maintaining a continuing stake in the appreciation of the ongoing (postacquisition) enterprise.

Given the elaborate regime for ISOs—and (by comparison) the very liberal rules for NSOs—ISOs and NSOs must be separately considered in an analysis of assumptions and substitutions of options.

[1] Assuming/Substituting ISOs

Where the target has outstanding ISOs, one huge concern will be preserving the qualified ISO status of those options. Some option plans contain hidden traps that would disqualify the ISO treatment. For example, the target's plan may provide that ISOs vest automatically on a change in control. This could cause a large number of options to lose ISO status because of the annual dollar cap ($100,000) mentioned earlier.

It is also important to ensure that the assumption does not result in a "modification" of the ISOs. *Modification* here is a technical term with (perhaps not surprisingly) negative consequences. A modification may occur if the option terms change, giving the employee additional benefits. The reason the determination of whether an ISO is modified is so important is what happens if it is treated as modified: the option is treated as *reissued* as of the date of the modification.[164]

This reissuance treatment means the option will be retested as of that moment to see if it satisfies all of the ISO requirements. Recall the long list of requirements that must be met for an option to qualify as an

ISO (see subsection [B]). It is a fairly odious list. For a variety of reasons, especially the fair market value of the underlying shares in the context of a merger or acquisition, the options may well exceed the option exercise price and thus preclude ISO treatment if this retesting must occur.

[2] Specialized Meaning of "Corporate Transaction"

Still, there may be a silver lining here. If an ISO is substituted or assumed in a "corporate transaction," that substitution or assumption is *not* treated as a modification (1) as long as the new option satisfies a "spread test" and a "ratio test"; and (2) as long as it does not provide additional benefits that were not provided under the old option. Before defining the spread and ratio tests, let's look at what constitutes a corporate transaction.

Two conditions must be met before a transaction will be considered a corporate transaction. First, the transaction must involve one of the following: a merger or consolidation, an acquisition of property or stock by any corporation, a spinoff, split-up or split-off, a reorganization, or any partial or complete liquidation.[165] Note that it is irrelevant whether the transaction qualifies as a tax-qualified reorganization under § 368 of the Code. The second requirement is that the transaction must result in a significant number of employees being transferred to a new employer, or discharged. (And, yes, there can be debates about the relative meaning of the term "significant number of employees" here!)

[3] Spread and Ratio Tests

Assuming that a corporation transaction (as previously defined) has occurred, the assumption or substitution of the ISO will be fine, as long as both the "spread" and "ratio" tests are met. The spread test is met if the aggregate spread of the new option (immediately after the substitution or assumption) is not more than the aggregate spread of the old option immediately before the substitution or assumption. This spread is the excess of the aggregate fair market value of the shares subject to the option over the aggregate option price for those shares.[166]

The "ratio" test is met by doing a share-by-share comparison. The ratio of the option price to the fair market value of the shares subject to the new option immediately after the substitution or assumption must be no more favorable to the optionee than the ratio of the option price to the fair market value of the shares subject to the old option (immediately before the substitution or assumption). This spread test is only regulatory (it does not appear in the Code itself). Examples in the Regulations help explain and illustrate both the spread and the tests.[167]

Predictably, there are some determinations to be made in assessing whether these tests have been met. For both tests, the parties may adopt "any reasonable method" to determine the fair market value of the stock subject to the option. Stock listed on an exchange can be based on the last sale before the transaction or the first sale after the transaction, as long as the sale clearly reflects the fair market value. Or, an average selling price may be used during a longer period. The fair market value can also be based on the stock value assigned for purposes of the deal (as long as it is an arm's-length deal).

Even if one gets over the corporate transaction hurdle, the spread hurdle, and the ratio hurdle, someone must also analyze the transaction to determine whether the new option provides any "additional benefits" to the optionholders. If it does, the ISOs assumed or substituted will be a problem. The new option must not provide the optionholder with additional time to exercise or more favorable terms for paying the exercise price. Significantly, though, shortening the period during which the option may be exercised, or accelerating vesting, are *not* treated as additional benefits. The acceleration-of-vesting exception is an important one and is widely used.

[4] Cancelling ISOs

Although the rules regarding assumption of ISOs are complex (actually, more complex than the preceding brief summary indicates) and a variety of issues can come up in that context, cancelling ISOs turns out to be remarkably simple. The tax consequences on a cancellation of ISOs are governed by § 83 of the Code. If the ISO does not have a readily ascertainable fair market value at the time it was granted, § 83 requires that the cash or property received for cancellation of the option be treated the same as if the cash or property had been transferred pursuant to the exercise of the option.[168]

Thus, if the cash or property received on cancellation is fully vested, the optionholder would recognize income on the cancellation of the option equal to this amount (less any amount paid by the optionholder to acquire the option, typically nothing). This income constitutes wages subject to withholding for income and employment taxes, and will generate a corresponding deduction to the company.

When the property received in exchange for the option (on its cancellation) is not substantially vested (if restricted stock is used, for example), the cancellation transaction will not be taxable until the property becomes substantially vested. Again, these are the rules set out in (and in the regulations underlying) § 83. Consequently, it should be possible for the employee to elect to take the property even before substantial vesting into income by making a § 83(b) election.

[5] Treatment of NSOs in Transactions

The treatment of NSOs in a transaction, as with the initial issuance of NSOs, is a good deal simpler than for ISOs. If a buyer wishes to assume the target's NSOs, one looks to § 83 to determine the tax consequences to both the optionholders and the company. Recall that § 83 generally does not apply to the grant of an option without an ascertainable fair market value. If an employee exchanges an NSO that does not have a fair market value, in an arm's-length transaction, the question is what he or she gets. Section 83 will apply to the transfer of the money (or other property) received in exchange.

Thus, if the new NSO received in exchange for the old NSO does not have a readily ascertainable fair market value, the employee will not recognize income in the exchange, nor will the company get a deduction. Of course, NSOs may have some value when they are issued, but this value generally is not readily ascertainable unless the option is actively traded on an established market. Assuming it is not actively traded on an established market, it will not have a readily ascertainable value unless all of the following exist for the option:

- it is transferable
- it is immediately exercisable in full
- it (or the property subject to the option) is not subject to any restriction or condition, other than a lien or other condition to secure payment, that has a significant effect on the fair market value of the option
- its fair market value is readily ascertainable in accordance with the Regulations.[169] Most NSOs do not satisfy all four of these conditions, so they do not have a readily ascertainable fair market value.

Unlike ISOs, with an NSO there is no need to focus on whether the assumption or substitution of the NSO results in a modification.

There is simply no qualified status to interrupt. Thus, the holder of an NSO should not recognize income even when the terms of the new option are different from the terms of the old. This is a somewhat murky area, though.

For example, suppose the new option has an exercise price that is nominal in relation to the fair market value of the underlying shares. Here, the optionholder may have to recognize the income on the transaction. If the buyer chooses to give the optionholder an alternative, to convert the option into an option in the buyer, or to take cash (or other property) for the option now, the situation is also easier with NSOs than with ISOs. Someone choosing cash will recognize income in an amount equal to the amount of cash received, less any amount paid for the option (but the amount paid is most typically zero). An optionholder who elects not to take cash should not be taxed.

[6] Cancellation of NSOs

One area in which the rules for ISOs and NSOs are remarkably parallel concerns cancellation. Although most of the complexity associated with the treatment of options (either ISOs or NSOs) in merger and acquisition transactions involves assumptions and substitutions, not too much can go wrong when it comes to a cancellation. If the NSOs are simply canceled in the deal, the employee looks to § 83 to determine how he or she will be taxed. Remarkably, this is the same set of rules that will apply when an ISO is cancelled. Thus, the preceding discussion concerning cancellation of ISOs applies to cancellation of NSOs as well.

[7] Accounting Treatment Change

Finally, there can be accounting issues on a modification. Under Financial Accounting Standards Board Interpretation No. 44 (FIN 44), *Accounting for Certain Transactions Involving Stock Compensation*, an assessment as to whether the proposed modification changes the life of the employee stock options, through an extension of the exercise period or a renewal of the exercise period, would have to be made. The assessment should also determine whether the modification changes the exercise price of the employee stock options or the number of shares the employee is entitled to receive.

A modification that does not affect the life of the stock option, the exercise price, or the number of shares to be issued has no accounting consequence. In most cases, a modification of this type would not affect the life of the stock option, the exercise price, or the number of shares to be issued. Accordingly, a new measurement date would not be deemed to have occurred.

[156] *See* Reg. § 1.83-7(a). *See also* Rev. Rul. 78-175, 1978-1 C.B. 304.
[157] *See* Reg. § 1.83-7(a).
[158] I.R.C. § 83(a).
[159] 101 T.C. 225 (1993).
[160] 480 F.2d 66 (9th Cir. 1973).
[161] 79 T.C. 864 (1982), *aff'd*, 734 F.2d 478 (9th Cir. 1984).
[162] *See* FASB.
[163] *See, e.g.*, United States v. Gilmore, 372 U.S. 39 (1963).
[164] *See* I.R.C. § 424(h)(1); Reg. § 1.425-1(e)(2).
[165] *See* I.R.C. § 424(a); Reg. § 1.425-1(a)(1)(ii).
[166] *See* I.R.C. § 424(a)(1); Reg. § 1.425-1(a)(1)(i).
[167] *See* Reg. § 1.425-1(a)(4).
[168] *See* Reg. § 1.83-7(a).
[169] *See* Reg. § 1.83-7(b).

CHAPTER 3

DETERMINATION OF EMPLOYEE/INDEPENDENT CONTRACTOR STATUS FOR FEDERAL TAX PURPOSES

§ 3.01 GENERAL EMPLOYEE/INDEPENDENT CONTRACTOR ISSUES

[C] Class of Employees Irrelevant

Page 75, add at end of carryover paragraph:

In *Deja Vu-Lynnwood, Inc. v. United States,*[9.1] the Ninth Circuit, in an unpublished memorandum reversing a district court's opinion, has awarded litigation costs to a nightclub that treated its exotic dancers as nonemployees for tax purposes. Deja Vu-Lynnwood, Inc. avoided paying employment taxes for its exotic dancers by treating them as "tenants" who rented space in the nightclub and not employees. The IRS determined the dancers were employees and assessed back taxes. The club challenged the assessments in district court. After the Ninth Circuit issued a decision in *Marlar, Inc. v. United States,*[9.2] the government conceded its case against the club. *Marlar* held that a club treating its female dancers as tenants was not subject to employment taxes. The club moved for litigation costs, which the district court denied.[9.3]

The Ninth Circuit awarded attorney's fees to the club, holding that the government's reasons for pursuing the assessments were not substantially justified. The court explained that the free legal services the club offered to dancers, after they were charged with criminal violations for their work at the clubs, were not reportable payments under § 6041 because the services were provided only at the club's discretion. The court also held that the government should have realized that because the club did not have to report cash payments given to dancers by club patrons, it would not have to report the patrons' use of cash to buy drinks for the dancers.

The court concluded that the government was not substantially justified in contending that the club failed to make the necessary filings to obtain safe-harbor relief under § 503 of the Revenue Act of 1978. One judge dissented, stating that he believed that the government was substantially justified in arguing that the dancers' drink credits were payments under § 6041 and thus the club had failed to satisfy § 530's filing requirements.

[9.1] No. 99-35832 (9th Cir. Oct. 26, 2001).

[9.2] *See* Doc 98-25330 (9 pages) or 98 TNT 154-7.

[9.3] Deja Vu-Lynnwood, Inc. v. United States, No. C96-1721C (W.D. Wash. Sept. 17, 1997) (for a summary, *see* Tax Notes, Oct. 6, 1997, p. 67; for the full text, *see* Doc 97-27081 (6 pages) or 97 TNT 187-15).

§ 3.02 FACTORS BEARING ON EMPLOYEE STATUS

[B] IRS 20-Factor Test

[6] Continuing Relationship

Page 80, add at end of section:

The issue whether there is a "continuing relationship" with a worker also comes up when the person has worked in one capacity and then switches to another. Workers who have been employees and then are terminated as employees (either by the employee or the company) only to be rehired as "independent contractors" bear special scrutiny. Although there is nothing inherently wrong with altering a relationship, special considerations come into play. Before we talk about danger zones, it is worth noting that it is quite common for a company executive to terminate employment (as in the context of the sale of a business) but to stay on as a "consultant" for some limited period of time. As long as the consultant's duties change substantially, as they typically will in the context of a business sale, there is rarely a problem with the independent contractor classification. The usual factors would be relevant, including whether the executive can work for other companies, whether the employment is full-time, whether the relationship is full-time, whether the executive can work on his or her own time, where the work is performed, and so forth.

A primary danger zone, however, occurs when a company terminates workers who have been treated as employees and then rehires them as "independent contractors" without making significant and important changes in the relationship. A good example of this problem is the recent case of *Barmes v. IRS*.[27.1] In this case, the company operated a gift shop and discharged its employees, only continuing its operations with workers purportedly classified as independent contractors. The facts are somewhat complicated, and it is apparent that the business owners had a plan (which turned out not to be successful) to alleviate payroll tax burdens. The gift shop began operations back in 1972. In 1995, ownership of the business was transferred to an entity denominated by a trust. The employees were discharged at the same time. A few months later, in 1996, the business continued its operations under the ownership of the trust, with workers being called "independent contractors."

Interestingly, in May 1996 the IRS informed the company (the trust now) that they no longer had to file quarterly withholding returns.

However, the IRS began investigating whether the business still had employees for purposes of withholding. At the end of 1996, the IRS sent notices of deficiency to the business for the first two quarters of 1996.

A great deal of legal wrangling ensued, and ultimately the IRS issued notices and the taxpayer and IRS wound up in district court. The plaintiffs (taxpayers) must have felt fairly confident, because they filed an action seeking declaratory judgment that the tax liens stemming from the 1996 tax assessments were invalid. Predictably, the IRS filed counterclaims, seeking to determine that the amounts were proper. The motion for summary judgment was granted in favor of Mrs. Barmes but denied with respect to her husband. The government, on the other hand, fared even better. Their motion for summary judgment on the issue of the procedural correctness of the assessments was granted.

[27.1] Dkt. No. TH 97-287-C-T/F (S.D. Ind. Mar. 8, 2000), *reported in* 76 Daily Tax Rep. K-17 (Apr. 19, 2000).

§ 3.03 ADDITIONAL CLASSIFICATION ISSUES AND EXAMPLES

[A] Statutory Employees

Page 87, add after first paragraph:

In two cases decided the same day, the Tax Court held that a veterinarian[52.1] and the operator of a drywall construction business[52.2] were employees. In each case, the worker performed substantial services for the S corporation and received remuneration for those services and was held to be an employee for federal employment tax purposes.

John Yeagle owed 99 percent of Yeagle Drywall Co.'s stock, and his wife owned the remaining 1 percent of the S corporation. Yeagle performed many services for the business, including soliciting business, ordering supplies, entering into oral and written agreements, overseeing finances, collecting money owed, and hiring and firing independent contractors. The business did not make regular payments to Yeagle, but he withdrew money and paid his personal expenses from the business's bank account at his discretion. The business did not treat Yeagle as an employee or file quarterly employment tax returns.

The business treated individuals who performed services for it, other than Yeagle, as independent contractors and issued Forms 1099-MISC to them. The business did not issue those forms or W-2s to Yeagle.

The business reported net income that it paid to Yeagle as Yeagle's share of the business's income. The Yeagles reported John's share of the business's income as nonpassive income from an S corporation on their individual returns. On audit, the IRS determined that Yeagle was the business's employee for employment tax purposes and that the business was not entitled to safe harbor relief from the taxes under the Revenue Act of 1978.

The Tax Court sustained the IRS's determination, holding that Yeagle was the business's employee for federal employment tax purposes. The court explained that Yeagle was an officer who performed substantial services for Yeagle Drywall Co. and received remuneration for those services. The court rejected the business's characterization of its payments to Yeagle as a distribution of the business's net income. The court concluded that the business distributed all its net income to the Yeagles and that those payments were clearly remuneration for services.

Thus, the payments were wages subject to federal employment taxes. The court further held that the business was not entitled to safe harbor relief under the Revenue Act because it had no reasonable basis for not treating Yeagle as an employee. The court rejected each of the business's arguments for meeting the safe haven requirements as described in § 530 of the Act. Finally, the court accepted the parties' stipulated amount of taxes due from the business.[52.3]

In *Veterinary Surgical Consultants, P.C. v. Commissioner*,[52.4] the company argued that the worker received distributions of corporate net income, not wages. The worker performed substantial services for the company, and was paid. The court found that regardless of how the employer chose to characterize the payments, the payments represented remuneration for services rendered.

Doctor Kenneth K. Sadanaga, DVM, worked full-time for Bristol-Meyers Squibb Co. in the years 1994, 1995, and 1996. He earned $91,212.18, $95,891.15, and $102,031.14, respectively, in those years, on which income and payroll taxes were duly withheld. But Sadanaga, a veterinarian, also offered consulting and surgical services to other veterinarians through his S corporation, Veterinary Surgical Consultants, P.C. (VSC). His practice was limited to surgical oncology, general surgery, plastic and reconstruction surgery, emergency surgery, neurosurgery, and selected orthopedic procedures. VSC's sole income came from Sadanaga's services. However, VSC did not pay him a salary. Instead, Sadanaga drew money from the VSC bank account "at his discretion."

These amounts were treated as "distributions other than dividend distributions paid from accumulated earnings and profits," which Sadanaga reported on Schedule E of his Form 1040 the VSC net income of $83,995.50, $173,030.39, and $161,483.35 for, respectively, 1994, 1995, and 1996.

VSC argued that Sadanaga paid the maximum FICA tax required by law each year as an employee of Bristol-Myers Squibb. "This argument is simply a 'red herring,'" wrote Judge Jacobs. The court continued, "For Federal employment tax purposes, the taxable wage base applies separately to each employer." Judge Jacobs pointed out that when an employee receives wages from more than one employer, and the total of the wages exceeds the FICA wage base, the employee may be entitled to a credit or refund for the employee portion of the FICA tax attributable to the combined wages in excess of that base.

Finally, VSC contended even if Sadanaga was an employee, the corporation's treatment was protected by the safe harbor of § 530. VSC had been consistent in its treatment of Sadanaga and had filed all necessary federal tax and information returns on the basis that he was not an employee. Furthermore, he had properly reported all of the S corporation income. Those are the two basic requirements for § 530 treatment, said the corporation.

Furthermore, the taxpayer had no reasonable basis for not treating Dr. Sadanaga as an employee. The court pointed out that a taxpayer is treated as having a reasonable basis for not treating an individual as an employee if the taxpayer's treatment of the individual was in reasonable reliance on judicial precedent, published rulings, technical advice with respect to the taxpayer, a letter ruling to the taxpayer, or long-standing recognized practice of a significant segment of the industry in which the individual was engaged.

VSC argued that it reasonably relied on cases and rulings, citing *Antonio R. Durando v. United States.*[52.5] The *Durando* opinion stated that it is "improper to treat income earned by a corporation through its trade or business as though it were earned directly by its shareholders, even when, as here, the shareholders' services help to produce that income." However, the court found that the issue in this case is whether the distributions paid to Dr. Sadanaga are wages paid to Dr. Sadanaga as an employee.

The court distinguished all of the cases and rulings VSC cited to establish reasonableness. He concluded that "respondent's position is supported by the plain language of the statute, the applicable Treasury regulations, published revenue rulings, and cases interpreting the applicable statutes." The conclusion resulted in VSC being liable for the employer FICA tax (both the 6.2-percent old age tax and the 1.45-percent health insurance tax on Sadanaga's drawings up to the FICA maximum in each year, and the 1.45-percent health insurance tax on the balance in excess of that). Sadanaga was also liable for the 1.45-percent health insurance tax on the total amount after offsetting his liability for the old age portion of the FICA with the credit for excess FICA as the result of having two employers.

[52.1] Veterinary Surgical Consultants, P.C. v. Commissioner, 117 T.C. No. 14 (2001).

[52.2] Yeagle Drywall Co. v. Commissioner, T.C. Memo 2001-284 (Oct. 15, 2001).

[52.3] *Id.*

[52.4] 117 T.C. No. 14 (2001).

[52.5] 70 F.3d 548, 552 (9th Cir. 1995).

Page 88, add after first paragraph:

Ways and Means Committee member Philip M. Crane, R-Ill., introduced H.R. 3877, which would clarify one of the criteria for determining whether agent-drivers and commission-drivers are independent contractors or statutory employees. Crane noted that the bill is intended to overturn a 1991 IRS general counsel memorandum interpreting § 3121(d) and would "reflect Congressional intent that an investment in facilities [by an agent-driver] can include an investment in a distribution right or territory," in contrast to an investment in education, training, and experience. "Thus, an independent contractor driver who is engaged in distributing meant, vegetable, bakery, beverage (other than milk) products, or laundry or dry-cleaning services and who has a substantial investment in his or her distribution right or territory will not be treated as a statutory employee," he added.

[E] Examples of Employee/Independent Contractor Classification

[12] Construction Workers

Page 111, add after Practice Tip:

In *Richard Mulzet v. R.L. Reppert, Inc.*,[107.1] a United States District Court has held that a construction worker was an independent contractor subject to self-employment taxes. When Richard Mulzet began working as a drywaller for R.L. Reppert, Inc., a manager told him he could not work as an employee because he did not have a driver's license. Mulzet signed an independent contractor agreement and continued working for R.L. Reppert. He did substantially the same job as an R.L. Reppert employee with whom he worked, and he used his own tools. Mulzet filed tax returns indicating that he worked as an independent contractor for R.L. Reppert and other companies and that he owed self-employment taxes. Mulzet filed suit, arguing that he was an employee.

United States District Judge Ronald L. Buckwalter held that Mulzet was an independent contractor. The court noted several factors that favored treating Mulzet as an employee. However, the court decided his contractor status was based on the evidence, which included his signed independent contractor agreement, the fact that he could work for other contractors at the same time, and the fact that he submitted weekly invoices and was not paid overtime.

[107.1] 88 A.F.T.R.2d ¶ 2001-5639; No. 00-4769 (Dec. 4, 2001).

[16] Drivers

Page 120, add at end of section:

In the recent case of *Lub's Enterprises, Inc. v. United States,*[123.1] drivers working for a company that provided services for manufacturers of vehicles and vehicle-leasing companies were held to be employees and not independent contractors. Even more significantly, this company was not held entitled to § 530 safe harbor relief.

The facts are quite specific in the case and may merit review by readers. The company, Lub's Enterprises, Inc., operated the vehicle transportation business and did so for various manufacturers of vehicles as well as for leasing companies. The Lub's drivers took the vehicles to various locations for these Lub's clients. In 1995, three years were under audit, 1992, 1993, and 1994, and the IRS determined that many of the Lub's workers had been incorrectly classified as independent contractors. The revenue agent found that all of the Lub's workers did substantially similar work, but Lub's treated some of them as employees and some of them as contractors. (Recall that this inconsistent treatment is one of the clear danger signs in this area, spelling a risk of reclassification.)

In the district court action, Lub's had paid the IRS the taxes sought but sued for a refund. The government sought summary judgment on the question whether Lub's was entitled to safe harbor relief under § 530 of the Revenue Act of 1978. The court agreed that the workers performed substantially similar work yet were classified in different ways. The court also found that Lub's inconsistently treated several of its workers as both employees and independent contractors. The court therefore found no material issue of fact as to whether Lub's was entitled to the § 530 relief.

Going on to the merits (after having dispensed with the safe harbor relief provision), the court also granted summary judgment for the government on the question whether the Lub's workers designated as independent contractors were properly treated as such under the employment tax laws. Predictably, the court concluded that they should have been treated as employees.

[123.1] Dkt. No. 97 CV 4718 (N.D. Ill. Jan. 24, 2000), *reported in* 20 Daily Tax Rep. K-5 (Jan. 31, 2000).

[26] Insurance Agents

Page 138, add at end of carryover paragraph:

In technical assistance, the IRS concluded that an employee did not qualify as a "full time life insurance salesman." The employee became disabled and drew benefits from his employer's long-term disability plan. To maintain his benefits, the employer treated the employee as a full-time insurance agent. The IRS originally determined that the employee should be treated as a statutory employee, but the employer asked for reconsideration. Noting that the substance of the arrangement should control, the IRS concluded that the employee was not a full-time agent of the employer.[168.1]

[168.1] ITA 200203005 (Sept. 19, 2001).

[32] Physicians

Page 144, add at end of first sentence:

In *North Louisiana Rehabilitation Center, Inc. v. United States*,[185.1] the United States District Court ordered the IRS to refund a rehabilitation hospital's employment taxes, finding that the Service should have determined that the hospital physicians were employees, not independent contractors. North Louisiana Rehabilitation Center, Inc. contracted with physicians to serve as medical and program directors. The corporation treated the physicians as independent contractors and therefore did not pay employment taxes or withhold income taxes from their pay. On audit, the IRS assessed employment and unemployment taxes against the corporation, which amended its returns, paid part of the taxes, and sought a refund.

The IRS failed to act on the refund claim, and the corporation filed a refund suit. The district court rejected the corporation's claim that a magistrate judge erred by allowing the government to withdraw its admission that all of the corporation's medical directors were treated as independent contractors. The court held that the withdrawal of the admission would not prejudice the corporation in its case.

Next the court granted summary judgment to the corporation, finding that it was entitled to relief under the safe harbor provision of § 530 of the Revenue Act of 1978. The court explained that the corporation showed that all medical and program directors were treated as independent contractors, that the corporation filed all required tax returns and consistently treated the physicians as contractors, and that the corporation relied reasonably and in good faith on the advice of in-house and outside counsel in deciding to treat the physicians as contractors. Thus, the court found that the corporation was entitled to a $7,000 judgment.

[185.1] 88 A.F.T.R.2d ¶ 2001-5589; No. 00-0445 (Nov. 8, 2001).

[37] Truck Drivers and Helpers

Page 154, add at end of section:

In the recent case of *Abillo v. Intermodal Container Service, Inc.,*[226] truck owner/operators and other drivers working for two major shipping companies in Los Angeles were held to be properly classified as independent contractors. Key to the judge's finding that the independent contractor relationship was present are the facts that the owner/operators chose their own schedules; worked as much or as little as they wished; started whenever they wanted to; and rejected trucking loads without any discipline or reprisal from the company. The court also noted that there was no evidence that the company supervised drivers while they were on the road. Finally, the drivers were free to hire substitute drivers when they were ill or on vacation. The relationship was also nonexclusive, allowing the drivers to work for other companies. Interestingly, there were a few inconsistencies between the actual relationship (based on testimony) and the written contract. Finding the testimony controlling, the three-judge panel ruled on January 14, 2000, that the actual working relationship is more instructive than the contract language. This is an important finding that might be useful in a future court proceeding!

[226] Dkt. No. BC 17450 (Cal. Sup. Ct. Jan. 14, 2000), *reported in* 14 Daily Tax Rep. G-8 (Jan. 21, 2000).

CHAPTER 4
EMPLOYER LIABILITY FOR MISCLASSIFICATION AND IRS AUDITS

§ 4.02 THE IRS AUDIT

[B] Who and Why the IRS Audits

Page 294, add after carryover paragraph:

As noted, there has been significant controversy concerning the adult entertainment industry. The most recent episode in this continuing fight was *303 West 42nd Street Enterprises, Inc. v. IRS.*[16.1] This case involved a jury verdict in favor of the government in a suit by an adult entertainment operator disputing liability for employment taxes assessed by the IRS. The jury verdict in favor of the government (holding the exotic dancers to be employees) was upheld, despite the taxpayer's motion to set aside the verdict. Furthermore, the taxpayer's motion for a new trial was denied.

This was a fairly well publicized case, with the business being a shop in New York City operating under the name "Show World." Show World operated one-on-one fantasy booths, where a single performer (usually female) provided a private erotic show requested by a patron. There was glass partition in the center of the booth, preventing physical contact, and the patron kept the show going by depositing tokens into a slot.

The court went through the arrangement between Show World and the dancers, but basically it was a 60/40 split, the performer receiving 40 percent and the company receiving 60 percent. Cleverly, the written agreements between the dancers and the company were drafted as a lease, Show World being the landlord and the performer being considered a renter of the booth. A variety of aspects of the arrangement were examined by the IRS and the court. But this was not the first time this matter had come before the court. In 1996, the court granted summary judgment in favor of the IRS. Show World appealed, and the court of appeals found no error in the analysis of the 20 common-law factors used to determine employment status. Furthermore, the court of appeals did not find erroneous the conclusion that relief under § 530 of the Revenue Act of 1978 required proof of a single industry-wide practice. The court remanded the case for a finding whether there was an industry-wide practice. In April 2000, a jury found in favor of the government. Show World, in this case, was asking the court to set aside the jury verdict and to order yet another trial. The court refused, leaving in effect the decision that the exotic dancers were truly employees and not independent contractors.

[16.1] 93 CIV. 4483 (LBS) (S.D.N.Y. May 22, 2000), *reported in* 110 Daily Tax Report K-12 (June 7, 2000).

§ 4.03 IRS PENALTY ASSESSMENT

Page 297, add at end of section:

In *Ewens and Miller, Inc. v. Commissioner,*[28.1] the Tax Court, in an issue of first impression, held that (1) the court had jurisdiction over additions to tax relating to employment taxes; (2) a bakery's workers were employees; and (3) the bakery was not entitled to relief under § 530 of the Revenue Act of 1978. Ewens and Miller, Inc. (EAM) was a corporation and filed a petition for redetermination of worker classification. EAM had four categories of employees: bakery workers, cash payroll workers, route distributors/salespeople, and outside sales workers. In 1991 EAM issued a memo to the staff saying that in 1992 it would subcontract its operation to outside groups, and employees who wanted to continue with EAM would be required to accept responsibility for their own payroll taxes. EAM filed Forms 941, Employer's Quarterly Federal Tax Return, for the four quarters of 1992 and reported no wages subject to withholding.

The Form 941 for the fourth quarter reported that EAM had no employees and that it was going out of business. The IRS determined that the bakery workers, cash payroll workers, route distributors, and outside sales workers were employees for employment tax purposes in 1992 and that EAM was not entitled to § 530 relief under the Tax Reform Act of 1978, Pub. L. 95-600, for any workers. The IRS also determined that EAM was liable for penalties under § 6656.

The Tax Court noted that until the amendment of § 7436(a), the court lacked jurisdiction to hear the case, and properly granted the IRS's motion to dismiss as to the employment taxes and related penalties. The court noted that the amendment of § 7436(a) was retroactive to August 5, 1997, but that it did not specifically provide the court with jurisdiction to decide the proper amount of additions to tax and penalties related to employment tax arising from worker classification or § 530 determinations. Judge Vasquez concluded that § 7436(e) does not exclude additions to tax or penalties from the definition of employment tax, and thus the court had jurisdiction over additions to tax and penalties in chapter 68 of subtitle F of the Code, including the proper additions to tax and penalties relating to taxes imposed by subtitle C for worker classification or § 530 treatment determinations.

In determining if the workers were common law employees, the Tax Court looked to the seven factors in *Weber v. Commissioner*:[28.2] (1) degree of control; (2) investment in facilities; (3) opportunity for profit or loss; (4) right to discharge; (5) integral part of business; (6) permanence of the relationship; and (7) the relationship the parties thought they created. The court concluded that the cash payroll workers, the bakery workers, and the outside sales workers were common law employees. The court held that the route distributors were not common law employees.

The court next concluded that the route distributors were statutory employees. The court noted that agent drivers or commission drivers engaged in distributing bakery products were includable in the term "employee" under § 3121(d)(3)(A). The court concluded that the route distributors fit under the definition of agent driver because the drivers did not have a substantial investment in the facilities; EAM controlled the customers they served and their compensation.

Finally, the court considered if EAM was entitled to relief under § 530 of the Revenue Act of 1978 and concluded that it was not. The court noted that EAM lacked a reasonable basis for not treating an individual as an employee, and it failed to prove that its treatment was based on reasonable reliance on a judicial precedent, published ruling, technical advice, or letter ruling to the taxpayer, or a past IRS audit of the employer. Before 1992, EAM treated all of its production workers as employees and some of its route distributors and outside sales workers as employees.

The court dismissed EAM's argument that it relied on a long-standing industry practice of "co-packing," noting that EAM failed to present evidence on the industry practice of co-packing and the treatment of co-packers as independent contractors. Thus, Judge Vasquez concluded that EAM had no reasonable basis for treating its employees as independent contractors and it was not entitled to § 530 relief.

[28.1] 117 T.C. No. 22; No. 13069-99 (Dec. 11, 2001).

[28.2] 103 T.C. 378 (1994), *aff'd per curiam*, 60 F.3d 1104 (4th Cir. 1995).

§ 4.04 SAFE HARBORS FROM RECLASSIFICATION

[C] Conditions to Protection

[1] Reasonable Basis

[a] Precedent

Page 303, add at end of section:

In *In re Arndt,*[50.1] the U.S. District Court for the Middle District of Florida held that an accountant's advice about classifying workers as not independent contractors was not "technical advice" within the meaning of § 530 of the Revenue Act of 1978. This would seem to be straightforward and obvious, since tax practitioners know what technical advice from the IRS is. However, the employer in this case relied on the advice of an accountant, and it may have seemed to make perfect sense that the accountant's advice could be used as the subject of reliance for purposes of § 530. However, the court in this case held otherwise.

The question, of course, is really what should the rule be. There is some evidence that reliance on an accountant's advice in classifying workers should satisfy the "reasonable basis" requirement of § 530. Courts have interpreted an accountant's advice to fit within § 530 either as "technical advice" under § 530(a)(2)(A) or as an "other reasonable basis" within the meaning of § 530(a)(1).

For example, in *J&J Cab Service v. United States,*[50.2] the court found that advice of an accountant about classifying workers as independent contractors did constitute "technical advice" to the taxpayer. Thus, the taxpayer's reliance on that advice gave rise to a reasonable basis for the taxpayer's treatment of workers as something other than employees. The case involved a taxpayer in the livery business that contracted with cab drivers to operate the taxpayer's cabs for a daily or weekly payment. The court granted the § 530 relief, without much analysis about whether the accountant's advice constituted the kind of technical advice that the statute seems to require.

Similarly, there is a lack of analysis (but a favorable result) in *Hospital Resource Personnel, Inc. v. United States.*[50.3] There, the circuit court neither reversed nor affirmed a lower court finding that an accountant's advice constituted "technical advice" under § 530. This taxpayer operated a nurse registry from which hospitals filled staffing needs. The taxpayer classified the nurses as independent contractors based on the advice of its accountants. For the lower court opinion, see *Hospital Resource Personnel, Inc. v. United States.*[50.4]

The IRS claimed that the nurses were employees, but the district court found that the taxpayer was reasonable in relying on the advice of the accountants. The circuit court upheld the finding that the taxpayer had a reasonable basis for classifying the nurses as independent contractors, because the taxpayer had satisfied the reasonableness requirement through the "published ruling" and "judicial precedent" portions of § 530(a)(2)(A). Unfortunately, the appellate court opinion contains no discussion about the term "technical advice" nor specific discussion about whether accountant advice (or lawyer, for that matter) should so qualify.

In re Arndt does specifically conclude that an accountant's advice was not "technical advice." Indeed, in *In re Arndt,* both the bankruptcy court (which considered the matter first) and the district court so concluded.[50.5] The *Arndt* courts (both courts) seemed to rely on two decisions that are not wholly clear. The two cases are *In re Compass Marine*[50.6] and *Henderson v. United States.*[50.7] Both of these cases obviously involved taxpayers who treated workers as independent contractors. In the *Henderson* case, the advice of an accountant was held not to be technical advice, but there was no reasoning behind the court's conclusion.

It is somewhat disturbing that the purpose of § 530—protecting taxpayers who have had a reasonable basis for treating workers in a certain fashion—certainly would be served by allowing advice of an accountant or lawyer to constitute an "other reasonable basis" for reliance. By its very nature, this "other reasonable basis" provision would seem to be a catchall, certainly one that should encompass reliance on an accountant (or attorney) giving advice.

At present, the *In re Arndt* court holding that good-faith reliance on an accountant's advice does not establish a reasonable basis for classifying workers as independent contractors stands as an impediment to the expansion of § 530 relief.[50.8]

[50.1] 201 B.R. 853 (M.D. Fla. 1996).

[50.2] 1995 WL 214326, Dkt. No. 1:93-CV-234 (W.D.N.C. Jan. 3, 1995).

[50.3] 68 F.3d 421 (11th Cir. 1995).

[50.4] 860 F. Supp. at 1558, *aff'd in part and rev'd in part,* 68 F.3d 421 (11th Cir. 1995).

[50.5] For the bankruptcy court opinion, *see In re Arndt,* 158 B.R. 863 (Bankr. M.D. Fla. 1993); for the district court opinion, *see In re Arndt,* 201 B.R. 853 (M.D. Fla. 1996).

[50.6] 146 B.R. 138 (Bankr. E.D. Pa. 1992).

[50.7] 1992 WL 104326, Dkt. No. 1:90-CV-1064 (W.D. Mich. Feb. 18, 1992).

[50.8] For more complete discussion of this issue, *see* Note, *Employer Who Misclassified Employees Based on Accountant's Advice Was Not Entitled to Section 530 Relief:* In re Arndt, 50 Tax Law. No. 4, at 855 (Summer 1997).

CHAPTER 4 FORMS

Page 319, add after Form 4-3:

FORM 4-4
SAMPLE OPINION LETTER
REGARDING CLASSIFICATION OF WORKERS

Date ____________

Leroy Litigator
Litigator & Settle, P.C.
123 Court Street
Cleveland, OH 12345

Re: *Smith et al. v. Bosco et al.*

Dear Mr. Litigator:

You have asked for my opinion of the status (as independent contractors or employees) of the "Independent Agents" whom your firm represents as a class in the above-referenced litigation. These Independent Agents now sell or previously sold insurance for Bad Life and Accident Insurance Company ("Bad") and/or Large People's Insurance Company ("Large People's"). Bad's parent company is Bosco Corporation ("Bosco").

As you are aware, I have significant experience in tax law and in the substantive area of independent contractor vs. employee characterization for tax purposes. I am the coauthor and editor of the two-volume treatise *Legal Guide to Independent Contractor Status* (3d ed., Aspen Law & Business, 1999). A complete *curriculum vitae* is attached to this letter as Exhibit A.

DOCUMENTS REVIEWED

In connection with rendering my opinion, you have provided me with the following documents:

1. IRS Form SS-8 from Bad completed with respect to Tom Jones, stamped as received by the Internal Revenue Service on February 22, 1995;

2. IRS Response to Form SS-8 dated June 12, 1996, addressed to Bad regarding Tom Jones;
3. Transmittal letter to Form SS-8 (item no. 1 above), including copy of the Independent Agent's Contract between Tom Jones and Bad;
4. Transcript of July 16, 1996, deposition of Liza Smith;
5. Transcript of June 28, 1996, deposition of Grendle Brown (Branch Manager in Oklahoma City);
6. Independent Agent's Contract from Large People's;
7. Broker's Commission Agreement from Bad;
8. Group of correspondence between Bad and Internal Revenue Service consisting of Bad's request for a ruling to determine employment tax status of Mack Spott, including October 5, 1992, letter to Bad from IRS; April 30, 1992, letter from Bart Katt of Bad to IRS; April 8, 1992, letter from IRS to Bad regarding John Doe and additional statements in response to Questionnaire for Determination of Employee Work Status; April 9, 1990, Independent Agent's Contract between Bad and John Doe; draft memo regarding Independent Agent's Contract status (undated and no author specified); October 13, 1993, letter from IRS to Bad's Bart Katt regarding John Doe; June 15, 1993, letter from IRS to Bad's Bart Katt; June 13, 1993, letter from IRS to Mack Spott; October 12, 1993, letter from IRS to Mack Spott; Form SS-8 (handwritten and undated, bearing Bates stamp no. D022177 through D022180);
9. April 8, 1992, letter from IRS to Bad regarding Mack Spott;
10. April 28, 1992, letter from IRS to Bad regarding John Doe;
11. Packet of correspondence and documents between the IRS and Bad relating to an unnamed employee (name has been redacted), including March 14, 1994, determination letter to Bad regarding status of worker, and various correspondence dating between 1992 and 1994 concerning the status of this worker;
12. Bad financial statements prepared by KPMG Peat Marwick for the years ending December 31, 1992, and December 31, 1993;
13. Independent Agent's Contract between Bad and Brad Smith dated December 3, 1994;
14. Career Agent's Contract between Bad and Gretchen Young dated February 28, 1967;

15. First Amended Complaint in the captioned action dated August 28, 1995, filed with the U.S. District Court for the Western District of Missouri, Southern Division;
16. Transcript of deposition of Bart Katt taken May 30, 1996;
17. Agent Contract Cancellation Data Sheet for Robin Earl dated August 8, 1989, terminating his agency because of multiple appointments with other companies;
18. Bad Standard Operating Procedure for Miami branch;
19. Bad memo dated January 18, 1991, to Jay Bennett from Jeff Tweedy;
20. Bad memo dated September 22, 1993, to Gary Louris from Jeff Tweedy;
21. Bad memo dated February 4, 1993, to all Branch Managers from Jeff Tweedy;
22. June 9, 1995, Large People's memo to all Agents/Unit Managers—Dayton Branch from Karen Grotberg;
23. February 23, 1990, Glove memo to Dave Alvin regarding termination of Frank Black;
24. Agent Minimum Standards (undated);
25. January 22, 1997, memo to all Agents and Unit Managers from Lena Lovich;
26. Bad memo to Directors of Agencies and all Branch Managers dated October 19, 1988, from Jeff Tweedy;
27. Memo from Bad Branch Manager Don Bone regarding Unit System guidelines, etc., mailed November 1, 1990;
28. January 3, 1990, Bad memo from Dave Alvin to all branches;
29. Undated, unsigned memo entitled "The Concept 'Hire to Fire' ";
30. October 25, 1988, Bad memo from Dave Alvin to all Managers assigned to Dave Alvin;
31. Bad memo dated January 26, 1990, from Dave Alvin to all Branch Managers;
32. Plaintiffs' Third Supplemental Answers to Defendants' Second Set of Interrogatories Directed to Plaintiffs.

STATEMENT OF FACTS

All of the documents listed above deal with the operation of Bad and Bad's branch office system of insurance agents. In 1994, Bad was acquired by Large People's, and this branch office system was continued in that name. Large People's maintains approximately 50,000–55,000 general agents to market insurance products offered by Bad/Large People's. General agents are treated as independent contractors by Bad/Large People's, and they are free to sell insurance products offered by Bad/Large People's as well as other insurance companies. The status of these "general agents" is not in question in the above-captioned litigation.

In addition to general agents, Bad/Large People's maintains approximately 1,000–1,200 "Independent Agents," whom Bad/Large People's also treat as independent contractors. Accordingly, Bad/Large People's pays them without any income or employment tax withholding and provides them no pension or health benefits. In contrast to the treatment of general agents, however, Independent Agents (the status of which is the primary subject of this litigation) are prohibited from selling any insurance products other than those underwritten by Bad/Large People's. These Independent Agents are "captive" agents who can sell only Bad/Large People's insurance products.

Independent Agents can be terminated for a variety of reasons, including selling any insurance products other than Bad/Large People's products, failing to attend the required meetings in their respective branch offices, failing to meet sales goals, etc. Required sales meetings are held regularly, in some branch offices as frequently as daily.

Independent Agents are expected to attend various meetings and conferences in addition to regular sales meetings at which their travel expenses, accommodations, and meal expenses are paid for by Bad/Large People's. Independent Agents are provided office space, telephones, office supplies, and various other office amenities by Bad/Large People's, and they do not separately pay Bad/Large People's for these services and facilities. Any advertising done by Independent Agents must be approved by the home office before it is used. Independent Agents also receive training in sales and regularly are given sales leads.

DISCUSSION OF APPLICABLE LAW

The Treasury Regulations make clear that the common law "right to control" standard is generally controlling in determining whether a worker is an independent contractor or an employee. Treas. Reg.

§§ 31.3121(d)-1(c)(1), 31.3401. The Treasury Regulations recognize that this right-to-control standard is based on a variety of factors:

> Generally [the employment] relationship exists when the person for whom services are performed has the right to control and direct the individual who performs the services, not only as to the result to be accomplished. That is, an employee is subject to the will and control of the employer not only as to what shall be done but how it shall be done. In this connection, it is not necessary that the employer actually direct or control the manner in which the services are performed; it is sufficient if he has the right to do so. The right to discharge is also an important factor indicating that the person possessing that right is an employer. Other factors characteristic of an employer, but not necessarily present in every case, are the furnishing of tools and the furnishing of a place to work, to the individual who performs the services. In general, if an individual is subject to the control or direction of another merely as to the result to be accomplished by the work and not as to the means and methods for accomplishing the result, he is an independent contractor. An individual performing services as an independent contractor is not as to such services an employee under the usual common law rules. Individuals such as physicians, lawyers, dentists, veterinarians, construction contractors, public stenographers, and auctioneers, engaged in the pursuit of an independent trade, business, or profession, in which they offer their services to the public, are independent contractors and not employees.

Id. § 31.3121(d)-1(c)(2).

IRS 20 FACTORS

The 20 factors deemed important by the Internal Revenue Service ("IRS") are set forth in Revenue Ruling 87-41, 1987-1 C.B. 296. The ruling indicates that special scrutiny is required in applying these factors to ensure that formal aspects of an arrangement designed to achieve a particular status do not obscure the true substance of the relationship. I have set out below a brief description of these criteria.

Instructions. A worker required to comply with other persons' instructions about when, where, and how work is performed is ordinarily an employee. This control factor is present if the person or persons for whom the services are performed have the right to require compliance with instructions. Rev. Rul. 68-598, 1968-2 C.B. 464; Rev. Rul. 66-381, 1966-2 C.B. 449.

Training. When the employer either requires an experienced employee to help train a worker, corresponds with the worker, or requires the worker to attend meetings, it indicates that the person for whom the services are performed wants them performed in a particular method or manner. Rev. Rul. 70-630, 1970-2 C.B. 229. If the work to be performed requires special training and skills, the person performing the services is more likely to be classified as an independent contractor.

Integration. Integration of the worker's services into the retaining company's business operations generally shows that the worker is subject to direction and control. When the success or continuation of a business depends to an appreciable degree upon the performance of certain services, the workers who perform those services must necessarily be subject to a certain amount of control by the business. *United States v. Silk,* 331 U.S. 704 (1947). If the work constitutes an integral part of the employer's business, this tends to support employee classification for the service providers. Conversely, if the work is only tangential to the employer's business, the service providers are more likely to be classified as independent contractors.

Services Rendered Personally. If the services must be rendered personally, the employer is implicitly interested in the methods used to accomplish the work as well as in the results. Rev. Rul. 55-695, 1955-2 C.B. 410. This tends to support employee classification. If the person providing the services has the right to delegate all or a portion of the work to others, independent contractor status may be indicated. Conversely, the lack of a right to delegate duties may support employee classification.

Hiring, Supervising, and Paying Assistants. If the entity for which the services are performed hires, supervises, and pays assistants, that factor generally shows control over the workers. However, if one worker hires, supervises, and pays assistants pursuant to a contract under which the worker agrees to provide materials and labor, and under which the worker is responsible only for the attainment of a result, this factor indicates independent contractor status. *Compare* Rev. Rul. 63-115, 1963-1 C.B. 178, *with* Rev. Rul. 55-593, 1955-2 C.B. 610.

Continuing Relationship. A continuing relationship between the worker and the employer indicates an employer-employee relationship. A continuing relationship may exist if work is performed at frequent although irregular intervals. *United States v. Silk,* 331 U.S. 704 (1947). The duration of the engagement will also be considered. A long-term engagement is more consistent with employee classification, while a short-term or one-time engagement is more consistent with independent contractor classification.

Set Hours of Work. The establishment of set hours of work by the person or persons for whom the services are performed indicates control. Rev. Rul. 73-591, 1973-2 C.B. 337.

Full-Time Required. If the worker must devote substantially all of his or her time to the business of the employer, the employer has control over the amount of time the worker spends working and impliedly restricts the worker from doing other gainful work. An independent contractor, on the other hand, is free to set hours and seek other work. Rev. Rul. 56-694, 1956-2 C.B. 694.

Performing Work on Employer's Premises. If the work is performed on the premises of the employer, this suggests control over the worker, especially if the work could be done elsewhere. Work done off the premises of the employer, such as at an independent office of the worker, indicates some freedom from control. Rev. Rul. 56-20, 1956-2 C.B. 496. This fact by itself does not mean that the worker is not an employee. The importance of this factor depends on the nature of the service involved and the extent to which an employer generally requires that employees perform such services on the employer's premises. Control over the place of work is also indicated when the person or persons for whom the services are performed have the right to compel the worker to travel a designated route, to canvass a territory within a certain time, or to work at specific places. Rev. Rul. 56-694, 1956-2 C.B. 694.

Order or Sequence Set. If a worker must perform services in the order or sequence set by the employer, the worker is not free to follow the worker's own pattern of work. Often, because of the nature of an occupation, employers do not set the order of the services or set the order infrequently. It is still sufficient to show control if the employer merely retains the right to do so. Rev. Rul. 56-694, 1956-2 C.B. 694.

Oral or Written Reports. A requirement that the worker submit regular or written reports to the employer indicates a degree of control. Rev. Rul. 70-309, 1970-1 C.B. 199; Rev. Rul. 68-248, 1968-1 C.B. 431.

Payment by Hour, Week, Month. Payment by the hour, week, or month generally points to an employer-employee relationship, provided that this method of payment is not just a convenient way of paying a lump sum agreed upon as the cost of a job. Payment made by the job or on a straight commission basis generally indicates that the worker is an independent contractor. Rev. Rul. 56-15, 1956-1 C.B. 451.

Payment of Business or Travel Expenses. If the employer ordinarily pays the worker's business or travel expenses, the worker is ordinarily an employee. To control expenses, an employer generally retains the right to regulate and direct the worker's business activities. Rev. Rul. 55-144, 1955-1 C.B. 483.

Furnishing Tools and Materials. If the employer furnishes significant tools, materials, and equipment, it tends to show an employer-employee relationship. Rev. Rul. 71-524, 1971-2 C.B. 346. A worker who provides his or her own equipment is more likely to be characterized as an independent contractor. One must define "equipment" for purposes of this factor and analyze how much equipment is required in each position. A person who is provided an office, a desk, a telephone, and secretarial service but who provides a calculator and other minimal equipment will probably not be considered to be providing equipment for purposes of this rule. Rev. Rul. 74-412, 1974-2 C.B. 332 (an architect who was furnished office and desk space, and secretarial and telephone service, but who provided his own drafting instruments and reference guides was held to be an employee).

Significant Investment. If the worker invests in facilities used in performing services that are not typically maintained by employees (such as an office rented at fair value from an unrelated party), that factor tends to indicate that the worker is an independent contractor. Conversely, lack of investment in facilities indicates dependence on the employer for such facilities and, accordingly, the existence of an employer-employee relationship. Rev. Rul. 74-412, 1974-2 C.B. 332. For a recent case involving significant investment by a contractor, see *Youngs v. Commissioner,* T.C. Memo 1995-94 (1995). *See also* Priv. Ltr. Rul. 9446009 (involving an x-ray technician ruled an employee based on lack of financial investment).

Realization of Profit or Loss. A worker who can realize a profit or suffer a loss (in addition to the profits or losses ordinarily realized by employees) is generally an independent contractor. The worker who cannot is an employee. Rev. Rul. 70-309, 1970-1 C.B. 199. If the worker risks economic loss due to significant investments or a *bona fide* liability for expenses, such as salary payments to unrelated employees, that factor indicates that the worker is an independent contractor. The risk that a worker will not receive payment for services, however, is common to both independent contractors and employees and thus does not constitute a sufficient economic risk to support treatment as an independent contractor.

Working for More Than One Firm at a Time. If a worker performs more than *de minimis* services for more than one entity at the same time, this factor generally indicates that the worker is an independent contractor. Rev. Rul. 70-572, 1970-2 C.B. 221. However, a worker who performs services for more than one person may be an employee of each of the persons, especially if such persons are part of the service arrangement. Conversely, if the person is free to offer services to the general public and to accept as many assignments as possible from others, this supports independent contractor characterization. Rev. Rul. 57-380, 1957-2 C.B. 634 (anesthetists performing services for many doctors and hospitals held to be independent contractors).

Making Services Available to General Public. When workers make their services available to the general public on a regular and consistent basis, it indicates independent contractor status. Rev. Rul. 56-660, 1956-2 C.B. 693.

Right to Discharge. The right to discharge a worker indicates that the worker is an employee, and the person possessing the right is an employer. An employer exercises control through the threat of dismissal, which causes the worker to obey the employer's instructions. An independent contractor, on the other hand, cannot be fired as long as the independent contractor produces a result that meets the contract specifications. Rev. Rul. 75-41, 1975-1 C.B. 323.

The lack of an ability to discharge the worker without cause is more consistent with independent contractor status. *Jones v. United States,* 79-1 U.S.T.C. (CCH) ¶ 9120 (E.D. Tex. 1978), *rev'd and remanded on another issue,* 613 F.2d 1311, 80-1 U.S.T.C. (CCH) ¶ 9291 (5th Cir. 1980). Consequently, some employers attempt to make their workers appear to be independent contractors by executing contracts that either limit the employer's right to discharge the worker except for specified reasons or provide for a notice period before discharge.

Right to Terminate. If workers have the right to end their relationship with the employer at any time they wish without incurring liability, that factor tends to indicate an employer-employee relationship. Rev. Rul. 70-309, 1970-1 C.B. 199.

ADDITIONAL FACTORS AND ANALYSIS

Parties' Understanding. Although not listed in the IRS Revenue Ruling, several other factors mentioned in the case law deserve mention. One is the intention of the parties. If the two parties believe they have an employment relationship, or if they believe they have an independent contractor relationship, this belief will generally be entitled to some weight in making the employee/independent contractor determination if other criteria are equivocal.

Custom in Industry. Another factor not enumerated in the IRS criteria but often mentioned in the case law is the custom in the industry. If the custom in the trade or industry is for the service to be provided by employees, employee status may be indicated. Conversely, if the custom is for the work to be performed by independent contractors, that status may be indicated.

Analysis of 20-Factor Test. The IRS 20 factors established by Revenue Ruling 87-41 can be grouped into broader categories for purposes of analysis. The major category is the *control over the worker on the job,* that is, control over when or where the worker works and what the worker does (training, instructions, and sequence). A second category concerns the *employer's dependence on the job.* The integration of the worker's role into the business reflects the dependence of the employer on the job, especially when the job is full-time.

A third category includes factors reflecting the *worker's dependence on the employer.* Workers who can advertise to the public or work for more than one business at a time are less dependent upon a particular employer and thus more apt to be seen as independent contractors. An exclusive arrangement that a worker not perform services for any other company (such as the exclusivity to which the Bad/Large People's Independent

Agents are subject) certainly mandates the worker's dependence. Having a continual working relationship and working on the employer's premises tend to also indicate dependence on the employer.

Several of the 20 factors deal with the *assumption of risks.* In order to make a profit, workers must take risks. In particular, the worker must have a significant investment in facilities or in his or her tools to be considered an independent contractor. Commissions are the riskiest form of pay. While acceptance of work on a commission basis may indicate a profit motive, such an arrangement can be seen as avoidance of risk by the employer because the employer is protected from lack of performance. By forcing workers to work on a straight commission basis, the employer is exerting control over the employment situation and thereby structuring it. Two other factors in this risk category are that independent contractors hire their own help and pay their own business expenses.

A fifth category deals with *ending the relationship* by discharge or termination. If an employer has the right to immediately discharge the worker, the worker generally has the corresponding right to immediate termination. If the parties have a contract that only allows the worker to be discharged for just cause, this reflects the employer's control over the work environment. *See* O'Neill & Nelsestuen, "Employee or Independent Contractor Status?: Conflicting Letter Rulings Continue Controversy," *Tax Notes,* May 17, 1993, at 961.

As is apparent from a review of the 20 factors enumerated by the IRS in Revenue Ruling 87-41, and the several additional factors mentioned above based on the case law, there is no one factor that by itself will dictate employee or independent contractor status. The determination whether a worker constitutes an employee or an independent contractor must be based on a case-by-case evaluation of all of these factors and their relative importance. Ultimately, what these factors seek to discern is whether the company has control over the worker in the manner and extent of the worker's services.

In the case of the Bad/Large People's Independent Agents, the control factors present are overwhelming, including day-to-day control by the company and over the manner in which sales activities are conducted, the extent of required sales meetings, the company's payment for expenses (and provision of tools and supplies), the company's insistence that the work relationship be an exclusive and full-time one, and many other manifestations of a pattern of control of Independent Agents that is systematic, consistent, and severe. The Independent Agents are routinely given sales leads. The regular required sales meetings, which in some Bad/Large People's offices are mandated daily, are a clear indication that these Independent Agents are subject to the control and direction of Bad/Large People's.

STATUTORY EMPLOYEES

For completeness, I have briefly described below the nature of the term "statutory employees" and its application. I do not view this category of employees as controlling here because, as described below in my opinion, I believe Independent Agents constitute employees under the common law and Revenue Ruling 87-41.

Most employee vs. independent contractor questions are resolved based on the common law standards. However, certain classes of workers are automatically classified as employees even though they do not meet the common law criteria. A statutory employee, like a common law employee, is subject to FICA coverage even though the common law control tests are not met. However, income tax withholding is optional. Wages and taxes withheld are reported on Form W-2. Statutory employees include

1. Agent drivers or commission drivers engaged in distributing meat, vegetable, or bakery products, beverages (other than milk), or laundry or dry cleaning.
2. Full-time life insurance salespersons.
3. Home workers performing work according to furnished specifications on materials provided, which products are required to be returned to the principal.
4. Full-time traveling or city salespersons soliciting orders from wholesalers or retailers for merchandise for resale or for supplies used in their business operations.

In addition to belonging to one of the specific occupations listed above, a statutory employee must satisfy all the following requirements:

1. Any contract of service contemplates that the worker will personally perform substantially all the work.

2. The worker has no substantial investment in the facilities.*
3. There is a continuing work relationship with the business for which the work is performed.

A statutory employee does not have any authority to delegate substantial portions of the work to another person. In addition, the work relationship must be regular and recurring. Thus, a single job transaction will not meet the definition of a statutory employee.

As noted above, one generally reaches the classification of statutory employees only after concluding that a worker does not constitute an employee under the common law criteria. Since I believe the common law criteria (as formulated in Revenue Ruling 87-41) are met with respect to the Bad/Large People's Independent Agents, I do not believe resort to the statutory employee category is necessary in order to conclude that the Independent Agents are employees.

Although insurance agents may be evaluated under common law criteria for employee status, the statutory provision may apply even if the common law criteria are absent. Under the statutory provision (I.R.C. § 3121 (d)(3)(B)), an insurance agent is an employee (1) if he or she is a full-time life insurance salesperson under a contract that contemplates that substantially all services will be performed personally; (2) if the agent does not have a substantial investment in the facilities used in performing services (other than facilities for transportation, such as a car); and (3) if the arrangement is part of a continuing relationship rather than being in the nature of a single transaction. The Treasury Regulations include the definition of a full-time life insurance salesperson as someone whose entire or principal business activity is devoted to the solicitation of life insurance or annuity contracts primarily for one life insurance company. Treas. Reg. § 31.3121(d)-1(d)(3)(ii). For an example of a ruling determining that an insurance agent was a statutory employee, see Revenue Ruling 90-93, 1990-2 C.B. 33.

Because of this definition of a full-time life insurance salesperson, it is possible that an insurance agent will not fit within this definition on one

* Vehicles used for the transportation of the worker (salesperson) or for transporting goods sold (delivery) are specifically excluded from the definition of facilities. Items such as office furniture and fixtures as well as the actual office are included. I.R.C. § 3121(d)(3).

of several bases. One possibility is that the agent sells property and casualty insurance as well as life insurance, so that the sale of life insurance does not constitute the entire or principal business activity. Another possibility is that the agent will sell policies for more than one company. However, in one revenue ruling (Rev. Rul. 54-312, 1954-2 C.B. 327), the IRS determined that an insurance agent was a statutory employee, although he sold accident and health policies in addition to life insurance, because his sales of the accident and health policies were only incidental. Of course, falling outside this statutory definition of what constitutes an employee does not obviate the common law criteria for employee status.

For example, in one revenue ruling (Rev. Rul. 69-288, 1969-1 C.B. 258), a contract gave the insurance agent the exclusive right to sell the company's policies within a specified geographic area. The agent operated his own office at his own expense and received commissions on insurance policies sold. The contracts did not allow the company to interfere with the activities of the agent. Based on the lack of control and the financial independence of the agent, the IRS held him to be an independent contractor.

In several court cases, the status of insurance agents was examined with emphasis on the degree of control that the companies had over the activities of the agents. In one case, *Reserve National Insurance Co. v. United States,* 74-1 U.S.T.C. (CCH) ¶ 9486 (W.D. Okla. 1974), the insurance company retained agents who had their own licenses to sell insurance. The arrangement was exclusive, with the agents agreeing that they would not sell for other companies without the company's written consent. Nevertheless, if an agent sold for another company without seeking such consent, the company would not terminate the arrangement but would merely withhold leads from the offending agent.

The agents were paid solely on commission, although they occasionally received prizes or bonuses for reaching certain goals. They paid their own expenses and were not furnished office space. They were not required to work any particular hours and were not even required to attend the periodic sales meetings that were held. They did, however, have to submit daily reports of the insurance applications they had solicited.

These insurance agents were also free to do their own advertising, subject to the approval of the company. The training the agents received was for minimal familiarity with company policies. Although the contracts recited the fact that the agents should abide by the company's rules and regulations, the court interpreted this to refer to underwriting requirements dealing with the policies themselves. Under these circumstances, the insurance agents were held to be independent contractors rather than employees.

There are several important differences between the facts in *Reserve National* and the facts pertaining to the Bad/Large People's Independent Agents. The exclusivity requirement is strictly enforced (by termination) by Bad/Large People's, whereas it was not by Reserve National. Many other aspects of control are also present with Bad/Large People's but not with Reserve National.

The overall degree of control imposed by a company on its workers is highly relevant to this determination. In *Butts v. Commissioner,* 49 F.3d 713 (11th Cir. 1995), the Tax Court had held that an insurance agent for Allstate was an independent contractor even though he had previously been an employee, the taxpayer received certain pension and employee benefits, and other employee-type factors existed. The court found that the taxpayer paid his own operating expenses and had a high degree of control over his day-to-day business affairs. The Eleventh Circuit Court of Appeals affirmed.

IRS EXAMINATION CRITERIA

Although Revenue Ruling 87-41 still constitutes prevailing law on employee characterization issues, recent internal IRS documents shed light on current audit techniques and standards applied in this area. In March 1996, the Internal Revenue Service released an important and lengthy document dealing with how its examiners should evaluate whether workers are properly classified as employees or independent contractors. *See* "IRS Drafts Guidelines for Determining If Workers Are Independent Contractors," *Wall St. J.,* Mar. 6, 1996, at B-2; IRS Document No. 9300 (9-94), Catalog No. 21066S. In August 1996, the IRS finalized its training manual, revising it again in October 1996.

The IRS training manual does not change the law or alter the fundamental 20-factor test. In an attempt to assist its field agents in resolving this factual question, the guidelines are divided into three broad categories of evidence that IRS agents are to consider when faced with worker classification problems. The three broad categories are "behavioral control," "financial control," and the "relationship of the parties."

The guidelines dealing with behavioral control ask whether the worker is controlled as to the manner in which the work is accomplished. The guidelines state that in determining the answer to this question, several factors must be examined, including instructions provided to the worker. These include time and place, tools or equipment to use, routines or patterns to use, order or sequence to follow, etc.

The principal question regarding financial control is whether the service recipient has the right to direct and control the business aspects of how its workers perform services. A number of factors are relevant here, including whether the worker must make a significant investment in the work, whether the worker incurs expenses for which he is not reimbursed, and the like. Another important financial factor is whether the worker can realize a profit or loss on the arrangement.

The focus of the IRS training manual on the "relationship of the parties" factor is not only the intent of the parties as expressed in a written contract but also the entire facts and circumstances under which a worker performs services. In this amorphous "relationship of the parties" area, the rights to terminate and/or discharge are viewed as very significant. These items are discussed above in connection with the 20 factors established by Revenue Ruling 87-41. The exclusivity and permanency of the relationship is also of prime importance.

As is apparent from the foregoing discussion, the formulation in the IRS training manual of the employee/independent contractor distinction is somewhat different from the 20 factors set forth in Revenue Ruling 87-41. The grouping of the various factors into three primary inquiries (behavioral control, financial control, and the relationship of the parties) suggests a more overarching analysis focused not on any one factor. In an audit circumstance, applying these factors to the facts described above and applicable to the Bad/Large People's Independent Agents, I believe that the Internal Revenue Service would view the Bad/Large People's Independent Agents as employees. The IRS training manual is clear that a written contract which states that persons are independent contractors is certainly not sufficient for such persons to be so treated if the reality of the relationship (and particularly behavioral and financial control by the company) suggests otherwise.

CONCLUSION

In view of the 20 factors enumerated by the Internal Revenue Service in Revenue Ruling 87-41, and in view of the clear control and direction received by the Independent Agents as shown in the above-listed documents with which I have been provided, based upon and subject to the foregoing, I am of the opinion that the Independent Agents are employees for federal income tax and federal employment tax purposes. Based on an evaluation of the documents with which I have been provided, and based on my knowledge of the conduct of the insurance industry, the Bad/Large People's Independent Agents are employees, despite being characterized as independent contractors in their written agreements.

In rendering this opinion, I am mindful that some of the above-listed documents show that Bad/Large People's solicited Internal Revenue Service guidance on the status of its Independent Agents, receiving confirmation from the Internal Revenue Service that such individuals were properly treated as independent contractors. It is my opinion that such guidance from the Internal Revenue Service is irrelevant to this inquiry, as the statements submitted to the Internal Revenue Service were inaccurate in their description of the relationship between Bad/Large People's and these Independent Agents.

If you have further questions, please don't hesitate to contact me.

Very truly yours,

FORM 4-5
SAMPLE APPEAL LETTER REGARDING RESPONSIBLE PERSON PENALTY

Date ____________

Mr. X
Settlement Officer
Internal Revenue Service
123 Collection Street
San Francisco, CA 94105

Re: *John Taxpayer, SSN #123-45-6789*

Dear Mr. X:

I am writing this letter regarding my client, John Taxpayer, and the Section 6672 penalty liability asserted against him, which is now on appeal. The law clearly shows that John Taxpayer was not a responsible person who willfully failed to pay over withholding taxes, within the meaning of Internal Revenue Code § 6672 and the regulations thereunder, with respect to Bad Risk Company (the "Company"). Mr. Taxpayer was not a responsible person because he did not have ultimate authority for Company policy or disbursement of funds. In addition, Mr. Taxpayer had no knowledge of the Company's failure to pay withholding taxes and had no reason to believe that the Company was not paying them. Being neither a responsible person nor willful, Mr. Taxpayer is not liable for the Section 6672 penalty that the IRS is currently assessing against him.

1. **Mr. Taxpayer Was Not a Responsible Person Because He Had No Authority Whatsoever over the Company's Tax Matters.** Mr. Taxpayer was merely an operations manager, employed to exercise his expertise in contracting and painting, and had no supervisory role in financial or payroll operations. As such, Mr. Taxpayer was not given a position from which he could control or exercise discretion over the Company's broader financial decisions, including income and payroll tax matters. The broader corporate financial decisions remained in the purview of those with responsibility for the overall success and operation of the Company and those with expertise in financial matters: Jane Black-

gard (the President), Bud Blackgard, (the Vice President), and Steve Finance (the Controller). Hired only as an operations manager, Mr. Taxpayer possessed no authority over the Company's broader financial decisions or tax liabilities and tax payments. For example, the Blackgards frequently used corporate funds for unspecified or personal purposes, withdrawing money from the corporate account without giving Mr. Taxpayer a reason for its withdrawal, and Mr. Taxpayer had no authority to stop them. Also, Mr. Taxpayer signed only those checks that Mr. Blackgard authorized him to sign, signed them only at the direction of Mr. Blackgard, never signed a tax return concerning the Company, and never had contact with the IRS regarding the Company. Without such ultimate authority regarding financial decisions and tax matters, case law shows that Mr. Taxpayer could not have been a responsible person.

In Barton v. United States, 988 F.2d 58 (8th Cir. 1993), the court held that Barton was not a responsible person because his authority did not extend to tax matters. As vice president of his corporation, Barton had extensive powers, including check-writing authority. However, despite the fact that Barton's signature was often the authorization for payment of payroll, the court ruled that it could not find Barton a responsible person because he lacked tax-paying authority and had absolutely no responsibility for payment of employee withholding funds to the IRS. The court of appeals found that the IRS and district court had clearly erred by placing unwarranted reliance on Barton's corporate titles, restricted management and supervisory powers, and bounded signature authority. "The Government could not reasonably rely on this kind of evidence in the face of uncontradicted evidence that Barton had no authority over tax matters." *Barton,* 988 F.2d at 60. Mr. Taxpayer, like Barton, had entirely no authority over tax matters and absolutely no responsibility over payment of withholding taxes. Mr. Blackgard and Mr. Finance at all times retained complete authority and responsibility over tax matters and tax payments.

In *Evans v. United States,* 188 B.R. 598 (Bankr. D. Neb. 1995), the court held that Evans was not a responsible person, despite the fact that Evans possessed all of the powers that one normally associates with responsible person status. Evans's daily duties included signing checks, hiring and firing personnel, corresponding with the IRS concerning his employer's tax liability, and managing production. Nevertheless, the court found that Evans lacked one power that is essential to responsible person status:

"Mr. Evans had no significant control over corporate financial decision-making, particularly as it pertained to tax matters." *Evans,* 188 B.R. at 603. The president and chief financial officer had control over financial decision-making. In sum, "Mr. Evans's title as general manager and duties overseeing production for the Corporation constituted, *at best, only circumstantial evidence* that he was a 'controlling person'." *Evans,* 188 B.R. at 603 (emphasis added). *Evans,* like *Barton,* clearly shows that the focus of any inquiry into authority over financial decision-making must give due consideration to authority over tax matters in particular. They show that an employee can have some limited financial decision-making ability, but if it is clearly limited and obviously does not extend to tax matters, the employee is not a responsible person.

Mr. Taxpayer had powers consistent with employment as a field operations manager but had absolutely no authority or responsibility over tax matters. Because of this, *Barton* and *Evans* are persuasive authority that the IRS's current position against Mr. Taxpayer is simply wrong. Mr. Taxpayer had previous experience in construction and its costs, not accounting. Any financial expertise that Mr. Taxpayer possessed related only to the cost of construction projects, not accounting. "The courts recognize the normal division of and limitations on authority exercised by various representatives of a particular business." *Godfrey v. United States,* 748 F.2d 1568 (Fed. Cir. 1984) (holding that an outside director, despite being the person considered to be the ultimate financial and managerial authority, was not a responsible person) (citing *Bauer v. United States,* 543 F.2d 142 (Ct. Cl. 1976)). Also, in *Evans,* the court was compelled by testimony that Evans, although a very accomplished production manager, would have been incompetent with regard to tax matters because of his lack of knowledge, background, and experience in tax matters. *Evans,* 188 B.R. at 605. In short, the IRS must give due regard to Mr. Taxpayer's limited authority as an operations manager, not a financial expert.

Consistent with the division of authority within the Company, Mr. Taxpayer had some authority, but it was clearly circumscribed. Mr. Taxpayer approved proposals and prepared project cost data. Unlike financial accounting, these tasks were within the range of Mr. Taxpayer's experience and past duties. Likewise, Mr. Taxpayer had only authority to hire personnel required to run his field operations, and to sign payroll and vendors' checks related to his field operations. At all times these managerial responsibilities and powers were strictly delineated. Mr. Taxpayer could not hire or fire personnel or direct checks that did not fall directly within his purview. Tax matters did not fall within his purview. In short, Mr. Taxpayer had no authority or responsibility regarding the Company's overall financial condition in general, or taxes in particular. Therefore, Mr. Taxpayer could not have been a responsible person.

In light of Mr. Taxpayer's clearly delineated duties, the IRS's current position falls within the line of cases holding that exercise of circumscribed financial responsibilities without concurrent status, duty, and authority to pay taxes will not lead to responsible person status. For example, in *Williams v. IRS,* 179 B.R. 929 (Bankr. D. Mont. 1995), the court held that a shareholder and manager with extensive powers was not a responsible person. Williams was vice president, treasurer, and office manager. His powers and knowledge were commensurate with his titles. Williams often wrote checks, supervised employees, and negotiated contracts. Also, Williams knew of delinquent withholding taxes. However, the court found Williams's authority to be so effectively confined that he could not be a responsible person. The court concluded that the facts showed that "Plaintiff had authority to direct the running of the daily operations, but the record shows Plaintiff never was granted the authority to pay delinquent taxes, even if Plaintiff wanted to." *Williams,* 179 B.R. at 933.

Similarly, in *Alsheskie v. United States,* 31 F.3d 837 (9th Cir. 1994), the court held that Alsheskie was not a responsible person despite the fact that he had sole authority to sign checks, hired and fired employees, signed contracts, and managed daily operations. The court stated that Alsheskie "did not have 'significant control' over what bills to pay or not to pay." *Alsheskie,* 31 F.3d at 839. These two cases clearly show that even if a person possesses broad powers, if those powers do not confer authority over tax matters, as with Mr. Taxpayer, the person is not a responsible person. Mr. Taxpayer simply could not pay the withholding taxes even if he wanted to. The authority and responsibility to manage the Company's greater financial affairs rested with the Blackgards and Mr. Finance. As evidenced by Mr. Taxpayer's lack of ability to stop the Blackgards' reckless treatment of the corporate accounts, Mr. Taxpayer lacked the authority and responsibility for their payment.

The Blackgards controlled the ultimate direction of Company business and funds. As owners of the Company, they abused the corporate form, respecting few of its formalities. For example, Bud Blackgard was the chief executive of the Company but had the title of vice president in order to receive preferential affirmative action treatment. Jane Blackgard was merely a figurehead. As a minority group member, she was given the title of president in order to obtain preference for government loans and contracts. Also, as stated above, the Blackgards frequently treated the corporate accounts as personal accounts. They withdrew money at will, without feeling a need to inform anyone else of the purpose for the withdrawal. On the other hand, Mr. Taxpayer did not even have access to the Company's financial information, including the Company's tax returns and their contents; the financial information stayed solely in the purview of Mr. Blackgard and Mr. Finance. Unaware of the Company's overall financial situation and powerless to stop the Blackgards' reckless treatment of the corporate accounts, Mr. Taxpayer attempted to maintain order as best he could despite his limited authority.

However, not having knowledge or significant control over the Company's financial situation, Mr. Taxpayer could do little. In fact, Mr. Taxpayer was not even aware of the Company's failure to pay withholding taxes, until the IRS informed him of the failure, several years later, in late 1996. The Blackgards and Mr. Finance possessed ultimate authority regarding financial and, particularly, tax matters. They delegated no responsibility and passed on no knowledge regarding them to Mr. Taxpayer. Under these circumstances, Mr. Taxpayer could not have been "responsible" for them.

In *Abramson v. United States,* 48 B.R. 809 (E.D.N.Y. 1985), the court held that the person who was the corporation's treasurer and secretary was not a responsible person. Abramson had power to write checks and did so, but it was clear that ultimate authority rested with others. The court found important that "Abramson had neither input nor influence over Weiss' [the president's] plan, which was already in place, to keep Hargil [the corporation] in business by preferring creditors over the United States." *Abramson,* 48 B.R. at 812. Also, Abramson only signed checks "if there w[as] an adequate balance in the account, and then only in response to a bill." *Id.* Further, the court emphasized that no one believed Abramson to have had responsibility to interact with the IRS or to pay withholding taxes.

Mr. Taxpayer faced the same situation as Abramson. Mr. Taxpayer only signed checks when they were within his range of responsibilities, only if there was money in the account, and was kept ignorant of any plan by Mr. Blackgard and Mr. Finance not to pay withholding taxes. Mr. Taxpayer had no input and was not preferring anyone, let alone creditors. Mr. Taxpayer was paying everyone whom he knew the Company owed and whom he had the authority to pay.

Again, in *Terrell v. United States,* 65 B.R. 365 (Bankr. N.D. Ala. 1986), the court held that the person who acted as shareholder, treasurer, and secretary was not a responsible person. Terrell performed virtually all financial tasks, including doing general accounting, signing checks to pay payroll and suppliers, signing tax returns, and negotiating with the IRS. Despite these powers, the court found that Terrell was not a responsible person, because the president effectively controlled the corporation. The court stated that "[Terrell] did not have the requisite 'significant' input into the financial decisions of the employer corporations and in no way participated in the decisions concerning payment of creditors." *Terrell,* 65 B.R. at 369.

As in *Abramson* and *Terrell,* Mr. Blackgard clearly confined Mr. Taxpayer's authority to pay expenses and manage field operations. Mr. Taxpayer never possessed sufficient discretion that could be construed as authority to pay withholding taxes. Merely incidental to Mr. Taxpayer's management of field operations, he signed payroll and vendor checks. At no point did Mr. Taxpayer have ultimate authority regarding the overall disbursement of funds. He did only what he was told, including hiring a limited range of employees and signing particular checks as directed. In short, Mr. Taxpayer never had responsibility or authority to pay over withholding taxes; and, without such, he could not have been a responsible person.

2. **Mr. Taxpayer Was Not Willful Because He Did Not Know That Withholding Taxes Were Not Being Paid and Believed That They Were Being Paid.** Even if you should determine that Mr. Taxpayer was "responsible," a finding that would be entirely inconsistent with the case law described above, you would also need to find that my client acted willfully. Although the above authorities make any inquiry into willfulness irrelevant, I will address this matter for completeness. The law is quite clear that Mr. Taxpayer, not having intentionally or recklessly failed to pay over withholding taxes, was not willful in regards to their nonpayment. Mere negligence will not suffice to show intent or recklessness;

some degree of fault on the part of a responsible person must also exist. *Godfrey,* 748 F.2d at 1577.

In *United States v. Leuschner,* 336 F.2d 246 (9th Cir. 1964), the court held that Leuschner, the director and general manager of the corporation, was not willful. A controller maintained the books and the tax records, and paid the bills and taxes. Leuschner ran the field operations, and signed checks and tax returns that the controller presented to him. Under these circumstances, the court found that Leuschner did not willfully fail to pay over withholding taxes. Despite Leuschner's broad responsibilities to manage the business, the court found that, because Leuschner did not know that withholding taxes were not being paid and believed that they were, Leuschner's conduct amounted to negligence and no more. *Leuschner,* 336 F.2d at 247.

In *Dudley v. United States,* 428 F.2d 1196 (9th Cir. 1970), the court held that Dudley, the president of the corporation, was not willful. The court stated that Dudley could not have been willful because "the evidence is clear that Dudley had no knowledge that the taxes had not been paid" and "[h]e believed that they had been paid." *Dudley,* 428 F.2d at 1200. This, the court explained, is negligence, and negligence is never willfulness. In short, both *Leuschner* and *Dudley* show that one who lacks knowledge of nonpayment and believes that payment has occurred carries no degree of fault and is not willful.

As mentioned above, Mr. Taxpayer did not know that withholding taxes had not been paid until 1996 and had no reason to believe that they had not been paid. Mr. Taxpayer's beliefs were justified. Mr. Taxpayer was not aware of any other wrongdoing within the Company that would have led him to believe that withholding taxes were not being paid. The limited checks that Mr. Taxpayer was allowed to endorse were clearing. Mr. Taxpayer had no authority whatsoever over broader financial decisions and tax matters in particular. Mr. Taxpayer was told to do only his specified tasks, and Mr. Taxpayer did only those.

Thus, *Leuschner* and *Dudley* show that Mr. Taxpayer was at most negligent. Mr. Taxpayer's lack of knowledge that withholding taxes were not being paid and belief that they were being paid amounts to negligence, not willfulness, and such negligence fails to import liability under Section 6672.

3. **Conclusion.** Thus, there are two independent bases upon which you can conclude that, in light of Mr. Taxpayer's clearly delineated authority at Bad Risk Company, Inc., and the above law, the imposition of a Section 6672 penalty against Mr. Taxpayer is wholly unwarranted. First, Mr. Taxpayer was only a manager with absolutely no duty or authority to pay the withholding taxes. Second, at most, Mr. Taxpayer may have been negligent, but he in no way was willful or reckless. Therefore, liability as a responsible person cannot attach.

I would also like to remark that the amount being assessed against Mr. Taxpayer is incorrect. The IRS claims that Mr. Taxpayer is liable for $209,126.09. That represents the total amount of unpaid withholding taxes for three quarters beginning from March 30, 1993, to December 31, 1993. However, Mr. Taxpayer had signature authority on the Company's general account only from June 8, 1993, to December 3, 1993, and on the Company's Payroll Account only from June 23, 1993, to December 3, 1993. For the periods in which Mr. Taxpayer did not have signature authority, there is absolutely no foundation on which to claim that Mr. Taxpayer was a responsible person. Therefore, Mr. Taxpayer cannot be liable for the withholding taxes not paid during those periods. Therefore, the assessment against Mr. Taxpayer is too large. Mr. Taxpayer is not liable for most of the withholding taxes for the quarter ending June 30, 1993, and for part of the withholding taxes for the quarter ending December 31, 1993.

Finally, as my associate Maxwell Smart mentioned to you recently, we still do not have an update of the trust fund taxes paid by Bad Risk Company pursuant to its bankruptcy plan. Please forward that (or a transcript) at your convenience. This also constitutes a formal request for information concerning the status of other Section 6672 assessments relating to Bad Risk Company. In particular, as Ms. Y of the IRS had told me more than a year ago that the penalty assessment against Mr. Blackgard had not been timely appealed, and Mr. Blackgard apparently has quite sizable liquid assets, I am anxious to know the details of that assessment and the success of the Service's collection efforts.

Thank you for your cooperation. I look forward to hearing from you.

Very truly yours,

CHAPTER 5

INDEPENDENT CONTRACTOR STATUS UNDER FEDERAL LABOR AND EMPLOYMENT LAWS

§ 5.02 TITLE VII OF THE CIVIL RIGHTS ACT OF 1964

[B] Scope of Title VII Coverage

Page 326, add at end of page:

Similarly, notwithstanding the *Doe v. St. Joseph's Hospital* and *Alexander v. Rush N. Shore Medical Center* opinions, a trial court within the Seventh Circuit recently recognized Title VII liability in an "indirect" employment relationship. In *Jensen v. Illinois Department of Corrections,*[30.1] the court permitted claims brought by a nurse, where the defendant employer exerted significant control over the individual's daily work activities.

[30.1] Case No. 97 C 50198 (N.D. Ill. 1999); *also see* Mattice v. Memorial Hospital, 9 A.D. Cases (BNA) 988 (N.D. Ind. 1999) (ADA claims allowed to proceed where the plaintiff was "an employee" of another employer).

Page 327, add to third paragraph:

However, employers should be aware that the First Circuit recently affirmed a large award in favor of a contractor business and its employee based on a racially hostile work environment at the contracting employer's work site. The award was not based on Title VII, but rather on another portion of federal law, the Civil Rights Act of 1866, which prohibits racial discrimination in business contracts.[36.1] It is unlikely that this theory will extend to independent contractors generally because, in its opinion, the appellate court noted that the employer had not properly argued that the individual, as opposed to the contractor entity, was not covered by the Civil Rights Act of 1866.

[36.1] Danco, Inc. v. Wal-Mart Stores, Inc., 178 F.3d 8 (1st Cir. 1999), *citing* 42 U.S.C. § 1981.

[C] Tests to Determine Coverage

[2] Agency Test

Page 335, add to paragraph at end of section:

In *Salamon v. Our Lady of Victory Hospital,*[83.1] a federal trial court applied *Cilecek* and denied summary judgment where the plaintiff physician had shown "at least a modicum of hospital control over certain aspects of her practice." The court clarified that "a miniscule degree of hospital control over some aspect of the plaintiff's professional practice" probably would not be sufficient to establish "employee" status under Title VII.

[83.1] 1999-2 Trade Cases (CCH) ¶ 72,715 (W.D.N.Y. 1999).

§ 5.03 AGE DISCRIMINATION IN EMPLOYMENT ACT

[A] Scope of ADEA

Page 344, add at end of footnote 135:

But see Mattice v. Memorial Hosp., *supra;* Saxon v. Thompson Orthodontics, 71 F. Supp. 2d 1085, 1088 (D. Kan. 1999) (discussing *Hyland* and looking to actual control of business in determining whether shareholders in a professional corporation were "employees").

§ 5.05 FAIR LABOR STANDARDS ACT

[C] Application of Test

Page 365, add at end of footnote 228:

A court recently applied the same principles to find that migrant workers engaged on a pickle farm were employees under both the FLSA and the federal Migrant and Seasonal Agricultural Worker Protection Act. *See* Elizondo v. Podgorniak, 70 F. Supp. 2d 758 (E.D. Mich. 1999).

§ 5.06 NATIONAL LABOR RELATIONS ACT

[C] Application of Test

Page 385, add at end of section:

[5] Physicians

The National Labor Relations Board upheld a regional determination that physicians contracted to provide services to a New Jersey HMO were independent contractors, not employees. Though the Board applied the traditional common law agency test as applied in *Roadway Package System,* it de-emphasized (1) the lack of substantial control over physicians' daily conduct (reasoning that such lack of control results from physicians' specialized training), and (2) the absence of on-site supervision (finding that the HMO could monitor physicians' conduct by other means such as paperwork and computer-based reporting). The Board noted expressly that it *was not* precluding a future finding that other physicians employed by an HMO could be statutory employees.[318.1]

[318.1] *AmeriHealth Inc./AmeriHealth HMO,* 329 NLRB No. 76 (1999).

CHAPTER 6

PENSION AND EMPLOYEE BENEFIT CONSIDERATIONS

§ 6.03 LITIGATION INVOLVING CLASSIFICATION CHANGE FROM EMPLOYEE TO INDEPENDENT CONTRACTOR

Page 410, add at end of section:

There may never be a last chapter written in the *Microsoft* case. However, some of the subsequent history of the *Microsoft* case following the Ninth Circuit's landmark decision is worth noting.

On June 10, 1999, the Ninth Circuit issued its third opinion (in four years) in the *Vizcaino v. Microsoft Corp.* case.[39.1] The issue in this part of the litigation was whether the plaintiffs could properly expand the class of persons who were litigants in the case.

The Ninth Circuit granted *mandamus* to the class of the classified Microsoft common law employees who were seeking benefits. The Ninth Circuit found that the district court lacked the authority to redefine the class on remand from the circuit court's earlier decision. Thus, the Ninth Circuit had to consider the *Microsoft* case twice. In effect, in the second Ninth Circuit opinion, the court found that the district court, after the remand, lacked the authority to redefine the class.[39.2]

As noted in the main volume, the *Microsoft* case involved the workers who were denied employee benefits because they were considered (and agreed to be considered) independent contractors or employees of third-party employment agencies. The Ninth Circuit found that despite all of this, they met the definition of common law employees. In the Ninth Circuit (the first opinion), the court found that the class members were common law employees and thus were eligible for employee benefits. The appeals court remanded the case to the district court for a determination of individual eligibility and calculation of damages or benefits that were due them.

On remand to the district court after the Ninth Circuit opinion, the district court clarified the class definition, limiting those in the class to persons who worked as independent contractors during 1987 through 1990, and whose positions were reclassified by the IRS, or persons who were converted in status by Microsoft. This new definition by the district court excluded temporary employees who were hired later into a reclassified position. It also excluded all other common law employees. Not happy with this limitation, the plaintiffs argued that the court lacked the authority to change the class under Federal Rules of Civil Procedure Rule 23. The district court rejected this position, citing as authority for the position "unusual circumstances."

This led the plaintiffs to petition the Ninth Circuit for *mandamus,* asking that the order reducing the class in size be vacated. Predictably, Microsoft objected, arguing that *mandamus* was not warranted because class certification was a discretionary matter for the district court.

The Ninth Circuit, however, with U.S. District Judge William W. Schwarzer sitting by designation, held that the district court's orders did not conform to the circuit court's mandate. The court of appeals pointed out that a judgment on the merits is *res judicata* to the plaintiff class as defined before the judgment was made. The scope of the class was material, because substantial rights were involved for the entire certified class. In addition, the Ninth Circuit noted that the original class certification had not been conditional and that Microsoft had not originally challenged it.

Microsoft raised a variety of other arguments which the court addressed. In particular, though, the Ninth Circuit noted that it was not persuaded that the district court had the power under Federal Rules of Civil Procedure Rule 23 to change the composition of the class based on "unusual circumstances," a term that the Ninth Circuit found lacked any legal underpinning.

In conclusion, the Ninth Circuit directed the district court, once again, to determine whether an individual is a common law employee using the factors set forth in *Nationwide Mutual Insurance Co. v. Darden.*[39.3] That case analyzes the workers. The Ninth Circuit also directed the district court to analyze the relationship of the workers with Microsoft. The fact that a worker occupies a reclassified position is a significant consideration, said the Ninth Circuit.[39.4]

[39.1] *See* 1999 WL 374084 (9th Cir. June 10, 1999).

[39.2] *See* Vizcaino v. United States, No. 98-71388 (9th Cir. June 24, 1999), *reported in* Tax Notes Doc. No. 1999-22308, 1999 TNT 133-5.

[39.3] 503 U.S. 318 (1992).

[39.4] For a recent discussion of the Microsoft case and its progeny, *see* Neiburger, *Compensation and Fringe Benefits: The Effect of Work Misclassification on Employee Benefit Plans, or Surprise, I'm Your Employee—Now Give Me Those Stock Options!,* J. Corp. Tax'n 359 (2000).

APPENDICES

Appendix A	**IRS Employer's Tax Guide**
Appendix A-1	**IRS Employer's Supplemental Tax Guide**

Appendix A

IRS EMPLOYER'S TAX GUIDE

Page 487, replace Appendix A in main volume with the following new Appendix A:

Department of the Treasury
Internal Revenue Service

Publication 15
(Rev. January 2002)
Cat. No. 10000W

Circular E, Employer's Tax Guide

(Including 2002 Wage Withholding and Advance Earned Income Credit Payment Tables)

Get forms and other information faster and easier by:

Computer • www.irs.gov or **FTP** • ftp.irs.gov

FAX • 703-368-9694 (from your fax machine)

www.irs.gov/elec_svs/efile-bus.html

Contents

Changes To Note

Social security and Medicare tax for 2002. The social security wage base for 2002 is $84,900. There is no wage base limit for Medicare tax. The tax rate remains 6.2% for social security and 1.45% for Medicare tax.

Redesignation of Estimated Income Tax Payments as Employment Tax Deposits. If you determine that your income tax liability for your current tax year will be lower than the amount of estimated income tax payments you have already made, you may redesignate estimated income tax payments as employment tax deposits. You may use these redesignated payments to satisfy deposit liabilities for income tax withholding and social security, Medicare, railroad retirement and Federal unemployment taxes. To make this redesignation, call 1-866-562-5227. Be certain that your redesignation of these payments does not result in an underpayment of the estimated income tax for the tax year. You may be subject to a penalty for an underpayment of estimated income tax.

Calendar

The following is a list of important dates. Also see **Pub. 509,** Tax Calendars for 2002.

Note: *If any date shown below falls on a Saturday, Sunday, or Federal holiday, use the next business day. A statewide legal holiday delays a filing due date only if the IRS office where you are required to file is located in that state. For any due date, you will meet the "file" or "furnish" requirement if the form is properly addressed and mailed First-Class or sent by an IRS designated private delivery service on or before the due date. See* ***Private Delivery Services*** *on page 5 for more information on IRS designated private delivery services.*

By January 31

Furnish Forms 1099 and W-2. Furnish each employee a completed **Form W-2,** Wage and Tax Statement. Furnish each recipient a completed Form 1099 (e.g., **Form 1099-R,** Distributions From Pensions, Annuities, Retirement or Profit-Sharing Plans, IRAs, Insurance Contracts, etc., and **Form 1099-MISC,** Miscellaneous Income).

File Form 940 or 940-EZ. File **Form 940** or **Form 940-EZ,** Employer's Annual Federal Unemployment (FUTA) Tax Return. However, if you deposited all the FUTA tax when due, you have ten additional days to file.

File Form 945. File **Form 945,** Annual Return of Withheld Federal Income Tax, to report any nonpayroll income tax withheld in 2001. See **Nonpayroll Income Tax Withholding** on page 4 for more information.

By February 15

Request new Form W-4 from exempt employees. Ask for a new **Form W-4,** Employee's Withholding Allowance Certificate, from each employee who claimed exemption from withholding last year.

On February 16

Exempt Forms W-4 expire. Any Form W-4 previously given to you claiming exemption from withholding has expired. Begin withholding for any employee who previously claimed exemption from withholding, but has not given you a new Form W-4 for the current year. If the employee does not give you a new Form W-4, withhold tax as if he or she is single, with zero withholding allowances. (See section 9.)

By February 28

File Forms 1099 and 1096. File Copy A of all Forms 1099 with **Form 1096,** Annual Summary and Transmittal of U.S. Information Returns, with the IRS. For electronically filed returns, see **By March 31** below.

File Forms W-2 and W-3. File Copy A of all Forms W-2 with **Form W-3,** Transmittal of Wage and Tax Statements, with the Social Security Administration (SSA). For electronically filed returns, see **By March 31** below.

File Form 8027. File **Form 8027,** Employer's Annual Information Return of Tip Income and Allocated Tips, with the Internal Revenue Service. (See section 6.) For electronically filed returns, see **By March 31** below.

By March 31

File electronic Forms 1099, W-2, and 8027. File Copy A of electronic (not magnetic media) Forms 1099 with the IRS and W-2 with the Social Security Administration. File electronic (not magnetic media) Form 8027 with the IRS.

By April 30, July 31, October 31, and January 31

Deposit FUTA taxes. Deposit Federal unemployment (FUTA) tax due if it is more than $100.

File Form 941. File **Form 941,** Employer's Quarterly Federal Tax Return, and **deposit** any undeposited income, social security, and Medicare taxes. You may pay these taxes with Form 941 if your total tax liability for the quarter is less than $2,500 and the taxes are paid in full with a timely filed return. If you deposited all taxes when due, you have 10 additional days from the due dates above to file the return.

Before December 1

New Forms W-4. Remind employees to submit a new Form W-4 if their withholding allowances have changed or will change for the next year.

On December 31

Form W-5 expires. Form W-5, Earned Income Credit Advance Payment Certificate, expires. Eligible employees who want to receive advance payments of the earned income credit next year must give you a new Form W-5.

Page 2

Employer Responsibilities: The following list provides a brief summary of your basic responsibilities. Because the individual circumstances for each employer can vary greatly, their responsibilities for withholding, depositing, and reporting employment taxes can differ. Each item in this list has a page reference to a more detailed discussion in this publication.

New Employees: Page

- ☐ Verify work eligibility of employees 3
- ☐ Record employees' names and SSNs from social security cards. 3
- ☐ Ask employees for 2002 Form W-4 13

Each Payday:

- ☐ Withhold Federal income tax based on each employee's Form W-4 32
- ☐ Withhold employee's share of social security and Medicare taxes 15
- ☐ Include advance earned income credit in paycheck if employee requested it on Form W-5. 15
- ☐ Deposit:
  - Withheld income tax
  - Withheld and employer social security taxes
  - Withheld and employer Medicare taxes 17

  Note: *Due date of deposit depends on your deposit schedule (monthly or semiweekly).*

Quarterly (By April 30, July 31, October 31, and January 31): Page

- ☐ Deposit FUTA tax in an authorized financial institution if undeposited amount is over $100 . 27
- ☐ File Form 941 (pay tax with return if not required to deposit) 22

Annually (See Calendar for due dates):

- ☐ Remind employees to submit a new Form W-4 if they need to change their withholding. . . .13
- ☐ Ask for a new Form W-4 from employees claiming exemption from income tax withholding 13
- ☐ Reconcile Forms 941 with Forms W-2 and W-3 23
- ☐ Furnish each employee a Form W-2 2
- ☐ File Copy A of Forms W-2 and the transmittal Form W-3 with the SSA. 2
- ☐ Furnish each recipient a Form 1099 (e.g., Forms 1099-R and 1099-MISC). 2
- ☐ File Forms 1099 and the transmittal Form 1096. 2
- ☐ File Form 940 or 940-EZ 27
- ☐ File Form 945 for any nonpayroll income tax withholding 4

Important Reminders

Electronic Filing

Form 940 and **Form 941** may now be filed electronically. For more information, visit the IRS Web Site at **www.irs.gov/elec_svs/efile-bus.html** or call 1-800-829-1040.

Electronic Deposit Requirement

You must make electronic deposits of all depository taxes (such as employment tax, excise tax, and corporate income tax) using the Electronic Federal Tax Payment System (EFTPS) in 2002 if:

- The total deposits of such taxes in 2000 were more than $200,000 or
- You were required to use EFTPS in 2001.

If you are required to use EFTPS and fail to do so, you may be subject to a 10% penalty. If you are not required to use EFTPS, you may participate voluntarily. To get more information or to enroll in EFTPS, call 1-800-555-4477 or 1-800-945-8400.

See section 11 for more information.

Hiring New Employees

Eligibility for employment. You must verify that each new employee is legally eligible to work in the United States. This will include completing the Immigration and Naturalization Service (INS) **Form I-9,** Employment Eligibility Verification. You can get the form from INS offices or by calling 1-800-870-3676. Contact the INS at 1-800-375-5283, or visit the INS Web Site at **www.ins.usdoj.gov** for further information.

New hire reporting. You are required to report any new employee to a designated state new hire registry. Many states accept a copy of Form W-4 with employer information added. Call the Office of Child Support Enforcement at 202-401-9267 or access its Web Site at **www.acf.dhhs.gov/programs/cse/newhire** for more information.

Income tax withholding. Ask each new employee to complete the 2002 Form W-4. (See section 9.)

Name and social security number. Record each new employee's name and number from his or her social security card. Any employee without a social security card should apply for one. (See section 4.)

Paying Wages, Pensions, or Annuities

Income tax withholding. Withhold tax from each wage payment or supplemental unemployment compensation

Page 3

plan benefit payment according to the employee's Form W-4 and the correct withholding rate. (If you have nonresident alien employees, see section 9.) Withhold from periodic pension and annuity payments as if the recipient is married claiming three withholding allowances, unless he or she has provided **Form W-4P,** Withholding Certificate for Pension or Annuity Payments, either electing no withholding or giving a different number of allowances, marital status, or an additional amount to be withheld. Do not withhold on direct rollovers from qualified plans. See section 9 and **Pub. 15-A,** Employer's Supplemental Tax Guide. Pub. 15-A includes information on withholding on pensions and annuities.

Information Returns

You may have to file information returns to report certain types of payments made during the year. For example, you must file **Form 1099-MISC,** Miscellaneous Income, to report payments of $600 or more to persons not treated as employees (e.g., independent contractors) for services performed for your trade or business. For details about filing Forms 1099 and for information about required electronic or magnetic media filing, see the **2002 General Instructions for Forms 1099, 1098, 5498, and W-2G** for general information and the separate specific instructions for each information return you file (for example, **2002 Instructions for Forms 1099-MISC**). Do not use Forms 1099 to report wages and other compensation you paid to employees; report these on Form W-2. See the separate **Instructions for Forms W-2 and W-3** for details about filing Form W-2 and for information about required magnetic media filing. If you file 250 or more Forms W-2 or 1099, you must file them on magnetic media or electronically.

Information reporting call site. The IRS operates a centralized call site to answer questions about reporting on Forms W-2, W-3, 1099, and other information returns. If you have questions related to reporting on information returns, call 1-866-455-7438.

Nonpayroll Income Tax Withholding

Nonpayroll income tax withholding must be reported on **Form 945,** Annual Return of Withheld Federal Income Tax. Form 945 is an annual tax return and the return for 2001 is due January 31, 2002. Separate deposits are required for payroll (Form 941) and nonpayroll (Form 945) withholding. Nonpayroll items include:

- Pensions, annuities, and IRAs.
- Military retirement.
- Gambling winnings.
- Indian gaming profits.
- Voluntary withholding on certain government payments.
- Backup withholding.

All income tax withholding reported on Forms 1099 or W-2G must be reported on Form 945. All income tax withholding reported on Form W-2 must be reported on Form 941, 943, or Schedule H (Form 1040).

Note: *Because distributions to participants from some nonqualified pension plans and deferred compensation plans are treated as wages and are reported on Form W-2, income tax withheld must be reported on Form 941, not Form 945. However, because distributions from such plans to a beneficiary or estate of a deceased employee are not wages and are reported on Forms 1099-R, income tax withheld must be reported on Form 945.*

For details on depositing and reporting nonpayroll income tax withholding, see the separate **Instructions for Form 945.**

Backup withholding. You generally must withhold 30% of certain taxable payments if the payee fails to furnish you with his or her correct taxpayer identification number (TIN). This withholding is referred to as backup withholding.

Payments subject to backup withholding include interest, dividends, patronage dividends, rents, royalties, commissions, nonemployee compensation, and certain other payments you make in the course of your trade or business. In addition, transactions by brokers and barter exchanges and certain payments made by fishing boat operators are subject to backup withholding.

Note: *Backup withholding does not apply to wages, pensions, annuities, IRAs (including simplified employee pension (SEP) and SIMPLE retirement plans), section 404(k) distributions from an employee stock ownership plan (ESOP), medical savings accounts, long-term care benefits, or real estate transactions.*

You can use **Form W-9,** Request for Taxpayer Identification Number and Certification, to request payees to furnish a TIN and to certify that the number furnished is correct. You can also use Form W-9 to get certifications from payees that they are not subject to backup withholding or that they are exempt from backup withholding. The **Instructions for the Requester of Form W-9** includes a list of types of payees who are exempt from backup withholding. For more information, see **Pub. 1679,** A Guide to Backup Withholding.

Recordkeeping

Keep all records of employment taxes for at least 4 years. These should be available for IRS review. Records should include:

- Your employer identification number.
- Amounts and dates of all wage, annuity, and pension payments.
- Amounts of tips reported.
- Records of allocated tips.
- The fair market value of in-kind wages paid.
- Names, addresses, social security numbers, and occupations of employees and recipients.

- Any employee copies of Form W-2 that were returned to you as undeliverable.
- Dates of employment.
- Periods for which employees and recipients were paid while absent due to sickness or injury and the amount and weekly rate of payments you or third-party payers made to them.
- Copies of employees' and recipients' income tax withholding allowance certificates (Forms W-4, W-4P, W-4S, and W-4V).
- Dates and amounts of tax deposits you made and acknowledgment numbers for deposits made by EFTPS.
- Copies of returns filed, including 941TeleFile Tax Records and confirmation numbers.
- Records of fringe benefits provided, including substantiation.

Change of Address

To notify the IRS of a new business mailing address or business location, file **Form 8822,** Change of Address.

Private Delivery Services

You can use certain private delivery services designated by the IRS to mail tax returns and payments. If you mail by the due date using any of these services, you are considered to have filed on time. The most recent list of designated private delivery services was published in October 2001. The list includes only the following:

- Airborne Express (Airborne): Overnight Air Express Service, Next Afternoon Service, Second Day Service.
- DHL Worldwide Express (DHL): DHL "Same Day" Service, DHL USA Overnight.
- Federal Express (FedEx): FedEx Priority Overnight, FedEx Standard Overnight, FedEx 2 Day.
- United Parcel Service (UPS): UPS Next Day Air, UPS Next Day Air Saver, UPS 2nd Day Air, UPS 2nd Day Air A.M., UPS Worldwide Express Plus, and UPS Worldwide Express.

The private delivery service can tell you how to get written proof of the mailing date.

Private delivery services cannot deliver items to P.O. boxes. You must use the U.S. Postal Service to mail any item to an IRS P.O. box address.

Telephone Help

Tax questions. You can call the IRS with your tax questions. Check your telephone book for the local number or call 1-800-829-1040.

Help for people with disabilities. Telephone help is available using TTY/TDD equipment. You may call 1-800-829-4059 with your tax question or to order forms and publications. You may also use this number for assistance with unresolved tax problems.

Recorded tax information (TeleTax). The TeleTax service provides recorded tax information on topics that answer many individual and business Federal tax questions. You can listen to up to three topics on each call you make. Touch-tone service is available 24 hours a day, 7 days a week. TeleTax topics are also available using a personal computer (connect to **www.irs.gov**).

A list of employment tax topics is provided below. Select, by number, the topic you want to hear and call 1-800-829-4477. For the directory of all topics, listen to topic 123.

TeleTax Topics

Topic No.	Subject
751	Social security and Medicare withholding rates
752	Form W-2—Where, when, and how to file
753	Form W-4—Employee's Withholding Allowance Certificate
754	Form W-5—Advance earned income credit
755	Employer identification number (EIN)—How to apply
756	Employment taxes for household employees
757	Form 941—Deposit requirements
758	Form 941—Employer's Quarterly Federal Tax Return
759	Form 940 and 940-EZ—Deposit requirements
760	Form 940 and 940-EZ—Employer's Annual Federal Unemployment Tax Return
761	Tips—Withholding and reporting
762	Independent contractor vs. employee

Unresolved Tax Issues

If you have attempted to deal with an IRS problem unsuccessfully, you should contact the Taxpayer Advocate. The Taxpayer Advocate independently represents your interests and concerns within the IRS by protecting your rights and resolving problems that have not been fixed through normal channels.

While Taxpayer Advocates cannot change the tax law or make a technical tax decision, they can clear up problems that resulted from previous contacts and ensure that your case is given a complete and impartial review.

Your assigned personal advocate will listen to your point of view and will work with you to address your concerns. You can expect the advocate to provide:

- A "fresh look" at a new or on-going problem.
- Timely acknowledgement.
- The name and phone number of the individual assigned to your case.

- Updates on progress.
- Timeframes for action.
- Speedy resolution.
- Courteous service.

When contacting the Taxpayer Advocate, you should provide the following information:

- Your name, address, and employer identification number.
- The name and telephone number of an authorized contact person and the hours he or she can be reached.
- The type of tax return and year(s)
- A detailed description of the problem.
- Previous attempts to solve the problem and the office that had been contacted.
- A description of the hardship you are facing (if applicable).

You may contact a Taxpayer Advocate by calling a toll-free number, **1-877-777-4778.** Persons who have access to TTY/TDD equipment may call 1-800-829-4059 and ask for Taxpayer Advocate assistance. If you prefer, you may call, write, or fax the Taxpayer Advocate office in your area. See **Pub. 1546,** The Taxpayer Advocate Service of the IRS, for a list of addresses and fax numbers.

General Information

This publication explains your tax responsibilities as an employer. It explains the requirements for withholding, depositing, reporting, and paying employment taxes. It explains the forms you must give your employees, those your employees must give you, and those you must send to the IRS and SSA. This guide also has tax tables you need to figure the taxes to withhold for each employee for 2002.

Additional employment tax information is available in **Pub. 15-A,** Employer's Supplemental Tax Guide. Pub. 15-A includes specialized information supplementing the basic employment tax information provided in this publication. **Pub. 15-B,** Employer's Tax Guide to Fringe Benefits, contains information about the employment tax treatment and valuation of various types of noncash compensation.

Most employers must withhold (except FUTA), deposit, report, and pay the following employment taxes—

- Income tax.
- Social security and Medicare taxes.
- Federal unemployment tax (FUTA).

There are exceptions to these requirements. See section 15, **Special Rules for Various Types of Services and Payments.** Railroad retirement taxes are explained in the **Instructions for Form CT-1.**

Federal Government employers. The information in this guide applies to Federal agencies except for the rules requiring deposit of Federal taxes only at Federal Reserve banks or through the FedTax option of the Government On-Line Accounting Link Systems (GOALS). See the **Treasury Financial Manual (I TFM 3-4000)** for more information.

State and local government employers. Employee wages are generally subject to Federal income tax withholding, but not Federal unemployment (FUTA) tax. In addition, wages, with certain exceptions, are subject to social security and Medicare taxes. See section 15 for more information on the exceptions.

You can get information on reporting and social security coverage from your local IRS office. If you have any questions about coverage under a section 218 (Social Security Act) agreement, contact the appropriate state official. To find out the State Social Security Administrator, contact the National Conference of State Social Security Administrators Web Site at **www.ncsssa.org.**

1. Employer Identification Number (EIN)

If you are required to report employment taxes or give tax statements to employees or annuitants, you need an EIN.

The EIN is a nine-digit number the IRS issues. The digits are arranged as follows: 00-0000000. It is used to identify the tax accounts of employers and certain others that have no employees. **Use your EIN on all the items you send to the IRS and SSA.** For more information, get **Pub. 1635,** Understanding Your EIN.

If you have not asked for an EIN, request one on **Form SS-4,** Application for Employer Identification Number. Form SS-4 has information on how to apply for an EIN by mail or by telephone.

You should have only one EIN. If you have more than one and are not sure which one to use, please check with the Internal Revenue Service office where you file your return. Give the numbers you have, the name and address to which each was assigned, and the address of your main place of business. The IRS will tell you which number to use.

If you took over another employer's business, do not use that employer's EIN. If you do not have your own EIN by the time a return is due, write "Applied for" and the date you applied in the space shown for the number.

See **Depositing without an EIN** on page 21 if you must make a deposit and you do not have an EIN.

2. Who Are Employees?

Generally, employees are defined either under common law or under special statutes for certain situations.

Employee status under common law. Generally, a worker who performs services for you is your employee if you can control what will be done and how it will be done. This is so even when you give the employee freedom of action. What matters is that you have the right to control the details of how the services are performed. See **Pub. 15-A,** Employer's Supplemental Tax Guide, for more information on how to determine whether an individual providing services is an independent contractor or an employee.

Generally, people in business for themselves are not employees. For example, doctors, lawyers, veterinarians, construction contractors, and others in an independent trade in which they offer their services to the public are usually not employees. However, if the business is incorporated, corporate officers who work in the business are employees.

If an employer-employee relationship exists, it does not matter what it is called. The employee may be called an agent or independent contractor. It also does not matter how payments are measured or paid, what they are called, or if the employee works full or part time.

Statutory employees. If someone who works for you is not an employee under the common law rules discussed above, do not withhold Federal income tax from his or her pay. Although the following persons may not be common law employees, they may be considered employees by statute for social security, Medicare, and FUTA tax purposes under certain conditions.

1) An agent (or commission) driver who delivers food, beverages (other than milk), laundry, or dry cleaning for someone else.
2) A full-time life insurance salesperson.
3) A homeworker who works by guidelines of the person for whom the work is done, with materials furnished by and returned to that person or to someone that person designates.
4) A traveling or city salesperson (other than an agent-driver or commission-driver) who works full time (except for sideline sales activities) for one firm or person getting orders from customers. The orders must be for items for resale or use as supplies in the customer's business. The customers must be retailers, wholesalers, contractors, or operators of hotels, restaurants, or other businesses dealing with food or lodging.

See Pub. 15-A for details on statutory employees.

Statutory nonemployees. Direct sellers and qualified real estate agents are by law considered nonemployees. They are instead treated as self-employed for all Federal tax purposes, including income and employment taxes. See Pub. 15-A for details.

Treating employees as nonemployees. You will be liable for social security and Medicare taxes and withheld income tax if you do not deduct and withhold them because you treat an employee as a nonemployee. See Internal Revenue Code section 3509 for details.

Relief provisions. If you have a reasonable basis for not treating a worker as an employee, you may be relieved from having to pay employment taxes for that worker. To get this relief, you must file all required information returns (Form 1099-MISC) on a basis consistent with your treatment of the worker. You (or your predecessor) must not have treated any worker holding a substantially similar position as an employee for any periods beginning after 1977.

IRS help. If you want the IRS to determine whether a worker is an employee, file **Form SS-8,** Determination of Worker Status for Purposes of Federal Employment Taxes and Income Tax Withholding.

3. Family Employees

Child employed by parents. Payments for the services of a child under age 18 who works for his or her parent in a trade or business are not subject to social security and Medicare taxes if the trade or business is a sole proprietorship or a partnership in which each partner is a parent of the child. If these services are for work other than in a trade or business, such as domestic work in the parent's private home, they are not subject to social security and Medicare taxes until the child reaches age 21. However, see **Covered services of a child or spouse** below. Payments for the services of a child under age 21 who works for his or her parent whether or not in a trade or business are not subject to Federal unemployment (FUTA) tax. Although not subject to FUTA tax, the wages of a child may be subject to income tax withholding.

One spouse employed by another. The wages for the services of an individual who works for his or her spouse in a trade or business are subject to income tax withholding and social security and Medicare taxes, but not to FUTA tax. However, the services of one spouse employed by another in other than a trade or business, such as domestic service in a private home, are not subject to social security, Medicare, and FUTA taxes.

Covered services of a child or spouse. The wages for the services of a child or spouse are subject to income tax withholding as well as social security, Medicare, and FUTA taxes if he or she works for:

1) A corporation, even if it is controlled by the child's parent or the individual's spouse,
2) A partnership, even if the child's parent is a partner, unless each partner is a parent of the child,
3) A partnership, even if the individual's spouse is a partner, or
4) An estate, even if it is the estate of a deceased parent.

Parent employed by child. The wages for the services of a parent employed by his or her child in a trade or business are subject to income tax withholding and social security and Medicare taxes. Social security and Medicare taxes do not apply to wages paid to a parent for services not in a trade or business, but they do apply to domestic services if:

1) The parent cares for a child who lives with a son or daughter and who is under age 18 or requires adult supervision for at least 4 continuous weeks in a calendar quarter due to a mental or physical condition, and
2) The son or daughter is a widow or widower, divorced, or married to a person who, because of a

physical or mental condition, cannot care for the child during such period.

Wages paid to a parent employed by his or her child are not subject to FUTA tax, regardless of the type of services provided.

4. Employee's Social Security Number (SSN)

You are required to get each employee's name and SSN and to enter them on Form W-2. (This requirement also applies to resident and nonresident alien employees.) You should ask your employee to show you his or her social security card. The employee is required to show the card if it is available. You may, but are not required to, photocopy the social security card if the employee provides it. If you do not provide the correct employee name and SSN on Form W-2, you may owe a penalty.

Any employee without a social security card can get one by completing **Form SS-5,** Application for a Social Security Card. You can get this form at Social Security Administration (SSA) offices or by calling 1-800-772-1213. Form SS-5 can also be obtained from the SSA Web Site at **www.ssa.gov.** The employee must complete and sign Form SS-5; it cannot be filed by the employer. If your employee applied for an SSN but does not have it when you must file Form W-2, enter "Applied for" on the form. When the employee receives the SSN, file **Form W-2c,** Corrected Wage and Tax Statement, to show the employee's SSN and furnish a copy to the employee.

Note: *Record the name and number of each employee exactly as they are shown on the employee's social security card. If the employee's name is not correct as shown on the card (for example, because of marriage or divorce), the employee should request a new card from the SSA. Continue to use the old name until the employee shows you the new social security card with the new name.*

If your employee was given a new social security card to show his or her correct name and number after an adjustment to his or her alien residence status, correct your records and show the new information on Form W-2. If you filed Form W-2 for the same employee in prior years under the old name and SSN, file Form W-2c to correct the name and number. Use a separate Form W-2c to correct each prior year and furnish a copy of each Form W-2c to the employee. Advise the employee to contact the local SSA office no earlier than 9 months after the Form W-2c is filed to ensure that the records were updated.

IRS individual taxpayer identification numbers (ITINs) for aliens. A resident or nonresident alien may request an ITIN for tax purposes if he or she does not have and is not eligible to get an SSN. Possession of an ITIN does not change an individual's employment or immigration status under U.S. law. Do not accept an ITIN in place of an SSN for employee identification or for work.

An individual with an ITIN who later becomes eligible to work in the United States must obtain an SSN.

Verification of social security numbers. The Social Security Administration (SSA) offers employers and authorized reporting agents two methods for verifying employee SSNs. Both methods match employee names and SSNs.

- **Telephone verification.** To verify up to five names and numbers, call 1-800-772-6270. To verify up to 50 names and numbers, contact your local social security office.
- **Large volume verification.** The **Enumeration Verification Service (EVS)** may be used to verify more than 50 employee names and SSNs. Preregistration is required for EVS or for requests made on magnetic media. For more information, call the EVS information line at 410-965-7140 or visit SSA's Web Site for Employers at **www.ssa.gov/employer.**

5. Wages and Other Compensation

Wages subject to Federal employment taxes include all pay you give an employee for services performed. The pay may be in cash or in other forms. It includes salaries, vacation allowances, bonuses, commissions, and fringe benefits. It does not matter how you measure or make the payments. Also, compensation paid to a former employee for services performed while still employed is wages subject to employment taxes. See section 6 for a discussion of tips and section 7 for a discussion of supplemental wages. Also, see section 15 for exceptions to the general rules for wages. **Pub. 15-A,** Employer's Supplemental Tax Guide, provides additional information on wages and other compensation. **Pub. 15-B,** Employer's Tax Guide to Fringe Benefits, provides information on other forms of compensation, including:

- Accident and health benefits
- Achievement awards
- Adoption assistance
- Athletic facilities
- De minimis (minimal) benefits
- Dependent care assistance
- Educational assistance
- Employee discounts
- Employee stock options
- Group-term life insurance coverage
- Lodging on your business premises
- Meals
- Moving expense reimbursements
- No-additional-cost services
- Retirement planning services
- Transportation (commuting) benefits

- Tuition reduction
- Working condition benefits

Employee business expense reimbursements. A reimbursement or allowance arrangement is a system by which you substantiate and pay the advances, reimbursements, and charges for your employees' business expenses. How you report a reimbursement or allowance amount depends on whether you have an accountable or a nonaccountable plan. If a single payment includes both wages and an expense reimbursement, you must specify the amount of the reimbursement.

These rules apply to all ordinary and necessary employee business expenses that would otherwise qualify for a deduction by the employee.

Accountable plan. To be an accountable plan, your reimbursement or allowance arrangement must require your employees to meet all three of the following rules.

1) They must have paid or incurred deductible expenses while performing services as your employees.
2) They must adequately account to you for these expenses within a reasonable period of time.
3) They must return any amounts in excess of expenses within a reasonable period of time.

Amounts paid under an accountable plan are not wages and are not subject to income tax withholding and payment of social security, Medicare, and Federal unemployment (FUTA) taxes.

If the expenses covered by this arrangement are not substantiated or amounts in excess of expenses are not returned within a reasonable period of time, the amount is treated as paid under a nonaccountable plan. This amount is subject to income tax withholding and payment of social security, Medicare, and FUTA taxes for the first payroll period following the end of the reasonable period.

A reasonable period of time depends on the facts and circumstances. Generally, it is considered reasonable if your employees receive the advance within 30 days of the time they incur the expense, adequately account for the expenses within 60 days after the expenses were paid or incurred, and they return any amounts in excess of expenses within 120 days after the expense was paid or incurred. Also, it is considered reasonable if you give your employees a periodic statement (at least quarterly) that asks them to either return or adequately account for outstanding amounts and they do so within 120 days.

Nonaccountable plan. Payments to your employee for travel and other necessary expenses of your business under a nonaccountable plan are wages and are treated as supplemental wages and subject to income tax withholding and payment of social security, Medicare, and FUTA taxes. Your payments are treated as paid under a nonaccountable plan if:

1) Your employee is not required to or does not substantiate timely those expenses to you with receipts or other documentation or
2) You advance an amount to your employee for business expenses and your employee is not required to or does not return timely any amount he or she does not use for business expenses.

See section 7 for more information on supplemental wages.

Per diem or other fixed allowance. You may reimburse your employees by travel days, or miles, or some other fixed allowance. In these cases, your employee is considered to have accounted to you if the payments do not exceed rates established by the Federal Government. The 2001 standard mileage rate for auto expenses was 34.5 cents per mile. The rate for 2002 is 36.5 cents per mile. The government per diem rates for meals and lodging in the continental United States are listed in **Pub. 1542,** Per Diem Rates. Other than the amount of these expenses, your employees' business expenses must be substantiated (for example, the business purpose of the travel or the number of business miles driven).

If the per diem or allowance paid exceeds the amounts specified, you must report the excess amount as wages. This excess amount is subject to income tax withholding and payment of social security, Medicare, and FUTA taxes. Show the amount equal to the specified amount (i.e., the nontaxable portion) in box 12 of Form W-2 using code L.

Wages not paid in money. If in the course of your trade or business you pay your employees in a medium that is neither cash nor a readily negotiable instrument, such as a check, you are said to pay them "in kind." Payments in kind may be in the form of goods, lodging, food, clothing, or services. Generally, the fair market value of such payments at the time they are provided is subject to income tax withholding and social security, Medicare, and FUTA taxes.

However, noncash payments for household work, agricultural labor, and service not in the employer's trade or business are exempt from social security, Medicare, and FUTA taxes. Withhold income tax on these payments only if you and the employee agree to do so. However, noncash payments for agricultural labor, such as commodity wages, are treated as cash payments subject to employment taxes if the substance of the transaction is a cash payment.

Moving expenses. Reimbursed and employer-paid qualified moving expenses (those that would otherwise be deductible by the employee) are not includible in an employee's income unless you have knowledge that the employee deducted the expenses in a prior year. Reimbursed and employer-paid nonqualified moving expenses are includible in income and are subject to employment taxes and income tax withholding. For more information on moving expenses, see **Pub. 521,** Moving Expenses.

Meals and lodging. The value of meals is not taxable income and is not subject to income tax withholding and social security, Medicare, and FUTA taxes if the meals are furnished for the employer's convenience and on the employer's premises. The value of lodging is not subject to income tax withholding and social security, Medicare, and FUTA taxes if the lodging is furnished for the employer's convenience, on the employer's premises, and as a condition of employment.

"For the convenience of the employer" means that you have a substantial business reason for providing the meals and lodging other than to provide additional compensation to the employee. For example, meals you provide at the place of work so an employee is available for emergencies during his or her lunch period are generally considered to be for your convenience.

However, whether meals or lodging are provided for the convenience of the employer depends on all the facts and circumstances. A written statement that the meals or lodging are for your convenience is not sufficient.

50% test. If over 50% of the employees who are provided meals on an employer's business premises receive these meals for the convenience of the employer, all meals provided on the premises are treated as furnished for the convenience of the employer. If this 50% test is met, the value of the meals is excludable for all employees and is not subject to income tax withholding or employment taxes.

For more information, see **Pub. 15-B,** Employer's Tax Guide to Fringe Benefits.

Health insurance plans. If you pay the cost of an accident or health insurance plan for your employees, which may include an employee's spouse and dependents, your payments are not wages and are not subject to social security, Medicare, and FUTA taxes, or income tax withholding. Generally, this exclusion applies to qualified long-term care insurance contracts. However, the cost of health insurance benefits must be included in the wages of S corporation employees who own more than 2% of the S corporation (2% shareholders).

Archer medical savings accounts. Your contributions to an employee's medical savings account (Archer MSA) are not subject to social security, Medicare, or FUTA taxes, or income tax withholding if it is reasonable to believe at the time of payment of the contributions that they will be excludable from the income of the employee. To the extent that it is **not** reasonable to believe they will be excludable, your contributions are subject to these taxes. Employee contributions to their Archer MSAs through a payroll deduction plan must be included in wages and are subject to social security, Medicare, and FUTA taxes, and income tax withholding.

Medical care reimbursements. Generally, medical care reimbursements paid for an employee under an employer's self-insured medical reimbursement plan are not wages and are not subject to social security, Medicare, and FUTA taxes, or income tax withholding. See Pub. 15-B for an exception for highly compensated employees.

Fringe benefits. You generally must include fringe benefits in an employee's gross income (but see ***Nontaxable fringe benefits*** next). The benefits are subject to income tax withholding and employment taxes. Fringe benefits include cars you provide, flights on aircraft you provide, free or discounted commercial flights, vacations, discounts on property or services, memberships in country clubs or other social clubs, and tickets to entertainment or sporting events. In general, the amount you must include is the amount by which the fair market value of the benefits is more than the sum of what the employee paid for it plus any amount the law excludes. There are other special rules you and your employees may use to value certain fringe benefits. See Pub. 15-B for more information.

Nontaxable fringe benefits. Some fringe benefits are not taxable if certain conditions are met. See Pub. 15-B for details. Examples are:

1) Services provided to your employees at no additional cost to you.
2) Qualified employee discounts.
3) Working condition fringes that are property or services the employee could deduct as a business expense if he or she had paid for it. Examples include a company car for business use and subscriptions to business magazines.
4) Minimal value fringes (including an occasional cab ride when an employee must work overtime, local transportation benefits provided because of unsafe conditions and unusual circumstances, and meals you provide at eating places you run for your employees if the meals are not furnished at below cost).
5) Qualified transportation fringes subject to specified conditions and dollar limitations (including transportation in a commuter highway vehicle, any transit pass, and qualified parking).
6) Qualified moving expense reimbursement. See page 9 for details.
7) The use of on-premises athletic facilities if substantially all the use is by employees, their spouses, and their dependent children.
8) Qualified tuition reduction, which an educational organization provides its employees for education. For more information, see **Pub. 520,** Scholarships and Fellowships.

However, do not exclude the following fringe benefits from the income of highly compensated employees unless the benefit is available to employees on a nondiscriminatory basis.

- No-additional-cost services (item 1 above).
- Qualified employee discounts (item 2 above).
- Meals provided at an employer operated eating facility (included in item 4 above).
- Reduced tuition for education (item 8 above).

For more information, including the definition of a highly compensated employee, see Pub. 15-B.

When fringe benefits are treated as paid. You may choose to treat certain noncash fringe benefits as paid by the pay period, or by the quarter, or on any other basis you choose as long as you treat the benefits as paid at least once a year. You do not have to make a formal choice of payment dates or notify the IRS of the dates you choose. You do not have to make this choice for all employees. You may change methods as often as you like, as long as you treat all benefits provided in a calendar year as paid by December 31 of the calendar year. See Pub.15-A for more information, including a discussion of the special accounting rule for fringe benefits provided during November and December.

Valuation of fringe benefits. Generally, you must determine the value of fringe benefits no later than January 31 of the next year. Prior to January 31, you may reasonably estimate the value of the fringe benefits for purposes of withholding and depositing on time.

Withholding on fringe benefits. You may add the value of fringe benefits to regular wages for a payroll period and figure withholding taxes on the total, or you may withhold Federal income tax on the value of the fringe benefits at the flat 27% supplemental wage rate.

You may choose not to withhold income tax on the value of an employee's personal use of a vehicle you provide. You must, however, withhold social security and Medicare taxes on the use of the vehicle. See Pub. 15-A for more information on this election.

Depositing taxes on fringe benefits. Once you choose payment dates for fringe benefits (discussed above), you must deposit taxes in the same deposit period you treat the fringe benefits as paid. To avoid a penalty, deposit the taxes following the general deposit rules for that deposit period.

If you determine by January 31 that you overestimated the value of a fringe benefit at the time you withheld and deposited for it, you may claim a refund for the overpayment or have it applied to your next employment tax return (see **Valuation of fringe benefits** above). If you underestimated the value and deposited too little, you may be subject to the failure to deposit penalty. See section 11 for information on deposit penalties.

If you deposited the required amount of taxes but withheld a lesser amount from the employee, you can recover from the employee the social security, Medicare, or income taxes you deposited on his or her behalf, and included in the employee's Form W-2. However, you must recover the income taxes before April 1 of the following year.

Sick pay. In general, sick pay is any amount you pay, under a plan you take part in, to an employee who is unable to work because of sickness or injury. These amounts are sometimes paid by a third party, such as an insurance company or employees' trust. In either case, these payments are subject to social security, Medicare, and FUTA taxes. Sick pay becomes exempt from these taxes after the end of 6 calendar months after the calendar month the employee last worked for the employer. The payments are also subject to income tax. See Pub. 15-A for more information.

6. Tips

Tips your employee receives from customers are generally subject to withholding. Your employee must report cash tips to you by the 10th of the month after the month the tips are received. The report should include tips you paid over to the employee for charge customers and tips the employee received directly from customers. No report is required for months when tips are less than $20. Your employee reports the tips on **Form 4070,** Employee's Report of Tips to Employer, or on a similar statement. The statement must be signed by the employee and must show the following:

- The employee's name, address, and SSN.
- Your name and address.
- The month or period the report covers.
- The total tips.

Both Forms 4070 and **4070-A,** Employee's Daily Record of Tips, are included in **Pub. 1244,** Employee's Daily Record of Tips and Report to Employer.

You must collect income tax, employee social security tax, and employee Medicare tax on the employee's tips. You can collect these taxes from the employee's wages or from other funds he or she makes available. (See **Tips treated as supplemental wages** in section 7 for further information.) Stop collecting the employee social security tax when his or her wages and tips for tax year 2002 reach $84,900; collect the income and employee Medicare taxes for the whole year on all wages and tips. You are responsible for the employer social security tax on wages and tips until the wages (including tips) reach the limit. You are responsible for the employer Medicare tax for the whole year on all wages and tips. File Form 941 to report withholding on tips.

If, by the 10th of the month after the month you received an employee's report on tips, you do not have enough employee funds available to deduct the employee tax, you no longer have to collect it. If there are not enough funds available, withhold taxes in the following order:

1) Withhold on regular wages and other compensation.
2) Withhold social security and Medicare taxes on tips.
3) Withhold income tax on tips.

Show these tips and any uncollected social security and Medicare taxes on Form W-2 and on lines 6c, 6d, 7a, and 7b of Form 941. Report an adjustment on line 9 of Form 941 for the uncollected social security and Medicare taxes. Enter the amount of uncollected social security and Medicare taxes in box 12 of Form W-2 with codes A and B. (See section 13 and the **Instructions for Forms W-2 and W-3**.)

If an employee reports to you in writing $20 or more of tips in a month, the tips are subject to FUTA tax.

Note: *You are permitted to establish a system for electronic tip reporting by employees. See Regulations section 31.6053-1.*

Allocated tips. If you operate a large food or beverage establishment, you must report allocated tips under certain circumstances. However, do not withhold income, social security, or Medicare taxes on allocated tips.

A large food or beverage establishment is one that provides food or beverages for consumption on the premises, where tipping is customary, and where there are normally more than 10 employees on a typical business day during the preceding year.

The tips may be allocated by one of three methods—hours worked, gross receipts, or good faith agreement. For information about these allocation methods, including the requirement to file Forms 8027 on magnetic media if 250 or more forms are filed, see the separate **Instructions for Form 8027.**

Tip Rate Determination and Education Program. Employers may participate in the Tip Rate Determination and Education Program. The program consists of two voluntary agreements developed to improve tip income reporting by helping taxpayers to understand and meet their tip reporting responsibilities. The two agreements are the **Tip Rate Determination Agreement** (TRDA) and the **Tip Reporting Alternative Commitment** (TRAC). To find out more about this program, or to identify the IRS Tip Coordinator for your state, call the IRS at 1-800-829-1040. To get more information about TRDA or TRAC agreements, access the IRS Web Site at **www.irs.gov** and search for Market Segment Understanding (MSU) agreements.

7. Supplemental Wages

Supplemental wages are compensation paid in addition to the employee's regular wages. They include, but are not limited to, bonuses, commissions, overtime pay, payments for accumulated sick leave, severance pay, awards, prizes, back pay and retroactive pay increases for current employees, and payments for nondeductible moving expenses. Other payments subject to the supplemental wage rules include taxable fringe benefits and expense allowances paid under a nonaccountable plan. How you withhold on supplemental payments depends on whether the supplemental payment is identified as a separate payment from regular wages.

Supplemental wages combined with regular wages. If you pay supplemental wages with regular wages but do not specify the amount of each, withhold income tax as if the total were a single payment for a regular payroll period.

Supplemental wages identified separately from regular wages. If you pay supplemental wages separately (or combine them in a single payment and specify the amount of each), the income tax withholding method depends partly on whether you withhold income tax from your employee's regular wages:

1) If you ***withheld*** income tax from an employee's regular wages, you can use one of the following methods for the supplemental wages:

   a) Withhold a flat 27% (no other percentage allowed).

   b) Add the supplemental and regular wages for the most recent payroll period this year. Then figure the income tax withholding as if the total were a single payment. Subtract the tax already withheld from the regular wages. Withhold the remaining tax from the supplemental wages.

2) If you ***did not withhold*** income tax from the employee's regular wages, use method **b** above. (This would occur, for example, when the value of the employee's withholding allowances claimed on Form W-4 is more than the wages.)

Regardless of the method you use to withhold income tax on supplemental wages, they are subject to social security, Medicare, and FUTA taxes.

Example 1. You pay John Peters a base salary on the 1st of each month. He is single and claims one withholding allowance. In January of 2002, he is paid $1,000. Using the wage bracket tables, you withhold $58 from this amount. In February 2002, he receives salary of $1,000 plus a commission of $2,000, which you include in regular wages. You figure the withholding based on the total of $3,000. The correct withholding from the tables is $394.

Example 2. You pay Sharon Warren a base salary on the 1st of each month. She is single and claims one allowance. Her May 1, 2002, pay is $2,000. Using the wage bracket tables, you withhold $208. On May 14, 2002, she receives a bonus of $2,000. Electing to use supplemental payment method **b**, you:

1) Add the bonus amount to the amount of wages from the most recent pay date ($2,000 + $2,000 = $4,000).

2) Determine the amount of withholding on the combined $4,000 amount to be $664 using the wage bracket tables.

3) Subtract the amount withheld from wages on the most recent pay date from the combined withholding amount ($664 – $208 = $456).

4) Withhold $456 from the bonus payment.

Example 3. The facts are the same as in Example 2, except that you elect to use the flat rate method of withholding on the bonus. You withhold 27% of $2,000, or $540, from Sharon's bonus payment.

Tips treated as supplemental wages. Withhold income tax on tips from wages or from other funds the employee makes available. If an employee receives regular wages and reports tips, figure income tax as if the tips were supplemental wages. If you have not withheld income tax from the regular wages, add the tips to the regular wages. Then withhold income tax on the total. If you withheld income tax from the regular wages, you can withhold on the tips by method **a** or **b** above.

Vacation pay. Vacation pay is subject to withholding as if it were a regular wage payment. When vacation pay is in addition to regular wages for the vacation period, treat it as a supplemental wage payment. If the vacation pay is for a time longer than your usual payroll period, spread it over the pay periods for which you pay it.

8. Payroll Period

The payroll period is a period of service for which you usually pay wages. When you have a regular payroll period, withhold income tax for that time period even if your employee does not work the full period.

When you do not have a regular payroll period, withhold the tax as if you paid wages for a daily or miscellaneous payroll period. Figure the number of days (including Sundays and holidays) in the period covered by the wage payment. If the wages are unrelated to a specific length of time (e.g., commissions paid on completion of a sale), count back the number of days from the payment period to the latest of:

1) The last wage payment made during the same calendar year,
2) The date employment began, if during the same calendar year, or
3) January 1 of the same year.

When you pay an employee for a period of less than 1 week, and the employee signs a statement under penalties of perjury that he or she is not working for any other employer during the same week for wages subject to withholding, figure withholding based on a weekly payroll period. If the employee later begins to work for another employer for wages subject to withholding, the employee must notify you within 10 days. You then figure withholding based on the daily or miscellaneous period.

---

9. Withholding From Employees' Wages

Income Tax Withholding

To know how much income tax to withhold from employees' wages, you should have a **Form W-4,** Employee's Withholding Allowance Certificate, on file for each employee. Ask all new employees to give you a signed Form W-4 when they start work. Make the form effective with the first wage payment. If a new employee does not give you a completed Form W-4, withhold tax as if he or she is single, with no withholding allowances.

You may establish a system to electronically receive Form W-4 from your employees. See Regulations section 31.3402(f)(5)-1(c) for more information.

A Form W-4 remains in effect until the employee gives you a new one. If an employee gives you a Form W-4 that replaces an existing Form W-4, begin withholding no later than the start of the first payroll period ending on or after the 30th day from the date you received the replacement Form W-4. For exceptions, see **Exemption from income tax withholding, Sending certain Forms W-4 to the IRS,** and **Invalid Forms W-4** later.

The amount of income tax withholding must be based on marital status and withholding allowances. Your employees may not base their withholding amounts on a fixed dollar amount or percentage. However, the employee may specify a dollar amount to be withheld **in addition** to the amount of withholding based on filing status and withholding allowances claimed on Form W-4.

Employees may claim **fewer** withholding allowances than they are entitled to claim. They may wish to claim fewer allowances to ensure that they have enough withholding or to offset other sources of taxable income that are not subject to adequate withholding.

Note: *A Form W-4 that makes a change for the next calendar year will not take effect in the current calendar year.*

See **Pub. 505,** Tax Withholding and Estimated Tax, for detailed instructions for completing Form W-4. Along with Form W-4, you may wish to order Pub. 505 and **Pub. 919,** How Do I Adjust My Tax Withholding?

When you receive a new Form W-4, do not adjust withholding for pay periods before the effective date of the new form. Also, do not accept any withholding or estimated tax payments from your employees in addition to withholding based on their Form W-4. If they require additional withholding, they should submit a new Form W-4 and, if necessary, pay estimated tax by filing **Form 1040-ES,** Estimated Tax for Individuals.

Exemption from income tax withholding. Generally, an employee may claim exemption from income tax withholding because he or she had no income tax liability last year and expects none this year. See the Form W-4 instructions for more information. However, the wages are still subject to social security and Medicare taxes.

A Form W-4 claiming exemption from withholding is valid for only one calendar year. To continue to be exempt from withholding in the next year, an employee must file a new Form W-4 by February 15 of that year. If the employee does not give you a new Form W-4, withhold tax as if the employee is single with zero withholding allowances.

Withholding on nonresident aliens. In general, if you pay wages to nonresident aliens, you must withhold income tax (unless excepted by regulations), social security, and Medicare taxes as you would for a U.S. citizen. However, income tax withholding from the wages of nonresident aliens is subject to the special rules shown in ***Form W-4*** below. You must also give a Form W-2 to the nonresident alien and file it with the SSA. The wages are subject to FUTA tax as well. However, see **Pub. 515,** Withholding of Tax on Nonresident Aliens and Foreign Entities, and **Pub. 519,** U.S. Tax Guide for Aliens, for exceptions to these general rules.

Form W-4. When completing Form W-4, nonresident aliens are required to:

- Not claim exemption from income tax withholding.
- Request withholding as if they are single, regardless of their actual marital status.
- Claim only one allowance (if the nonresident alien is a resident of Canada, Mexico, Japan, or Korea, he or she may claim more than one allowance).
- Request an additional income tax withholding amount, depending on the payroll period, as follows:

Payroll Period	Additional Withholding
Weekly	7.60
Biweekly	15.30
Semimonthly	16.60
Monthly	33.10
Quarterly	99.40
Semiannually	198.80
Annually	397.50
Daily or Miscellaneous (each day of the payroll period)	1.50

Note: *Nonresident alien students from India are not subject to the additional income tax withholding requirement.*

Nonwage withholding. In some cases, an Internal Revenue Code section or a U.S. treaty provision will exempt payments to a nonresident alien from wages. These payments are not subject to regular income tax withholding. Form W-2 is not required in these cases. Instead, the payments are subject to withholding at a flat 30% or lower treaty rate, unless exempt from tax because of a Code or U.S. tax treaty provision.

Report these payments and any withheld tax on **Form 1042-S,** Foreign Person's U.S. Source Income Subject to Withholding. Form 1042-S is sent to the IRS with **Form 1042,** Annual Withholding Tax Return for U.S. Source Income of Foreign Persons. You may have to make deposits of the withheld income tax, using **Form 8109,** Federal Tax Deposit Coupon, or EFTPS (see page 20). See Pub. 515 and the **Instructions for Form 1042-S** for more information.

Sending certain Forms W-4 to the IRS. Generally, you must send to the IRS copies of certain Forms W-4 received during the quarter from employees still employed by you at the end of the quarter. Send copies when the employee claims (a) more than 10 withholding allowances or (b) exemption from withholding and his or her wages would normally be more than $200 per week. Send the copies to the IRS office where you file your Form 941. You are not required to send any other Forms W-4 unless the IRS notifies you in writing to do so.

Send in Forms W-4 that meet either of the above conditions each quarter with Form 941. Complete boxes 8 and 10 on any Forms W-4 you send in. You may use box 9 to identify the office responsible for processing the employee's payroll information. Also send copies of any written statements from employees in support of the claims made on Forms W-4. Send these statements even if the Forms W-4 are not in effect at the end of the quarter. You can send them to the IRS more often if you like. If you do so, include a cover letter giving your name, address, EIN, and the number of forms included. In certain cases, the IRS may notify you in writing that you must submit specified Forms W-4 more frequently, separate from your Form 941.

Note: *Please make sure that the copies of Form W-4 you send to the IRS are clear and legible.*

If your Forms 941 are filed on magnetic media, this Form W-4 information also should be filed with the IRS on magnetic media. (See **Filing Form W-4 on magnetic media** below.) Magnetic media filers of Form 941 may send paper Forms W-4 to the IRS with a cover letter if they are unable to file them on magnetic media. If you file Form 941 by TeleFile, send your paper Forms W-4 to the IRS with a cover letter.

Note: Any Form W-4 you send to the IRS without a Form 941 should be mailed to the "Return without payment" address on the back of Form 941.

Base withholding on the Forms W-4 that you send in unless the IRS notifies you in writing to do otherwise. If the IRS notifies you about a particular employee, base withholding on the number of withholding allowances shown in the IRS notice. The employee will get a similar notice directly from the IRS. If the employee later gives you a new Form W-4, follow it only if (a) exempt status is not claimed or (b) the number of withholding allowances is equal to or lower than the number in the IRS notice. Otherwise, disregard it and do not submit it to the IRS. Continue to follow the IRS notice.

If the employee prepares a new Form W-4 explaining any difference with the IRS notice, he or she may either submit it to the IRS or to you. If submitted to you, send the Form W-4 and an explanation to the IRS office shown in the notice. Continue to withhold based on the notice until the IRS tells you to follow the new Form W-4.

Filing Form W-4 on magnetic media. Form W-4 information may be filed with the IRS on magnetic media. If you wish to file on magnetic media, you must submit **Form 4419,** Application for Filing Information Returns Magnetically/Electronically, to request authorization. See **Pub. 1245,** Specifications for Filing Form W-4, Employee's Withholding Allowance Certificate, Magnetically or Electronically. To get more information about magnetic media filing, call the IRS Martinsburg Computing Center at 304-263-8700.

Invalid Forms W-4. Any unauthorized change or addition to Form W-4 makes it invalid. This includes taking out any language by which the employee certifies that the form is correct. A Form W-4 is also invalid if, by the date an employee gives it to you, he or she indicates in any way that it is false. An employee who files a false Form W-4 may be subject to a $500 penalty.

When you get an invalid Form W-4, do not use it to figure withholding. Tell the employee it is invalid and ask for another one. If the employee does not give you a valid one, withhold taxes as if the employee were single and claiming no withholding allowances. However, if you have an earlier Form W-4 for this worker that is valid, withhold as you did before.

Amounts exempt from levy on wages, salary, and other income. If you receive a Notice of Levy on Wages, Salary, and Other Income (Forms 668-W, 668-W(c), or 668-W(c)(DO), you must withhold amounts as described in the instructions for these forms. **Pub. 1494,** Table for Figuring Amount Exempt From Levy on Wages, Salary, and Other Income (Forms 668-W(c) and 668-W(c)(DO)) 2002, shows the exempt amount.

Social Security and Medicare Taxes

The Federal Insurance Contributions Act (FICA) provides for a Federal system of old-age, survivors, disability, and hospital insurance. The old-age, survivors, and disability insurance part is financed by the social security tax. The hospital insurance part is financed by the Medicare tax. Each of these taxes is reported separately.

Generally, you are required to withhold social security and Medicare taxes from your employees' wages and you must also pay a matching amount of these taxes. Certain types of wages and compensation are not subject to social security taxes (see sections 5 and 15 for details). Generally, employee wages are subject to social security and Medicare taxes regardless of the employee's age or whether he or she is receiving social security benefits. (If the employee reported tips, see section 6.)

Tax rates and the social security wage base limit. These taxes have different tax rates and only the social security tax has a wage base limit. The wage base limit is the maximum wage that is subject to the tax for the year. Determine the amount of withholding for social security and Medicare taxes by multiplying each payment by the employee tax rate. There are no withholding allowances for social security and Medicare taxes.

The employee tax rate for social security is 6.2% (amount withheld). The employer tax rate for social security is also 6.2% (12.4% total). The 2001 wage base limit was $80,400. For 2002, the wage base limit is $84,900.

The employee tax rate for Medicare is 1.45% (amount withheld). The employer tax rate for Medicare tax is also 1.45% (2.9% total). There is no wage base limit for Medicare tax; all covered wages are subject to Medicare tax.

Successor employer. If you received all or most of the property used in the trade or business of another employer, or a unit of that employer's trade or business, you may include the wages the other employer paid to your employees when you figure the annual wage base limit for social security. See Regulations section 31.3121(a)(1)-1(b) for more information. Also see Rev. Proc. 96-60, 1996-2 C.B. 399, for the procedures used in filing returns in a predecessor-successor situation.

Example: Early in 2001, you bought all the assets of a plumbing business from Mr. Martin. Mr. Brown, who had been employed by Mr. Martin and received $2,000 in wages before the date of purchase, continued to work for you. The wages you paid Mr. Brown are subject to social security taxes on the first $78,400 ($80,400 less $2,000). Medicare tax is due on all wages you pay him during the calendar year.

International social security agreements. The United States has social security agreements with many countries that eliminate dual taxation and dual coverage. Compensation subject to social security and Medicare taxes may be exempt under one of these agreements. You can get more information and a list of agreement countries from SSA at **www.ssa.gov/international** or see **Pub. 15-A,** Employer's Supplemental Tax Guide.

Part-Time Workers

For income tax withholding and social security, Medicare, and Federal unemployment (FUTA) tax purposes, there are no differences among full-time employees, part-time employees, and employees hired for short periods. It does not matter whether the worker has another job or has the maximum amount of social security tax withheld by another employer. Income tax withholding may be figured the same way as for full-time workers. Or it may be figured by the part-year employment method explained in Pub. 15-A.

10. Advance Earned Income Credit (EIC) Payment

An employee who is eligible for the earned income credit (EIC) and has a qualifying child is entitled to receive EIC payments with his or her pay during the year. To get these payments, the employee must provide to you a properly completed **Form W-5,** Earned Income Credit Advance Payment Certificate, using either the paper form or using an approved electronic format. You are required to make advance EIC payments to employees who give you a completed and signed Form W-5. You may establish a system to electronically receive Form W-5 from your employees. See Announcement 99-3 (1999-1 C.B. 324) for information on electronic requirements for Form W-5.

Certain employees who do not have a qualifying child may be able to claim the EIC on their tax return. However, they **cannot** get advance EIC payments.

For 2002, the advance payment can be as much as $1,503. The tables that begin on page 56 reflect that limit.

Form W-5. Form W-5 states the eligibility requirements for receiving advance EIC payments. On Form W-5, an employee states that he or she expects to be eligible to claim the EIC and shows whether he or she has another Form W-5 in effect with any other current employer. The employee also shows the following:

- Whether he or she has a qualifying child.
- Whether he or she will file a joint return.
- If the employee is married, whether his or her spouse has a Form W-5 in effect with any employer.

An employee may have only one certificate in effect with a current employer at one time. If an employee is married and his or her spouse also works, each spouse should file a separate Form W-5.

Length of effective period. Form W-5 is effective for the first payroll period ending on or after the date the employee gives you the form (or the first wage payment made without regard to a payroll period). It remains in effect until the end of the calendar year unless the employee revokes it or files another one. Eligible employees must file a new Form W-5 each year.

Change of status. If an employee gives you a signed Form W-5 and later becomes ineligible for advance EIC payments, he or she must revoke Form W-5 within 10 days after learning about the change of circumstances. The employee must give you a new Form W-5 stating that he or she is no longer eligible for or no longer wants advance EIC payments.

If an employee's situation changes because his or her spouse files a Form W-5, the employee must file a new Form W-5 showing that his or her spouse has a Form W-5 in effect with an employer. This will reduce the maximum amount of advance payments you can make to that employee.

If an employee's spouse has filed a Form W-5 that is no longer in effect, the employee may file a new Form W-5 with you, but is not required to do so. A new form will certify that the spouse does not have a Form W-5 in effect and will increase the maximum amount of advance payments you can make to that employee.

Invalid Form W-5. The Form W-5 is invalid if it is incomplete, unsigned, or has an alteration or unauthorized addition. The form has been altered if any of the language has been deleted. Any writing added to the form other than the requested entries is an unauthorized addition.

You should consider a Form W-5 invalid if an employee has made an oral or written statement that clearly shows the Form W-5 to be false. If you receive an invalid form, tell the employee that it is invalid as of the date he or she made the oral or written statement. For advance EIC payment purposes, the invalid Form W-5 is considered void.

You are not required to determine if a completed and signed Form W-5 is correct. However, you should contact the IRS if you have reason to believe it has any incorrect statement.

How to figure the advance EIC payment. To figure the amount of the advance EIC payment to include with the employee's pay, you must consider:

- Wages, including reported tips, for the same period. Generally, figure advance EIC payments using the amount of wages subject to income tax withholding. If an employee's wages are not subject to income tax withholding, use the amount of wages subject to withholding for social security and Medicare taxes.
- Whether the employee is married or single.
- Whether a married employee's spouse has a Form W-5 in effect with an employer.

Note: *If during the year you have paid an employee total wages of at least $29,201 ($30,201 if married filing jointly), you must stop making advance EIC payments to that employee for the rest of the year.*

Figure the amount of advance EIC to include in the employee's pay by using the tables that begin on page 56. There are separate tables for employees whose spouses have a Form W-5 in effect. See page 33 for instructions on using the advance EIC payment tables. The amount of advance EIC paid to an employee during 2002 cannot exceed $1,503.

Paying the advance EIC to employees. An advance EIC payment is not wages and is not subject to withholding of income, social security, or Medicare taxes. An advance EIC payment does not change the amount of income, social security, or Medicare taxes you withhold from the employee's wages. You add the EIC payment to the employee's net pay for the pay period. At the end of the year, you show the total advance EIC payments in box 9 on Form W-2. Do not include this amount as wages in box 1.

Employer's returns. Show the total payments you made to employees on the advance EIC line of your Form 941. Subtract this amount from your total taxes (see the separate **Instructions for Form 941**). Reduce the amounts reported on line 17 of Form 941 or on appropriate lines of **Schedule B (Form 941),** Employer's Record of Federal Tax Liability, by any advance EIC paid to employees.

Generally, employers will make the advance EIC payment from withheld income tax and employee and employer social security and Medicare taxes. These taxes are normally required to be paid over to the IRS either through Federal tax deposits or with employment tax returns. For purposes of deposit due dates, advance EIC payments are treated as deposits of these taxes on the day you pay wages (including the advance EIC payment) to your employees. The payments are treated as deposits of these taxes in the following order: (1) Income tax withholding, (2) Withheld employee social security and Medicare taxes, and (3) The employer's share of social security and Medicare taxes.

Example: You have 10 employees, each entitled to an advance EIC payment of $10. The total amount of advance EIC payments you make for the payroll period is $100. The total amount of income tax withholding for the payroll period is $90. The total employee and employer social security and Medicare taxes for the payroll period is $122.60 ($61.30 each).

You are considered to have made a deposit of $100 advance EIC payment on the day you paid wages. The $100 is treated as if you deposited the $90 total income tax withholding and $10 of the employee social security and Medicare taxes. You remain liable for depositing the remaining $112.60 of the social security and Medicare taxes ($51.30 + $61.30 = $112.60).

Advance EIC payments more than taxes due. For any payroll period, if the total advance EIC payments are more than the total payroll taxes (withheld income tax and both employee and employer shares of social security and Medicare taxes), you may choose either to:

1) Reduce each employee's advance payment proportionally so that the total advance EIC payments equal the amount of taxes due or
2) Elect to make full payment of the advance EIC and treat the excess as an advance payment of employment taxes.

Example: You have 10 employees who are each entitled to an advance EIC payment of $10. The total amount of advance EIC payable for the payroll period is $100. The total employment tax for the payroll period is $90 (including income tax withholding and social security and Medicare taxes). The advance EIC payable is $10 more than the total employment tax. The $10 excess is 10% of the advance EIC payable ($100). You may—

Page 16

1) Reduce each employee's payment by 10% (to $9 each) so the advance EIC payments equal your total employment tax ($90) or
2) Pay each employee $10, and treat the excess $10 as an advance payment of employment taxes. Attach a statement to Form 941 showing the excess advance EIC payments and the pay period(s) to which the excess applies.

U.S. territories. If you are in American Samoa, the Commonwealth of the Northern Mariana Islands, Guam, or the U.S. Virgin Islands, consult your local tax office for information on the EIC. You cannot take advance EIC payments into account on Form 941-SS.

Required Notice to Employees

You must notify employees who have no income tax withheld that they may be able to claim a tax refund because of the EIC. Although you do not have to notify employees who claim exemption from withholding on **Form W-4,** Employee's Withholding Allowance Certificate, about the EIC, you are encouraged to notify any employees whose wages for 2001 were less than $32,121 that they may be eligible to claim the credit for 2001. This is because eligible employees may get a refund of the amount of EIC that is more than the tax they owe. For example, an employee who had no tax withheld in 2001 and owes no tax, but is eligible for a $791 EIC, can file a 2001 tax return to get a $791 refund.

You will meet this notification requirement if you issue the IRS Form W-2 with the EIC notice on the back of Copy B, or a substitute Form W-2 with the same statement. You may also meet the requirement by providing **Notice 797,** Possible Federal Tax Refund Due to the Earned Income Credit (EIC), or your own statement that contains the same wording.

If a substitute Form W-2 is given on time but does not have the required statement, you must notify the employee within 1 week of the date the substitute Form W-2 is given. If Form W-2 is required but is not given on time, you must give the employee Notice 797 or your written statement by the date Form W-2 is required to be given. If Form W-2 is not required, you must notify the employee by February 7, 2002.

11. Depositing Taxes

In general, you must deposit income tax withheld and both the employer and employee social security and Medicare taxes (minus any advance EIC payments) by mailing or delivering a check, money order, or cash to a financial institution that is an authorized depositary for Federal taxes. However, some taxpayers are required to deposit using the Electronic Federal Tax Deposit System (EFTPS). See **How To Deposit** on page 20 for information on electronic deposit requirements for 2002.

Payment with return. You may make a payment with Form 941 instead of depositing if:

- You accumulate less than a $2,500 tax liability (reduced by any advance earned income credit) during the quarter (line 13 of Form 941), and you pay in full with a timely filed return. (However, if you are unsure that you will accumulate less than $2,500, deposit under the appropriate rules so that you will not be subject to failure to deposit penalties.) Or
- You are a monthly schedule depositor (defined below) and make a payment in accordance with the **Accuracy of Deposits Rule** discussed on page 19. This payment may be $2,500 or more.

Caution: *Only monthly schedule depositors are allowed to make this payment with the return.*

Separate deposit requirements for nonpayroll (Form 945) tax liabilities. Separate deposits are required for nonpayroll and payroll income tax withholding. **Do not** combine deposits for Forms 941 and 945 tax liabilities. Generally, the deposit rules for nonpayroll liabilities are the same as discussed below, except that the rules apply to an annual rather than a quarterly return period. Thus, the $2,500 threshold for the deposit requirement discussed above applies to Form 945 on an annual basis. See the separate **Instructions for Form 945** for more information.

When To Deposit

There are two deposit schedules—***monthly*** or ***semi-weekly***—for determining when you deposit social security, Medicare, and withheld income taxes. These schedules tell you when a deposit is due after a tax liability arises (e.g., when you have a payday). Prior to the beginning of each calendar year, you must determine which of the two deposit schedules you are required to use. The deposit schedule you must use is based on the total tax liability you reported on Form 941 during a four-quarter ***lookback period*** discussed below. Your deposit schedule is **not** determined by how often you pay your employees or make deposits (see **Application of Monthly and Semi-weekly Schedules** on page 19).

These rules do not apply to Federal unemployment (FUTA) tax. See section 14 for information on depositing FUTA tax.

Lookback period. Your deposit schedule for a calendar year is determined from the total taxes (not reduced by any advance EIC payments) reported on your Forms 941 (line 11) in a four-quarter lookback period. The lookback period begins July 1 and ends June 30 as shown in Table 1 below. If you reported $50,000 or less of taxes for the lookback period, you are a monthly schedule depositor; if you reported more than $50,000, you are a semiweekly schedule depositor.

Table 1. **Lookback Period for Calendar Year 2002**

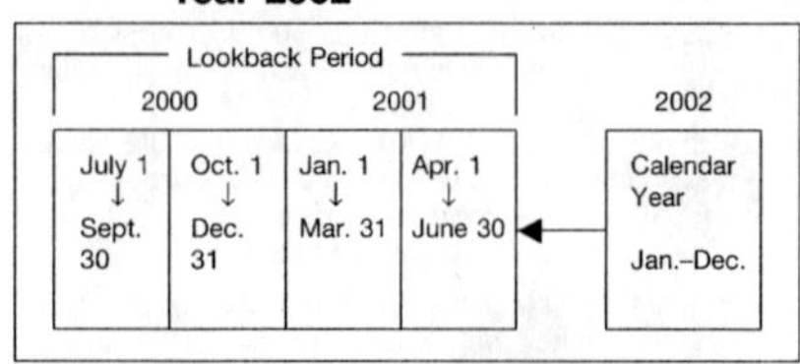

Adjustments and the lookback rule. Determine your tax liability for the four quarters in the lookback period based on the tax liability as **originally** reported on Form 941. If you made adjustments to correct errors on previously filed Forms 941, these adjustments do not affect the amount of tax liability for purposes of the lookback rule. If you report adjustments on your current Form 941 to correct errors on prior Forms 941, include these adjustments as part of your tax liability for the current quarter. If you filed Form 843 to claim a refund for a prior period overpayment, your tax liability does not change for either the prior period or the current period for purposes of the lookback rule.

Example: An employer originally reported a tax liability of $45,000 for the four quarters in the lookback period ending June 30, 2001. The employer discovered during January 2002 that the tax during one of the lookback period quarters was understated by $10,000 and corrected this error with an adjustment on the 2002 first quarter return. This employer is a monthly schedule depositor for 2002 because the lookback period tax liabilities are based on the amounts originally reported, and they were less than $50,000. The $10,000 adjustment is part of the 2002 first quarter tax liability.

Monthly Deposit Schedule

You are a monthly schedule depositor for a calendar year if the total taxes on Form 941 (line 11) for the four quarters in your lookback period were $50,000 or less. Under the monthly deposit schedule, deposit Form 941 taxes on payments made during a month by the 15th day of the following month.

Monthly schedule depositors should **not** file Form 941 on a monthly basis. Do not file **Form 941-M,** Employer's Monthly Federal Tax Return, unless you are instructed to do so by an IRS representative.

New employers. During the first calendar year of your business, your tax liability for each quarter in the lookback period is considered to be zero. Therefore, you are a monthly schedule depositor for the first calendar year of your business (but see the **$100,000 Next-Day Deposit Rule** on page 19).

Semiweekly Deposit Schedule

You are a semiweekly schedule depositor for a calendar year if the total taxes on Form 941 (line 11) during your lookback period were more than $50,000. Under the semiweekly deposit schedule, deposit Form 941 taxes on payments made on Wednesday, Thursday, and/or Friday by the following Wednesday. Deposit amounts accumulated on payments made on Saturday, Sunday, Monday, and/or Tuesday by the following Friday.

Table 2. **Semiweekly Deposit Schedule**

IF the payday falls on a . . .	**THEN deposit taxes by the following . . .**
Wednesday, Thursday, and/or Friday	Wednesday
Saturday, Sunday, Monday, and/or Tuesday	Friday

Deposit period. The term ***deposit period*** refers to the period during which tax liabilities are accumulated for each required deposit due date. For monthly schedule depositors, the deposit period is a calendar month. The deposit periods for semiweekly schedule depositors are Wednesday through Friday and Saturday through Tuesday.

Semiweekly deposit period spanning two quarters. If you have more than one pay date during a semiweekly period, and the pay dates fall in different calendar quarters, you will need to make separate deposits for the separate liabilities. For example, if you have a pay date on Saturday, March 30, 2002 (first quarter), and another pay date on Tuesday, April 2, 2002 (second quarter), two separate deposits would be required even though the pay dates fall within the same semiweekly period. Both deposits would be due Friday, April 5, 2002 (three banking days from the end of the semiweekly deposit period).

Summary of Steps To Determine Your Deposit Schedule

1. Identify your lookback period (see Table 1).
2. Add the total taxes (line 11 of Form 941) you reported during the lookback period.
3. Determine if you are a monthly or semiweekly schedule depositor:

If the total taxes you reported in the lookback period were . . .	**Then you are a . . .**
$50,000 or less	Monthly Schedule Depositor
More than $50,000	Semiweekly Schedule Depositor

Example of Monthly and Semiweekly Schedules

Rose Co. reported Form 941 taxes as follows:

2001 Lookback Period		2002 Lookback Period	
3rd Quarter 1999	$12,000	3rd Quarter 2000	$12,000
4th Quarter 1999	$12,000	4th Quarter 2000	$12,000
1st Quarter 2000	$12,000	1st Quarter 2001	$12,000
2nd Quarter 2000	$12,000	2nd Quarter 2001	$15,000
	$48,000		$51,000

Rose Co. is a monthly schedule depositor for 2001 because its tax liability for the four quarters in its lookback period (third quarter 1999 through second quarter 2000) was not more than $50,000. However, for 2002, Rose Co. is a semiweekly schedule depositor because the total taxes exceeded $50,000 for the four quarters in its lookback period (third quarter 2000 through second quarter 2001).

Deposits on Banking Days Only

If a deposit is required to be made on a day that is not a banking day, the deposit is considered timely if it is made by the close of the next banking day. In addition to Federal and state bank holidays, Saturdays and Sundays are treated as nonbanking days. For example, if a deposit is required to be made on a Friday and Friday is not a banking day, the deposit will be considered timely if it is made by the following Monday (if that Monday is a banking day).

Semiweekly schedule depositors have at least 3 banking days to make a deposit. That is, if any of the 3 weekdays after the end of a semiweekly period is a banking holiday, you will have one additional banking day to deposit. For example, if a semiweekly schedule depositor accumulated taxes for payments made on Friday and the following Monday is not a banking day, the deposit normally due on Wednesday may be made on Thursday (allowing 3 banking days to make the deposit).

Application of Monthly and Semiweekly Schedules

The terms "monthly schedule depositor" and "semiweekly schedule depositor" do **not** refer to how often your business pays its employees or even how often you are required to make deposits. The terms identify which set of deposit rules you must follow when an employment tax liability arises. The deposit rules are based on the dates wages are paid; **not** on when tax liabilities are accrued.

Monthly schedule example. Spruce Co. is a monthly schedule depositor with seasonal employees. It paid wages each Friday. During March it paid wages but did not pay any wages during April. Under the monthly deposit schedule, Spruce Co. must deposit the combined tax liabilities for the four March paydays by April 15. Spruce Co. does not have a deposit requirement for April (due by May 15) because no wages were paid and, therefore, it did not have a tax liability for April.

Semiweekly schedule example. Green Inc., which has a semiweekly deposit schedule, pays wages once each month on the last day of the month. Although Green Inc. has a semiweekly deposit schedule, it will deposit just once a month because it pays wages only once a month. The deposit, however, will be made under the semiweekly deposit schedule as follows: Green Inc.'s tax liability for the July 31, 2002 (Wednesday) payday must be deposited by August 7, 2002 (Wednesday). Under the semiweekly deposit schedule, liabilities for wages paid on Wednesday through Friday must be deposited by the following Wednesday.

$100,000 Next-Day Deposit Rule

If you accumulate a tax liability (reduced by any advance EIC payments) of $100,000 or more on any day during a ***deposit period,*** you must deposit the tax by the next banking day, whether you are a monthly or semiweekly schedule depositor.

For purposes of the $100,000 rule, do not continue accumulating tax liability after the end of a deposit period. For example, if a semiweekly schedule depositor has accumulated a liability of $95,000 on a Tuesday (of a Saturday-through-Tuesday deposit period) and accumulated a $10,000 liability on Wednesday, the $100,000 next-day deposit rule does not apply. Thus, $95,000 must be deposited by Friday and $10,000 must be deposited by the following Wednesday.

In addition, once you accumulate at least $100,000 in a deposit period, stop accumulating at the end of that day and begin to accumulate anew on the next day. For example, Fir Co. is a semiweekly schedule depositor. On Monday, Fir Co. accumulates taxes of $110,000 and must deposit this amount on Tuesday, the next banking day. On Tuesday, Fir Co. accumulates additional taxes of $30,000. Because the $30,000 is not added to the previous $110,000 and is less than $100,000, Fir Co. must deposit the $30,000 by Friday following the semiweekly deposit schedule.

If you are a monthly schedule depositor and accumulate a $100,000 tax liability on any day, you become a semiweekly schedule depositor on the next day and remain so for at least the rest of the calendar year and for the following calendar year.

Example: Elm Inc. started its business on April 1, 2002. On April 16, it paid wages for the first time and accumulated a tax liability of $40,000. On April 23, 2002, Elm Inc. paid wages and accumulated a liability of $60,000, bringing its accumulated tax liability to $100,000. Because this was the first year of its business, the tax liability for its lookback period is considered to be zero, and it would be a monthly schedule depositor based on the lookback rules. However, since Elm Inc. accumulated a $100,000 liability on April 23, it became a semiweekly schedule depositor on April 24. It will be a semiweekly schedule depositor for the remainder of 2002 and for 2003. Elm Inc. is required to deposit the $100,000 by April 24, the next banking day.

Accuracy of Deposits Rule

You are required to deposit 100% of your tax liability on or before the deposit due date. However, penalties will not be applied for depositing less than 100% if **both** of the following conditions are met:

1) Any deposit shortfall does not exceed the greater of $100 or 2% of the amount of taxes otherwise required to be deposited and
2) The deposit shortfall is paid or deposited by the shortfall makeup date as described below.

Makeup Date for Deposit Shortfall:

1) **Monthly schedule depositor.** Deposit the shortfall or pay it with your return by the due date of the Form 941 for the quarter in which the shortfall occurred. You may pay the shortfall with Form 941 even if the amount is $2,500 or more.
2) **Semiweekly schedule depositor.** Deposit by the earlier of:
   a) The first Wednesday or Friday that falls on or after the 15th of the month following the month in which the shortfall occurred or
   b) The due date of Form 941 (for the quarter of the tax liability).

For example, if a semiweekly schedule depositor has a deposit shortfall during July 2002, the shortfall makeup date is August 16, 2002 (Friday). However, if the shortfall occurred on the required October 2 (Wednesday) deposit due date for a September 25 (Wednesday) pay date, the return due date for the September 25 pay date (October 31) would come before the November 15 (Friday) shortfall makeup date. In this case, the shortfall must be deposited by October 31.

How To Deposit

The two methods of depositing employment taxes, including Form 945 taxes, are discussed below. See page 17 for exceptions explaining when taxes may be paid with the tax return instead of deposited.

Electronic deposit requirement. You must make electronic deposits of all depository taxes (such as employment tax, excise tax, and corporate income tax) using the Electronic Federal Tax Payment System (EFTPS) in 2002 if:

- The total deposits of such taxes in 2000 were more than $200,000 or
- You were required to use EFTPS in 2001.

If you are required to use EFTPS and fail to do so, you may be subject to a 10% penalty. If you are not required to use EFTPS, you may participate voluntarily. To get more information or to enroll in EFTPS, call 1-800-555-4477 or 1-800-945-8400.

Depositing on time. For deposits made by EFTPS to be on time, you must initiate the transaction at least one business day before the date the deposit is due.

Making deposits with FTD coupons. If you are not making deposits by EFTPS, use **Form 8109,** Federal Tax Deposit Coupon, to make the deposits at an authorized financial institution.

For new employers, the IRS will send you a Federal Tax Deposit (FTD) coupon book 5 to 6 weeks after you receive an employer identification number (EIN). (Apply for an EIN on Form SS-4.) The IRS will keep track of the number of FTD coupons you use and **automatically** will send you additional coupons when you need them. If you do not receive your resupply of FTD coupons, call 1-800-829-1040. You can have the FTD coupon books sent to a branch office, tax preparer, or service bureau that is making your deposits by showing that address on **Form 8109-C,** FTD Address Change, which is in the FTD coupon book. (Filing Form 8109-C will not change your address of record; it will change only the address where the FTD coupons are mailed.) The FTD coupons will be preprinted with your name, address, and EIN. They have entry boxes for indicating the type of tax and the tax period for which the deposit is made.

It is very important to clearly mark the correct type of tax and tax period on each FTD coupon. This information is used by the IRS to credit your account.

If you have branch offices depositing taxes, give them FTD coupons and complete instructions so they can deposit the taxes when due.

Please use only your FTD coupons. If you use anyone else's FTD coupon, you may be subject to the failure to deposit penalty. This is because your account will be underpaid by the amount of the deposit credited to the other person's account. See **Deposit Penalties** on page 21 for details.

How to deposit with an FTD coupon. Mail or deliver each FTD coupon and a single payment covering the taxes to be deposited to an authorized depositary. An authorized depositary is a financial institution (e.g., a commercial bank) that is authorized to accept Federal tax deposits. Follow the instructions in the FTD coupon book. Make the check or money order payable to the depositary. To help ensure proper crediting of your account, include your EIN, the type of tax (e.g., Form 941), and tax period to which the payment applies on your check or money order.

Authorized depositaries must accept cash, a postal money order drawn to the order of the depositary, or a check or draft drawn on and to the order of the depositary. You may deposit taxes with a check drawn on another financial institution only if the depositary is willing to accept that form of payment. Be sure that the financial institution where you make deposits is an authorized depositary. Deposits made at an unauthorized institution may be subject to the failure to deposit penalty.

If you prefer, you may mail your coupon and payment to Financial Agent, Federal Tax Deposit Processing, P.O. Box 970030, St. Louis, MO 63197. Make your check or money order payable to **Financial Agent.**

Depositing on time. The IRS determines whether deposits are on time by the date they are received by an authorized depositary. To be considered timely, the funds must be available to the depositary on the deposit due date before the institution's daily cutoff deadline. Contact your local depositary for information concerning check clearance and cutoff schedules. However, a deposit received by the authorized depositary after the due date will be considered timely if the taxpayer establishes that it was mailed in the United States at least 2 days before the due date.

Note: *If you are required to deposit any taxes more than once a month, any deposit of $20,000 or more must be made by its due date to be timely.*

Depositing without an EIN. If you have applied for an EIN but **have not** received it, and you must make a deposit, make the deposit with the IRS. **Do not** make the deposit at an authorized depositary. Make it payable to the "United States Treasury" and show on it your name (as shown on Form SS-4), address, kind of tax, period covered, and date you applied for an EIN. Send an explanation with the deposit. **Do not** use **Form 8109-B,** Federal Tax Deposit Coupon, in this situation.

Depositing without Form 8109. If you do not have the preprinted Form 8109, you may use Form 8109-B to make deposits. Form 8109-B is an over-the-counter FTD coupon that is not preprinted with your identifying information. You may get this form by calling 1-800-829-1040. Be sure to have your EIN ready when you call.

Use Form 8109-B to make deposits only if—

- You are a new employer and you have been assigned an EIN, but you have not received your initial supply of Forms 8109 or
- You have not received your resupply of preprinted Forms 8109.

Deposit record. For your records, a stub is provided with each FTD coupon in the coupon book. The FTD coupon itself will not be returned. It is used to credit your account. Your check, bank receipt, or money order is your receipt.

How to claim credit for overpayments. If you deposited more than the right amount of taxes for a quarter, you can choose on Form 941 for that quarter to have the overpayment refunded or applied as a credit to your next return. Do not ask the depositary or EFTPS to request a refund from the IRS for you.

Deposit Penalties

Penalties may apply if you do not make required deposits on time, make deposits for less than the required amount, or if you do not use EFTPS when required. The penalties do not apply if any failure to make a proper and timely deposit was due to reasonable cause and not to willful neglect. For amounts not properly or timely deposited, the penalty rates are:

2% - Deposits made 1 to 5 days late.
5% - Deposits made 6 to 15 days late.
10% - Deposits made 16 or more days late. Also applies to amounts paid within 10 days of the date of the first notice the IRS sent asking for the tax due.
10% - Deposits made at an unauthorized financial institution, paid directly to the IRS, or paid with your tax return (but see **Depositing without an EIN** above and **Payment with return** earlier for exceptions).
10% - Amounts subject to electronic deposit requirements but not deposited using EFTPS.
15% - Amounts still unpaid more than 10 days after the date of the first notice the IRS sent asking for the tax due or the day on which you receive notice and demand for immediate payment, whichever is earlier.

Order in which deposits are applied. Beginning in 2002, deposits generally are applied to the most recent tax liability within the quarter. For examples on how the IRS will apply deposits, see Rev. Proc. 2001-58 (2001-50 I.R.B. 579). Before 2002, deposits generally were applied first to the oldest tax liability. However, if you receive a failure-to-deposit penalty notice, you may designate how your payment is to be applied in order to minimize the amount of the penalty. Follow the instructions on the penalty notice you receive. For more information on designating deposits, see Revenue Procedure 99-10 (1999-1 C.B. 272). You can find Rev. Proc. 99-10 on page 11 of Internal Revenue Bulletin 1999-2 at **www.irs.gov.**

Example: Cedar Inc. is required to make a deposit of $1,000 on April 15 and $1,500 on May 15. It does not make the deposit on April 15. On May 15, Cedar Inc. deposits $2,000. Under the new rule, which applies deposits to the most recent tax liability, $1,500 of the deposit is applied to the May 15 deposit and the remaining $500 is applied to the April deposit. Accordingly, $500 of the April 15 liability remains undeposited. The penalty on this underdeposit will apply as explained above.

Trust fund recovery penalty. If income, social security, and Medicare taxes that must be withheld are not withheld or are not deposited or paid to the United States Treasury, the trust fund recovery penalty may apply. The penalty is the full amount of the unpaid trust fund tax. This penalty may apply to you if these unpaid taxes cannot be immediately collected from the employer or business.

The trust fund recovery penalty may be imposed on all persons who are determined by the IRS to be ***responsible*** for collecting, accounting for, and paying over these taxes, and who acted ***willfully*** in not doing so.

A ***responsible person*** can be an officer or employee of a corporation, a partner or employee of a partnership, an accountant, a volunteer director/trustee, or an employee of a sole proprietorship. A responsible person also may include one who signs checks for the business or otherwise has authority to cause the spending of business funds.

Willfully means voluntarily, consciously, and intentionally. A responsible person acts willfully if the person knows the required actions are not taking place.

Separate accounting when deposits are not made or withheld taxes are not paid. Separate accounting may be required if you do not pay over withheld employee social security, Medicare, or income taxes; deposit required taxes; make required payments; or file tax returns. In this case, you would receive written notice from the IRS requiring you to deposit taxes in a special trust account for the U.S. Government. You would also have to file monthly tax returns on **Form 941-M,** Employer's Monthly Federal Tax Return.

12. Filing Form 941

Each quarter, all employers who pay wages subject to income tax withholding (including withholding on sick pay and supplemental unemployment benefits) or social security and Medicare taxes must file **Form 941,** Employer's Quarterly Federal Tax Return. However, the following exceptions apply:

1) **Seasonal employers who no longer file for quarters when they regularly have no tax liability because they have paid no wages.** To alert the IRS that you will not have to file a return for one or more quarters during the year, mark the Seasonal employer box above line 1 on Form 941. The IRS will mail two Forms 941 to the seasonal filer once a year after March 1. The preprinted label will not include the date the quarter ended. You must enter the date the quarter ended when you file the return. Generally, the IRS will not inquire about unfiled returns if at least one taxable return is filed each year. However, you must mark the Seasonal employer box on every Form 941 you file. Otherwise, the IRS will expect a return to be filed for each quarter.
2) **Household employers reporting social security and Medicare taxes and/or withheld income tax.** If you are a sole proprietor and file Form 941 for business employees, you may include taxes for household employees on your Form 941. Otherwise, report social security and Medicare taxes and income tax withholding for household employees on **Schedule H (Form 1040),** Household Employment Taxes. See **Pub. 926,** Household Employer's Tax Guide, for more information.
3) **Employers reporting wages for employees in American Samoa, Guam, the Commonwealth of the Northern Mariana Islands, the U.S. Virgin Islands, or Puerto Rico.** If the employees are not subject to U.S. income tax withholding, use Form 941-SS. Employers in Puerto Rico use Form 941-PR.
4) **Agricultural employers reporting social security, Medicare, and withheld income taxes.** Report these on **Form 943,** Employer's Annual Tax Return for Agricultural Employees.

Form 941 e-file. The Form 941 e-file program allows a taxpayer to electronically file Form 941 using a personal computer, modem, and commercial tax preparation software. Contact the IRS at 1-800-829-1040 or visit the IRS Web Site at **www.irs.gov/elec_sys/efile-bus.html** for more information. See Pub. 1855 for technical specifications.

941TeleFile. You may be able to file Form 941 and pay any balance due by phone. If you receive TeleFile materials with your Form 941 package, check page TEL-1 of the 941TeleFile Instructions to see if you qualify for this method of filing. If you have questions related to filing Form 941 using TeleFile, call 1-800-829-8815. This phone number is for 941TeleFile information only and is not the number used to file the return.

Electronic and magnetic tape filing by reporting agents. Reporting agents filing Forms 941 for groups of taxpayers can file them electronically or on magnetic tape. See the reporting agent discussion in section 6 of **Pub. 15-A,** Employer's Supplemental Tax Guide, for more information.

Penalties. For each whole or part month a return is not filed when required (disregarding any extensions of the filing deadline), there is a penalty of 5% of the unpaid tax due with that return. The maximum penalty is 25% of the tax due. Also, for each whole or part month the tax is paid late (disregarding any extensions of the payment deadline), a penalty of 0.5% per month of the amount of tax generally applies. This penalty is 0.25% per month if an installment agreement is in effect. You must have filed your return on or before the due date of the return to qualify for the reduced penalty. The maximum for this penalty is also 25%. The penalties will not be charged if you have a reasonable cause for failing to file or pay.

Use of a reporting agent or other third-party payroll service provider does not relieve an employer of the responsibility to ensure that tax returns are filed and all taxes are paid or deposited correctly and on time.

Do not file more than one Form 941 per quarter. Employers with multiple locations or divisions must file only one Form 941 per quarter. Filing more than one return may result in processing delays and may require correspondence between you and the IRS. For information on making corrections to previously filed returns, see section 13.

Hints on filing. Do not report more than one calendar quarter on a return.

Use the preaddressed form mailed to you. If you do not have the form, get one from the IRS in time to file the return when due. If you use a form that is not preaddressed, show your name and EIN on it. Be sure they are exactly as they appeared on earlier returns. See the **Instructions for Form 941** for information on preparing the form.

Final return. If you go out of business, you must file a final return for the last quarter in which wages are paid. If you continue to pay wages or other compensation for quarters following termination of your business, you must file returns for those quarters. See the Instructions for Form 941 for details on how to file a final return.

Note: *If you are required to file a final Form 941, you are also required to furnish Form W-2 to your employees by the due date of the final Form 941. File Forms W-2 and W-3 with the SSA by the last day of the month that follows the due date of your final Form 941. See the* ***Instructions for Forms W-2 and W-3*** *for more information.*

Filing late Forms 941 for prior years. If you are filing an original return for a quarter in a prior year and you are using the current year form, you will have to modify Form 941. A form for a particular year generally can be used without modification for any quarter within that year. For example, a form with any 2002 revision date (e.g., January or October 2002) generally can be used without modification for any quarter of 2002.

In all cases, however, be sure to correctly fill out the "Date quarter ended" section at the top of the form. If you are modifying a form with preprinted information, change the date (the date is shown with the month and year the quarter ends; for example, JUN02 would be for the quarter ending June 30, 2002). Cross out any inapplicable tax rate(s) shown on the form and write in the rate from Table 3 below. You can get tax rates and wage base limits for years not shown in the table from the IRS.

The instructions on the form may be inappropriate for the year for which you are reporting taxes because of changes in the law, regulations, or procedures. The revision date (found under the form number at the top of the form) will tell you the year for which the form was developed. Contact the IRS if you have any questions.

Table 3. **Social Security and Medicare Tax Rates** *(For 3 prior years)*

Calendar Year	Wage Base Limit (each employee)	Tax Rate on Taxable Wages and Tips
2001–Social Security	$80,400	12.4%
2001–Medicare	All Wages	2.9%
2000–Social Security	$76,200	12.4%
2000–Medicare	All Wages	2.9%
1999–Social Security	$72,600	12.4%
1999–Medicare	All Wages	2.9%

Reconciling Forms W-2, W-3, and 941. When there are discrepancies between Forms 941 filed with the IRS and Forms W-2 and W-3 filed with the SSA, we must contact you to resolve the discrepancies.

To help reduce discrepancies—

1) Report bonuses as wages and as social security and Medicare wages on Forms W-2 and 941.
2) Report both social security and Medicare wages and taxes separately on Forms W-2, W-3, and 941.
3) Report social security taxes on Form W-2 in the box for social security tax withheld, not as social security wages.
4) Report Medicare taxes on Form W-2 in the box for Medicare tax withheld, not as Medicare wages.
5) Make sure the social security wage amount for each employee does not exceed the annual social security wage base limit.
6) Do not report noncash wages that are not subject to social security or Medicare taxes as social security or Medicare wages.
7) If you used an EIN on any Form 941 for the year that is different from the EIN reported on Form W-3, enter the other EIN on Form W-3 in the box for "Other EIN used this year."

To reduce the discrepancies between amounts reported on Forms W-2, W-3, and 941—

1) Be sure the amounts on Form W-3 are the total amounts from Forms W-2.
2) Reconcile Form W-3 with your four quarterly Forms 941 by comparing amounts reported for—
   a) Income tax withholding.
   b) Social security wages, social security tips, and Medicare wages and tips. Form W-3 should include Form 941 adjustments only for the current year (i.e., if the Form 941 adjustments include amounts for a prior year, do not report those prior year adjustments on the current-year Forms W-2 and W-3).
   c) Social security and Medicare taxes. The amounts shown on the four quarterly Forms 941, including current-year adjustments, should be approximately twice the amounts shown on Form W-3. This is because Form 941 includes both the employer and employee shares of social security and Medicare taxes.
   d) Advance earned income credit.

Do not report on Form 941 backup withholding or income tax withholding on nonpayroll payments such as pensions, annuities, and gambling winnings. Nonpayroll withholding must be reported on Form 945 (see the separate **Instructions for Form 945** for details). Income tax withholding required to be reported on Forms 1099 or W-2G must be reported on Form 945. Only taxes and withholding properly reported on Form W-2 should be reported on Form 941.

Amounts reported on Forms W-2, W-3, and 941 may not match for valid reasons. If they do not match, you should determine that the reasons are valid. Keep your reconciliation so you will have a record of why amounts did not match in case there are inquiries from the IRS or the SSA.

13. Reporting Adjustments on Form 941

There are two types of adjustments: current period adjustments and prior period adjustments to correct errors. See the instructions for Forms 941 and 941c for more information on how to report these adjustments.

Current Period Adjustments

In certain cases, amounts reported as social security and Medicare taxes on lines 6b, 6d, and 7b of Form 941 must be adjusted to arrive at your correct tax liability (e.g., excluding amounts withheld by a third-party payer or amounts you were not required to withhold). Current period

Page 23

adjustments are reported on line 9 of Form 941 and include the following:

Adjustment of tax on tips. If, by the 10th of the month after the month you received an employee's report on tips, you do not have enough employee funds available to withhold the employee's share of social security and Medicare taxes, you no longer have to collect it. Report the entire amount of these tips on lines 6c (social security tips) and 7a (Medicare wages and tips). Include as an adjustment in the "Other" space on line 9 the total uncollected employee share of the social security and Medicare taxes.

Adjustment of tax on group-term life insurance premiums paid for former employees. The employee share of social security and Medicare taxes on group-term life insurance over $50,000 **for a former employee** is paid by the former employee with his or her tax return and is not collected by the employer. However, include all social security and Medicare taxes for such coverage on lines 6b and 7b (social security and Medicare taxes), and back out the amount of the employee share of these taxes as an adjustment in the "Other" space on line 9. See Pub. 15-A for more information on group-term life insurance.

Note: *For the above adjustments, provide a brief supporting statement explaining the nature and amount of the adjustments (see the example of reporting current period adjustments below). Do not use Form 941c as the supporting statement for current period adjustments.*

Adjustment of tax on third-party sick pay. Report both the employer and employee shares of social security and Medicare taxes for sick pay on lines 6b and 7b of Form 941. Deduct on line 9 the social security and Medicare taxes withheld on sick pay by a third-party payer. Also enter the sick pay tax adjustment in the "Sick Pay" adjustment entry space. No additional statement for this adjustment is required. See section 7 of Pub. 15-B for more information.

Fractions of cents adjustment. If there is a small difference between net taxes (line 13) and total deposits (line 14), it may have been caused, all or in part, by rounding to the nearest cent each time you computed payroll. This rounding occurs when you figure the amount of social security and Medicare tax to be withheld from each employee's wages. If you pay your taxes with Form 941 instead of making deposits because your total taxes for the quarter are less than $2,500, you also may report a fractions-of-cents adjustment.

To determine if you have a fractions-of-cents adjustment, multiply the total wages and tips for the quarter subject to:

- Social security tax (reported on lines 6a and 6c) by 6.2% (.062).
- Medicare tax (reported on line 7a) by 1.45% (.0145).

Compare these amounts (the employee share of social security and Medicare taxes) with the total social security and Medicare taxes actually withheld from employees for the quarter (from your payroll records). The difference, positive or negative, is your fractions-of-cents adjustment. If the actual amount withheld is less, report a negative adjustment in parentheses in the entry space for "Fractions of cents." If the actual amount is more, report a positive adjustment. No supporting statement is required for this adjustment.

Example of reporting current period adjustments. Cedar Inc. was entitled to the following current period adjustments:

- **Third-party sick pay.** Cedar Inc. included taxes of $2,000 for sick pay on lines 6b and 7b for social security and Medicare taxes. However, the third-party payer of the sick pay withheld and paid the employee share ($1,000) of these taxes. Cedar Inc. is entitled to a $1,000 sick pay adjustment (negative).
- **Fractions of cents.** Cedar Inc. determined that the amounts withheld and deposited for social security and Medicare taxes during the quarter were a net $1.44 more than the employee share of the amount figured on lines 6b, 6d, and 7b (social security and Medicare taxes). This difference was caused by adding or dropping fractions of cents when figuring social security and Medicare taxes for each wage payment. It must report a positive $1.44 fractions-of-cents adjustment.
- **Life insurance premiums.** Cedar Inc. paid group-term life insurance premiums for policies in excess of $50,000 for former employees. The former employees must pay the employee share of the social security and Medicare taxes ($200) on the policies. However, Cedar Inc. must include the employee share of these taxes with the social security and Medicare taxes reported on lines 6b and 7b

Current Period Adjustment Example

7	Taxable Medicare wages and tips . . . . 7a $ × 2.9% (.029) =	7b		
8	Total social security and Medicare taxes (add lines 6b, 6d, and 7b). Check here if wages are not subject to social security and/or Medicare tax . . . . ▶ ☐	8		
9	Adjustment of social security and Medicare taxes (see instructions for required explanation) Sick Pay $ (1000.00) ± Fractions of Cents $ 1.44 ± Other $ (200.00) =	9	(1198	56)
10	Adjusted total of social security and Medicare taxes (line 8 as adjusted by line 9—see instructions).	10		
11	**Total taxes** (add lines 5 and 10)	11		

Page 24

of Form 941. It is entitled to a negative $200 adjustment.

Cedar Inc. reported these adjustments on line 9 of Form 941 as shown in the **Current Period Adjustment Example** on page 24. A brief supporting statement was filed with Form 941 explaining the life insurance adjustment.

Prior Period Adjustments

Generally, you can correct errors on prior quarter Forms 941 by making an adjustment on the Form 941 for the quarter during which the error was discovered. For example, if you made an error in reporting social security tax on your second quarter 2001 Form 941 and discovered the error during January 2002, correct the error by making an adjustment on your first quarter 2002 Form 941.

The adjustment increases or decreases your tax liability for the quarter in which it is reported (the quarter the error is discovered) and is interest free. The net adjustments reported on Form 941 may include any number of corrections for one or more previous quarters, including both overpayments and underpayments.

You are required to provide background information and certifications supporting prior quarter adjustments. File with Form 941 a **Form 941c,** Supporting Statement To Correct Information, or attach a statement that shows:

- What the error was.
- Quarter in which the error was made.
- The amount of the error for each quarter.
- Date on which you found the error.
- That you repaid the employee tax or received from each affected employee a written consent to this refund or credit, if the entry corrects an overcollection.
- If the entry corrects social security and Medicare taxes overcollected in an earlier year, that you received from the employee a written statement that he or she will not claim a refund or credit for the amount.

Do not file Form 941c or the equivalent supporting statement separately. The IRS will not be able to process your adjustments on Form 941 without this supporting information. See the instructions for Form 941c for more information.

Income tax withholding adjustments. Correct prior quarter income tax withholding errors by making an adjustment on line 4 of Form 941 for the quarter during which you discovered the error.

Note: *You may make an adjustment to correct income tax withholding errors only for quarters during the* ***same calendar year.*** *This is because the employee uses the amount shown on Form W-2 as a credit when filing the income tax return (Form 1040, etc.).*

You cannot adjust amounts reported as income tax withheld in a prior calendar year unless it is to correct an **administrative error.** An administrative error occurs if the amount you entered on Form 941 is not the amount you actually withheld. For example, if the total income tax actually withheld was incorrectly reported on Form 941 due to a mathematical or transposition error, this would be an administrative error. The administrative error adjustment corrects the amount reported on Form 941 to agree with the amount actually withheld from employees.

Social security and Medicare tax adjustments. Correct prior quarter social security and Medicare tax errors by making an adjustment on line 9 of Form 941 for the quarter during which you discovered the error. You may report adjustments on the current quarter Form 941 for previous quarters in the current and prior years.

Reporting prior quarter adjustments on the record of Federal tax liability. Adjustments to correct errors in prior quarters must be taken into account on either Form 941, line 17, Monthly Summary of Federal Tax Liability, or on **Schedule B (Form 941),** Employer's Record of Federal Tax Liability.

If the adjustment corrects an **underreported liability** in a prior quarter, report the adjustment on the entry space corresponding to the date the error was discovered. If the adjustment corrects an **overreported liability,** use the adjustment amount as a credit to offset subsequent liabilities until it is used up.

Example of reporting prior period adjustments: Elm Co., a monthly schedule depositor, discovered on January 9, 2002, that it overreported social security tax on a prior quarter return by $5,000. Its total tax liabilities for the first quarter of 2002 were: January—$4,500, February—$4,500, and March—$4,500. Elm Co. completed line 17 of Form 941 as shown in the **Prior Period Adjustment Example** on page 26.

The adjustment for the $5,000 overreported liability offset the January liability, so the $4,500 liability was not deposited and a -0- liability was reported on line 17, column (a). The remaining $500 of the $5,000 adjustment credit was used to partially offset the liability for February, so only $4,000 of the $4,500 liability was deposited and reported on line 17, column (b).

Note: *Do not make any changes to the record of Federal tax liability for current quarter adjustments. The amounts reported on the record reflect the* ***actual*** *amounts you withheld from employees' wages for social security and Medicare taxes. Because the current quarter adjustments make the amounts reported on lines 6b, 6d, and 7b of Form 941 equal the actual amounts you withheld (the amounts reported on the record), no additional changes to the record of Federal tax liability are necessary for these adjustments.*

Filing a claim for overreported prior period liabilities. If you discover an error on a prior quarter return resulting in a tax overpayment, you may file **Form 843,** Claim for Refund and Request for Abatement, for a refund. This form also can be used to request an abatement of an overassessment of employment taxes, interest, and/or penalties. You must file Form 941c, or an equivalent statement, with Form 843. See the separate **Instructions for Form 843.**

Collecting underwithheld taxes from employees. If you withheld no income, social security, or Medicare taxes

Prior Period Adjustment Example

- **All filers:** If line 13 is less than $1,000, you need not complete line 17 or Schedule B (Form 941).
- **Semiweekly schedule depositors:** Complete Schedule B (Form 941) and check here . . . . ▶ ☐
- **Monthly schedule depositors:** Complete line 17, columns (a) through (d), and check here. . . . . ▶ ☒

17 **Monthly Summary of Federal Tax Liability.** Do not complete if you were a semiweekly schedule depositor.			
(a) First month liability	**(b)** Second month liability	**(c)** Third month liability	**(d)** Total liability for quarter
-0-	4000.00	4500.00	8500.00

Sign Here Under penalties of perjury, I declare that I have examined this return, including accompanying schedules and statements, and to the best of my knowledge and belief, it is true, correct, and complete.

Signature ▶ **Print Your Name and Title ▶** **Date ▶**

For Privacy Act and Paperwork Reduction Act Notice, see back of Payment Voucher. Cat. No. 17001Z Form **941** (Rev. 10-2000)

or less than the right amount from an employee's wages, you can make it up from later pay to that employee. But you are the one who owes the underpayment. Reimbursement is a matter for settlement between you and the employee. Underwithheld income tax must be recovered from the employee *on or before the last day of the calendar year.* There are special rules for tax on tips (see section 6) and fringe benefits (see section 5).

Refunding amounts incorrectly withheld from employees. If you withheld more than the right amount of income, social security, or Medicare taxes from wages paid, give the employee the excess. Any excess income tax withholding must be reimbursed to the employee prior to the end of the calendar year. Keep in your records the employee's written receipt showing the date and amount of the repayment. If you do not have a receipt, you must report and pay each excess amount when you file Form 941 for the quarter in which you withheld too much tax.

Correcting filed Forms W-2 and W-3. When adjustments are made to correct social security and Medicare taxes because of a change in the wage totals reported for a previous year, you also may need to file **Form W-2c,** Corrected Wage and Tax Statement, and **Form W-3c,** Transmittal of Corrected Wage and Tax Statements.

Wage Repayments

If an employee repays you for wages received in error, do not offset the repayments against current-year wages unless the repayments are for amounts received in error in the current year.

Repayment of current-year wages. If you receive repayments for wages paid during a prior quarter in the current year, report adjustments on Form 941 to recover income tax withholding and social security and Medicare taxes for the repaid wages (as discussed earlier). Report the adjustments on Form 941 for the quarter during which the repayment occurred.

Repayment of prior year wages. If you receive repayments for wages paid during a prior year, report an adjustment on the Form 941 for the quarter during which the repayment was made to recover the social security and Medicare taxes. Instead of making an adjustment on Form 941, you may file a claim for these taxes using Form 843. You may not make an adjustment for income tax withholding because the wages were paid during a prior year.

You also must file Forms W-2c and W-3c with the SSA to correct social security and Medicare wages and taxes. **Do not** correct wages (box 1) on Form W-2c for the amount paid in error. Give a copy of Form W-2c to the employee.

Note: *The wages paid in error in the prior year remain taxable to the employee for that year. This is because the employee received and had use of those funds during that year. The employee is not entitled to file an amended return (Form 1040X) to recover the income tax on these wages. Instead, the employee is entitled to a deduction (or credit in some cases) for the repaid wages on his or her income tax return for the year of repayment.*

14. Federal Unemployment (FUTA) Tax

The Federal Unemployment Tax Act (FUTA), with state unemployment systems, provides for payments of unemployment compensation to workers who have lost their jobs. Most employers pay both a Federal and a state unemployment tax. A list of state unemployment tax agencies, including addresses and phone numbers, is available in **Pub. 926,** Household Employer's Tax Guide. Only the employer pays FUTA tax; it is not deducted from the employee's wages. For more information, see the **Instructions for Form 940.**

Note: *Services rendered after December 20, 2000, to a federally recognized Indian tribal government (or any subdivision, subsidiary, or business wholly owned by such an Indian tribe) are exempt from FUTA tax, subject to the tribe's compliance with state law. For more information, see Announcement 2001-16 and Code section 3309(d). You can find Announcement 2001-16 on page 715 of Internal Revenue Bulletin 2001-8, at* ***www.irs.gov.***

Use the following three tests to determine whether you must pay FUTA tax. Each test applies to a different category of employee, and each is independent of the others. If a test describes your situation, you are subject to FUTA tax on the wages you pay to employees in that category during the current calendar year.

1) General test.

You are subject to FUTA tax in 2002 on the wages you pay employees who are not farmworkers or

Page 26

household workers if in the current or preceding calendar year:

a) You paid wages of $1,500 or more in any calendar quarter in 2001 or 2002 or

b) You had one or more employees for at least some part of a day in any 20 or more different weeks in 2001 or 20 or more different weeks in 2002.

2) Household employees test.
You are subject to FUTA tax only if you paid total cash wages of $1,000 or more (for all household employees) in any calendar quarter in 2001 or 2002. A household worker is an employee who performs household work in a private home, local college club, or local fraternity or sorority chapter.

3) Farmworkers test.
You are subject to FUTA tax on the wages you pay to farmworkers if:

a) You paid cash wages of $20,000 or more to farmworkers during any calendar quarter in 2001 or 2002 or

b) You employed 10 or more farmworkers during at least some part of a day (whether or not at the same time) during any 20 or more different weeks in 2001 or 20 or more different weeks in 2002.

Computing FUTA tax. For 2001 and 2002, the FUTA tax rate is 6.2%. The tax applies to the first $7,000 you pay each employee as wages during the year. The $7,000 is the Federal wage base. Your state wage base may be different. Generally, you can take a credit against your FUTA tax for amounts you paid into state unemployment funds. This credit cannot be more than 5.4% of taxable wages. If you are entitled to the maximum 5.4% credit, the FUTA tax rate after the credit is 0.8%.

Successor employer. If you acquired a business from an employer who was liable for FUTA tax, you may be able to count the wages that employer paid to the employees who continue to work for you when you figure the $7,000 FUTA wage base. See the Instructions for Form 940.

Depositing FUTA tax. For deposit purposes, figure FUTA tax quarterly. Determine your FUTA tax liability by multiplying the amount of wages paid during the quarter by .008 (0.8%). Stop depositing FUTA tax on an employee's wages when he or she reaches $7,000 in wages for the calendar year. If any part of the wages subject to FUTA are exempt from state unemployment tax, you may have to deposit more than the tax using the 0.8% rate. For example, in certain states, wages paid to corporate officers, certain payments of sick pay by unions, and certain fringe benefits, are exempt from state unemployment tax.

If your FUTA tax liability for a quarter is $100 or less, you do not have to deposit the tax. Instead, you may carry it forward and add it to the liability figured in the next quarter to see if you must make a deposit. If your FUTA tax liability for any calendar quarter in 2002 is over $100 (including any FUTA tax carried forward from an earlier quarter), you must deposit the tax by electronic funds transfer (EFTPS) or in an authorized financial institution using **Form 8109,** Federal Tax Deposit Coupon. See section 11 for information on these two deposit methods.

Note: *You are not required to deposit FUTA taxes for household employees unless you report their wages on Form 941 or 943. See* ***Pub. 926,*** *Household Employer's Tax Guide, for more information.*

When to deposit. Deposit the FUTA tax by the last day of the first month after the quarter ends.

If your liability for the fourth quarter (plus any undeposited amount from any earlier quarter) is over $100, deposit the entire amount by the due date of Form 940 or Form 940-EZ (January 31). If it is $100 or less, you can either make a deposit or pay the tax with your Form 940 or 940-EZ by January 31.

Table 4. **When To Deposit FUTA Taxes**

Quarter	Ending	Due Date
Jan.–Feb.–Mar.	Mar. 31	Apr. 30
Apr.–May–June	June 30	July 31
July–Aug.–Sept.	Sept. 30	Oct. 31
Oct.–Nov.–Dec.	Dec. 31	Jan. 31

Reporting FUTA tax. Use **Form 940** or **940-EZ,** Employer's Annual Federal Unemployment (FUTA) Tax Return, to report this tax. The IRS will mail a preaddressed Form 940 or 940-EZ to you if you filed a return the year before. If you do not receive Form 940 or 940-EZ, you can get the form by calling 1-800-TAX-FORM (1-800-829-3676).

Form 940-EZ requirements. You may be able to use Form 940-EZ instead of Form 940 if (1) you paid unemployment taxes ("contributions") to only one state, (2) you paid state unemployment taxes by the due date of Form 940 or 940-EZ, and (3) all wages that were taxable for FUTA tax purposes were also taxable for your state's unemployment tax. For example, if you paid wages to corporate officers (these wages are subject to FUTA tax) in a state that exempts these wages from its unemployment taxes, you cannot use Form 940-EZ.

Household employees. If you did not report employment taxes for household employees on Form 941 or 943, report FUTA tax for these employees on **Schedule H (Form 1040),** Household Employment Taxes. See Pub. 926 for more information.

15. Special Rules for Various Types of Services and Payments

(Section references are to the Internal Revenue Code unless otherwise noted.)

Special Classes of Employment and Special Types of Payments	Treatment Under Employment Taxes		
	Income Tax Withholding	**Social Security and Medicare**	**Federal Unemployment**
Aliens, nonresident.	See page 13 and **Pub. 515,** Withholding of Tax on Nonresident Aliens and Foreign Corporations, and **Pub. 519,** U.S. Tax Guide for Aliens.		
Aliens, resident:			
1. Service performed in the U.S.	Same as U.S. citizen.	Same as U.S. citizen. (Exempt if any part of service as crew member of foreign vessel or aircraft is performed outside U.S.)	Same as U.S. citizen.
2. Service performed outside U.S.	Withhold	Taxable if (1) working for an American employer or (2) an American employer by agreement covers U.S. citizens and residents employed by its foreign affiliates.	Exempt unless on or in connection with an American vessel or aircraft and either performed under contract made in U.S., or alien is employed on such vessel or aircraft when it touches U.S. port.
Cafeteria plan benefits under section 125.	If employee chooses cash, subject to all employment taxes. If employee chooses another benefit, the treatment is the same as if the benefit were provided outside the plan. (See **Pub. 15-A** for more information.)		
Deceased worker:			
1. Wages paid to beneficiary or estate in same calendar year as worker's death. (See **Instructions for Form W-2** for details.)	Exempt	Taxable	Taxable
2. Wages paid to beneficiary or estate after calendar year of worker's death.	Exempt	Exempt	Exempt
Dependent care assistance programs (limited to $5,000; $2,500 if married filing separately).	Exempt to the extent it is reasonable to believe that amounts are excludable from gross income under section 129.		
Disabled worker's wages paid after year in which worker became entitled to disability insurance benefits under the Social Security Act.	Withhold	Exempt, if worker did not perform any service for employer during period for which payment is made.	Taxable
Employee business expense reimbursement:			
a. Accountable plan.			
1. Amounts not exceeding specified government rate for per diem or standard mileage.	Exempt	Exempt	Exempt
2. Amounts in excess of specified government rate for per diem or standard mileage.	Withhold	Taxable	Taxable
b. Nonaccountable plan. (See page 9 for details.)	Withhold	Taxable	Taxable
Family employees:			
1. Child employed by parent (or partnership in which each partner is a parent of the child).	Withhold	Exempt until age 18; age 21 for domestic service	Exempt until age 21
2. Parent employed by child.	Withhold	Taxable if in course of the son's or daughter's business. For domestic services, see section 3.	Exempt
3. Spouse employed by spouse. (See section 3 for more information.)	Withhold	Taxable if in course of spouse's business.	Exempt
Fishing and related activities.	See **Pub. 595,** Tax Highlights for Commercial Fishermen.		
Foreign governments and international organizations.	Exempt	Exempt	Exempt

Page 28

Special Classes of Employment and Special Types of Payments	Treatment Under Employment Taxes		
	Income Tax Withholding	Social Security and Medicare	Federal Unemployment
Foreign service by U.S. citizens:			
1. As U.S. government employee.	Withhold	Same as within U.S.	Exempt
2. For foreign affiliates of American employers and other private employers.	Exempt if at time of payment (1) it is reasonable to believe employee is entitled to exclusion from income under section 911 or (2) the employer is required by law of the foreign country to withhold income tax on such payment.	Exempt unless (1) an American employer by agreement covers U.S. citizens employed by its foreign affiliates or (2) U.S. citizen works for American employer.	Exempt unless (1) on American vessel or aircraft and work is performed under contract made in U.S. or worker is employed on vessel when it touches U.S. port or (2) U.S. citizen works for American employer (except in a contiguous country with which the U.S. has an agreement for unemployment compensation) or in the U.S. Virgin Islands.
Homeworkers (industrial, cottage industry):			
1. Common law employees.	Withhold	Taxable	Taxable
2. Statutory employees. (See page 7 for details.)	Exempt	Taxable if paid $100 or more in cash in a year.	Exempt
Hospital employees:			
1. Interns	Withhold	Taxable	Exempt
2. Patients	Withhold	Taxable (Exempt for state or local government hospitals.)	Exempt
Household employees:			
1. Domestic service in private homes. (Farmers see Circular A.)	Exempt (withhold if both employer and employee agree).	Taxable if paid $1,300 or more in cash in 2002. Exempt if performed by an individual under age 18 during any portion of the calendar year and is not the principal occupation of the employee.	Taxable if employer paid total cash wages of $1,000 or more (for **all** household employees) in any quarter in the current or preceding calendar year.
2. Domestic service in college clubs, fraternities, and sororities.	Exempt (withhold if both employer and employee agree).	Exempt if paid to regular student; also exempt if employee is paid less than $100 in a year by an income-tax-exempt employer.	Taxable if employer paid total cash wages of $1,000 or more (for **all** household employees) in any quarter in the current or preceding calendar year.
Insurance for employees:			
1. Accident and health insurance premiums under a plan or system for employees and their dependents generally or for a class or classes of employees and their dependents.	Exempt (except 2% shareholder-employees of S corporations).	Exempt	Exempt
2. Group-term life insurance costs. (See **Pub. 15-A** for more details.)	Exempt	Exempt, except for the cost of group-term life insurance that is includible in the employee's gross income. (Special rules apply for former employees.)	Exempt
Insurance agents or solicitors:			
1. Full-time life insurance salesperson.	Withhold only if employee under common law. (See page 6.)	Taxable	Taxable if (1) employee under common law and (2) not paid solely by commissions.
2. Other salesperson of life, casualty, etc., insurance.	Withhold only if employee under common law.	Taxable only if employee under common law.	Taxable if (1) employee under common law and (2) not paid solely by commissions.

Special Classes of Employment and Special Types of Payments	Treatment Under Employment Taxes		
	Income Tax Withholding	**Social Security and Medicare**	**Federal Unemployment**
Interest on loans with below-market interest rates (foregone interest and deemed original issue discount). (See **Pub. 15-A** for more information.)	Exempt (but deemed payments of compensation-related loans must be shown on Form W-2).	Exempt, unless loans are compensation related.	Exempt, unless loans are compensation related.
Leave-sharing plans: Amounts paid to an employee under a leave-sharing plan.	Withhold	Taxable	Taxable
Newspaper carriers and vendors: Newspaper carriers under age 18; newspaper and magazine vendors buying at fixed prices and retaining receipts from sales to customers. See Pub 15-A for information on statutory nonemployee status.	Exempt (withhold if both employer and employee voluntarily agree).	Exempt	Exempt
Noncash payments: 1. For household work, agricultural labor, and service not in the course of the employer's trade or business.	Exempt (withhold if both employer and employee voluntarily agree).	Exempt	Exempt
2. To certain retail commission salespersons ordinarily paid solely on a cash commission basis.	Optional with employer.	Taxable	Taxable
Nonprofit organizations.	See **Pub. 15-A.**		
Partners: Payments to members of general partnership.	Exempt	Exempt	Exempt
Railroads: Payments subject to the Railroad Retirement Act	Withhold	Exempt	Exempt
Religious exemptions.	See **Pub. 15-A.**		
Retirement and pension plans: 1. Employer contributions to a qualified plan.	Exempt	Exempt	Exempt
2. Elective employee contributions and deferrals to a plan containing a qualified cash or deferred compensation arrangement (e.g., 401(k)).	Generally exempt, but see section 402(g) for limitation.	Taxable	Taxable
3. Employer contributions to individual retirement accounts under simplified employee pension plan (SEP).	Generally exempt, but see section 402(g) for salary reduction SEP limitation.	Exempt, except for amounts contributed under a salary reduction SEP agreement.	
4. Employer contributions to section 403(b) annuities.	Generally exempt, but see section 402(g) for limitation.	Taxable if paid through a salary reduction agreement (written or otherwise).	
5. Employee salary reduction contributions to a SIMPLE retirement account.	Exempt	Taxable	Taxable
6. Distributions from qualified retirement and pension plans and section 403(b) annuities. (See **Pub. 15-A** for information on pensions, annuities, and employer contributions to nonqualified deferred compensation arrangements.)	Withhold, but recipient may elect exemption on Form W-4P in certain cases; mandatory 20% withholding applies to an eligible rollover distribution that is not a direct rollover; exempt for direct rollover. (See Pub. 15-A.)	Exempt	Exempt
Salespersons: 1. Common law employees.	Withhold	Taxable	Taxable
2. Statutory employees.	Exempt	Taxable	Taxable, except for full-time life insurance sales agents.
3. Statutory nonemployees (qualified real estate agents and direct sellers). (See page 7 for details.)	Exempt	Exempt	Exempt
Scholarships and fellowship grants: (includible in income under section 117(c)).	Withhold	Taxability depends on the nature of the employment and the status of the organization. See **Students** on page 31.	

Special Classes of Employment and Special Types of Payments	Treatment Under Employment Taxes		
	Income Tax Withholding	Social Security and Medicare	Federal Unemployment
Severance or dismissal pay.	Withhold	Taxable	Taxable
Service not in the course of the employer's trade or business, other than on a farm operated for profit or for household employment in private homes.	Withhold only if employee earns $50 or more in cash in a quarter and works on 24 or more different days in that quarter or in the preceding quarter.	Taxable if employee receives $100 or more in a calendar year.	Taxable only if employee earns $50 or more in cash in a quarter and works on 24 or more different days in that quarter or in the preceding quarter.
Sick pay. (See **Pub. 15-A** for more information.)	Withhold	Exempt after end of 6 calendar months after the calendar month employee last worked for employer.	
State governments and political subdivisions, employees of:			
1. Fees of public official.	Exempt	Taxable if certain transportation services or if covered by a section 218 (Social Security Act) agreement.	Exempt
2. Salaries and wages.	Withhold	Taxable (1) for services performed by employees who are either (a) covered under a section 218 agreement or (b) not a member of a public retirement system, and (2) (for Medicare tax only) for employees hired after 3/31/86 who are members of a public retirement system not covered by a section 218 social security agreement.	Exempt
3. Election workers.	Exempt	Taxable if paid $1,100 or more in 2001 (lesser amount if specified by a section 218 social security agreement); file Form W-2 for $600 or more.	Exempt
Students, scholars, trainees, teachers, etc.:			
1. Student enrolled and regularly attending classes, performing services for:			
a. Private school, college, or university	Withhold	Exempt	Exempt
b. Auxiliary nonprofit organization operated for and controlled by school, college, or university.	Withhold	Exempt unless services are covered by a section 218 (Social Security Act) agreement	Exempt
c. Public school, college, or university	Withhold	Exempt unless services are covered by a section 218 (Social Security Act) agreement	Exempt
2. Full-time student performing service for academic credit, combining instruction with work experience as an integral part of the program.	Withhold	Taxable	Exempt unless program was established for or on behalf of an employer or group of employers.
3. Student nurse performing part-time services for nominal earnings at hospital as incidental part of training.	Withhold	Exempt	Exempt
4. Student employed by organized camps.	Withhold	Taxable	Exempt
5. Student, scholar, trainee, teacher, etc., as nonimmigrant alien under section 101(a)(15)(F), (J), (M), or (Q) of Immigration and Nationality Act (i.e., aliens holding F-1, J-1, M-1, or Q-1 visas).	Withhold unless excepted by regulations.	Exempt if service is performed for purpose specified in section 101(a)(15)(F), (J), (M), or (Q) of Immigration and Nationality Act. However, these taxes may apply if the employee becomes a resident alien.	
Supplemental unemployment compensation plan benefits.	Withhold	Exempt	Exempt
Tips:			
1. If $20 or more in a month.	Withhold	Taxable	Taxable for all tips reported in writing to employer.
2. If less than $20 in a month. (See section 6 for more information.)	Exempt	Exempt	Exempt
Worker's compensation.	Exempt	Exempt	Exempt

Page 31

16. How To Use the Income Tax Withholding and Advance Earned Income Credit (EIC) Payment Tables

Income Tax Withholding

There are several ways to figure income tax withholding. The following methods of withholding are based on information you get from your employees on **Form W-4,** Employee's Withholding Allowance Certificate. See section 9 for more information on Form W-4.

Wage Bracket Method

Under the wage bracket method, find the proper table (on pages 36-55) for your payroll period and the employee's marital status as shown on his or her Form W-4. Then, based on the number of withholding allowances claimed on the Form W-4 and the amount of wages, find the amount of tax to withhold. If your employee is claiming more than 10 withholding allowances, see below.

Note: *If you cannot use the wage bracket tables because wages exceed the amount shown in the last bracket of the table, use the percentage method of withholding described below. Be sure to reduce wages by the amount of total withholding allowances in Table 5 before using the percentage method tables (pages 34-35).*

Adjusting wage bracket withholding for employees claiming more than 10 withholding allowances. The wage bracket tables can be used if an employee claims up to 10 allowances. More than 10 allowances may be claimed because of the special withholding allowance, additional allowances for deductions and credits, and the system itself.

To adapt the tables to more than 10 allowances:

1) Multiply the number of withholding allowances over 10 by the allowance value for the payroll period. (The allowance values are in Table 5, **Percentage Method—2002 Amount for One Withholding Allowance** later.)
2) Subtract the result from the employee's wages.
3) On this amount, find and withhold the tax in the column for 10 allowances.

This is a voluntary method. If you use the wage bracket tables, you may continue to withhold the amount in the "10" column when your employee has more than 10 allowances, using the method above. You can also use any other method described below.

Percentage Method

If you do not want to use the wage bracket tables on pages 36 through 55 to figure how much income tax to withhold, you can use a percentage computation based on Table 5 and the appropriate rate table. This method works for any number of withholding allowances the employee claims and any amount of wages.

Use these steps to figure the income tax to withhold under the percentage method:

1) Multiply one withholding allowance for your payroll period (see Table 5 below) by the number of allowances the employee claims.
2) Subtract that amount from the employee's wages.
3) Determine the amount to withhold from the appropriate table on pages 34 and 35.

Table 5. **Percentage Method—2002 Amount for One Withholding Allowance**

Payroll Period	One Withholding Allowance
Weekly	$57.69
Biweekly	115.38
Semimonthly	125.00
Monthly	250.00
Quarterly	750.00
Semiannually	1,500.00
Annually	3,000.00
Daily or miscellaneous (each day of the payroll period)	11.54

Example: An unmarried employee is paid $600 weekly. This employee has in effect a Form W-4 claiming two withholding allowances. Using the percentage method, figure the income tax to withhold as follows:

1. Total wage payment		$600.00
2. One allowance	$57.69	
3. Allowances claimed on Form W-4	2	
4. Multiply line 2 by line 3		$115.38
5. Amount subject to withholding (subtract line 4 from line 1)		$484.62
6. Tax to be withheld on $484.62 from Table 1—single person, page 34		$ 59.39

To figure the income tax to withhold, you may reduce the last digit of the wages to zero, or figure the wages to the nearest dollar.

Annual income tax withholding. Figure the income tax to withhold on annual wages under the Percentage Method for an annual payroll period. Then prorate the tax back to the payroll period.

Example: A married person claims four withholding allowances. She is paid $1,000 a week. Multiply the weekly wages by 52 weeks to figure the annual wage of $52,000. Subtract $12,000 (the value of four withholding allowances for 2002) for a balance of $40,000. Using the table for the annual payroll period on page 35, $4,432.50 is withheld. Divide the annual tax by 52. The weekly tax to withhold is $85.24.

Page 32

Alternative Methods of Income Tax Withholding

Rather than the Percentage or Wage Bracket Methods described on page 32, you can use an alternative method to withhold income tax. **Pub. 15-A,** Employer's Supplemental Tax Guide, describes these alternative methods and contains:

1) Formula tables for percentage method withholding (for automated payroll systems).
2) Wage bracket percentage method tables (for automated payroll systems).
3) Combined income, social security, and Medicare tax withholding tables.

Some alternative methods explained in Pub. 15-A are annualized wages, average estimated wages, cumulative wages, and part-year employment.

Advance Payment Methods for the Earned Income Credit (EIC)

To figure the advance EIC payment, you may use either the Wage Bracket Method or the Percentage Method explained below. You may use other methods for figuring advance EIC payments if the amount of the payment is about the same as it would be using tables in this booklet. See the tolerances allowed in the chart in section 10 of Pub. 15-A. See section 10 in this booklet for an explanation of the advance payment of the EIC.

The number of withholding allowances an employee claims on Form W-4 is not used in figuring the advance EIC payment. Nor does it matter that the employee has claimed exemption from income tax withholding on Form W-4.

Wage Bracket Method

If you use the wage bracket tables on pages 58 through 61, figure the advance EIC payment as follows.

Find the employee's gross wages before any deductions using the appropriate table. There are different tables for **(a)** single or head of household **(b)** married without spouse filing certificate **(c)** married with both spouses filing certificates. Determine the amount of the advance EIC payment shown in the appropriate table for the amount of wages paid.

Percentage Method

If you do not want to use the wage bracket tables to figure how much to include in an employee's wages for the advance EIC payment, you can use the percentage method based on the appropriate rate table on pages 56 and 57.

Find the employee's gross wages before any deductions in the appropriate table on pages 56 and 57. There are different tables for **(a)** single or head of household **(b)** married without spouse filing certificate **(c)** married with both spouses filing certificates. Find the advance EIC payment shown in the appropriate table for the amount of wages paid.

Whole-Dollar Withholding and Paying Advance EIC (Rounding)

The income tax withholding amounts in the wage bracket tables (pages 36-55) have been rounded to whole-dollar amounts.

When employers use the percentage method (pages 34-35) or an alternative method of income tax withholding, the tax for the pay period may be rounded to the nearest dollar.

The wage bracket tables for advance EIC payments (pages 58-61) have also been rounded to whole-dollar amounts. If you use the percentage method for advance EIC payments (pages 56-57), the payments may be rounded to the nearest dollar.

Tables for Percentage Method of Withholding

(For Wages Paid in 2002)

TABLE 1—WEEKLY Payroll Period

(a) SINGLE person (including head of household)—

If the amount of wages (after subtracting withholding allowances) is: / The amount of income tax to withhold is:

Not over $51 . . . . . $0

Over—	But not over—		of excess over—
$51	—$164	10%	—$51
$164	—$570	$11.30 plus 15%	—$164
$570	—$1,247	$72.20 plus 27%	—$570
$1,247	—$2,749	$254.99 plus 30%	—$1,247
$2,749	—$5,938	$705.59 plus 35%	—$2,749
$5,938	. . . . .	$1,821.74 plus 38.6%	—$5,938

(b) MARRIED person—

If the amount of wages (after subtracting withholding allowances) is: / The amount of income tax to withhold is:

Not over $124 . . . . $0

Over—	But not over—		of excess over—
$124	—$355	10%	—$124
$355	—$991	$23.10 plus 15%	—$355
$991	—$2,110	$118.50 plus 27%	—$991
$2,110	—$3,400	$420.63 plus 30%	—$2,110
$3,400	—$5,998	$807.63 plus 35%	—$3,400
$5,998	. . . . .	$1,716.93 plus 38.6%	—$5,998

TABLE 2—BIWEEKLY Payroll Period

(a) SINGLE person (including head of household)—

If the amount of wages (after subtracting withholding allowances) is: / The amount of income tax to withhold is:

Not over $102 . . . . $0

Over—	But not over—		of excess over—
$102	—$329	10%	—$102
$329	—$1,140	$22.70 plus 15%	—$329
$1,140	—$2,493	$144.35 plus 27%	—$1,140
$2,493	—$5,498	$509.66 plus 30%	—$2,493
$5,498	—$11,875	$1,411.16 plus 35%	—$5,498
$11,875	. . . . .	$3,643.11 plus 38.6%	—$11,875

(b) MARRIED person—

If the amount of wages (after subtracting withholding allowances) is: / The amount of income tax to withhold is:

Not over $248 . . . . $0

Over—	But not over—		of excess over—
$248	—$710	10%	—$248
$710	—$1,983	$46.20 plus 15%	—$710
$1,983	—$4,219	$237.15 plus 27%	—$1,983
$4,219	—$6,800	$840.87 plus 30%	—$4,219
$6,800	—$11,996	$1,615.17 plus 35%	—$6,800
$11,996	. . . . .	$3,433.77 plus 38.6%	—$11,996

TABLE 3—SEMIMONTHLY Payroll Period

(a) SINGLE person (including head of household)—

If the amount of wages (after subtracting withholding allowances) is: / The amount of income tax to withhold is:

Not over $110 . . . . $0

Over—	But not over—		of excess over—
$110	—$356	10%	—$110
$356	—$1,235	$24.60 plus 15%	—$356
$1,235	—$2,701	$156.45 plus 27%	—$1,235
$2,701	—$5,956	$552.27 plus 30%	—$2,701
$5,956	—$12,865	$1,528.77 plus 35%	—$5,956
$12,865	. . . . .	$3,946.92 plus 38.6%	—$12,865

(b) MARRIED person—

If the amount of wages (after subtracting withholding allowances) is: / The amount of income tax to withhold is:

Not over $269 . . . . $0

Over—	But not over—		of excess over—
$269	—$769	10%	—$269
$769	—$2,148	$50.00 plus 15%	—$769
$2,148	—$4,571	$256.85 plus 27%	—$2,148
$4,571	—$7,367	$911.06 plus 30%	—$4,571
$7,367	—$12,996	$1,749.86 plus 35%	—$7,367
$12,996	. . . . .	$3,720.01 plus 38.6%	—$12,996

TABLE 4—MONTHLY Payroll Period

(a) SINGLE person (including head of household)—

If the amount of wages (after subtracting withholding allowances) is: / The amount of income tax to withhold is:

Not over $221 . . . . $0

Over—	But not over—		of excess over—
$221	—$713	10%	—$221
$713	—$2,471	$49.20 plus 15%	—$713
$2,471	—$5,402	$312.90 plus 27%	—$2,471
$5,402	—$11,913	$1,104.27 plus 30%	—$5,402
$11,913	—$25,729	$3,057.57 plus 35%	—$11,913
$25,729	. . . . .	$7,893.17 plus 38.6%	—$25,729

(b) MARRIED person—

If the amount of wages (after subtracting withholding allowances) is: / The amount of income tax to withhold is:

Not over $538 . . . . $0

Over—	But not over—		of excess over—
$538	—$1,538	10%	—$538
$1,538	—$4,296	$100.00 plus 15%	—$1,538
$4,296	—$9,142	$513.70 plus 27%	—$4,296
$9,142	—$14,733	$1,822.12 plus 30%	—$9,142
$14,733	—$25,992	$3,499.42 plus 35%	—$14,733
$25,992	. . . . .	$7,440.07 plus 38.6%	—$25,992

Page 34

Tables for Percentage Method of Withholding (Continued)

(For Wages Paid in 2002)

TABLE 5—QUARTERLY Payroll Period

(a) SINGLE person (including head of household)—

If the amount of wages (after subtracting withholding allowances) is: Not over $663 . . . . The amount of income tax to withhold is: $0

Over—	But not over—		of excess over—
$663	—$2,138	10%	—$663
$2,138	—$7,413	$147.50 plus 15%	—$2,138
$7,413	—$16,205	$938.75 plus 27%	—$7,413
$16,205	—$35,738	$3,312.59 plus 30%	—$16,205
$35,738	—$77,188	$9,172.49 plus 35%	—$35,738
$77,188	. . . . . .	$23,679.99 plus 38.6%	—$77,188

(b) MARRIED person—

If the amount of wages (after subtracting withholding allowances) is: Not over $1,613 . . . . The amount of income tax to withhold is: $0

Over—	But not over—		of excess over—
$1,613	—$4,613	10%	—$1,613
$4,613	—$12,888	$300.00 plus 15%	—$4,613
$12,888	—$27,425	$1,541.25 plus 27%	—$12,888
$27,425	—$44,200	$5,466.24 plus 30%	—$27,425
$44,200	—$77,975	$10,498.74 plus 35%	—$44,200
$77,975	. . . . . .	$22,319.99 plus 38.6%	—$77,975

TABLE 6—SEMIANNUAL Payroll Period

(a) SINGLE person (including head of household)—

If the amount of wages (after subtracting withholding allowances) is: Not over $1,325 . . . . The amount of income tax to withhold is: $0

Over—	But not over—		of excess over—
$1,325	—$4,275	10%	—$1,325
$4,275	—$14,825	$295.00 plus 15%	—$4,275
$14,825	—$32,410	$1,877.50 plus 27%	—$14,825
$32,410	—$71,475	$6,625.45 plus 30%	—$32,410
$71,475	—$154,375	$18,344.95 plus 35%	—$71,475
$154,375	. . . . . .	$47,359.95 plus 38.6%	—$154,375

(b) MARRIED person—

If the amount of wages (after subtracting withholding allowances) is: Not over $3,225 . . . . The amount of income tax to withhold is: $0

Over—	But not over—		of excess over—
$3,225	—$9,225	10%	—$3,225
$9,225	—$25,775	$600.00 plus 15%	—$9,225
$25,775	—$54,850	$3,082.50 plus 27%	—$25,775
$54,850	—$88,400	$10,932.75 plus 30%	—$54,850
$88,400	—$155,950	$20,997.75 plus 35%	—$88,400
$155,950	. . . . . .	$44,640.25 plus 38.6%	—$155,950

TABLE 7—ANNUAL Payroll Period

(a) SINGLE person (including head of household)—

If the amount of wages (after subtracting withholding allowances) is: Not over $2,650 . . . . The amount of income tax to withhold is: $0

Over—	But not over—		of excess over—
$2,650	—$8,550	10%	—$2,650
$8,550	—$29,650	$590.00 plus 15%	—$8,550
$29,650	—$64,820	$3,755.00 plus 27%	—$29,650
$64,820	—$142,950	$13,250.90 plus 30%	—$64,820
$142,950	—$308,750	$36,689.90 plus 35%	—$142,950
$308,750	. . . . . .	$94,719.90 plus 38.6%	—$308,750

(b) MARRIED person—

If the amount of wages (after subtracting withholding allowances) is: Not over $6,450 . . . . The amount of income tax to withhold is: $0

Over—	But not over—		of excess over—
$6,450	—$18,450	10%	—$6,450
$18,450	—$51,550	$1,200.00 plus 15%	—$18,450
$51,550	—$109,700	$6,165.00 plus 27%	—$51,550
$109,700	—$176,800	$21,865.50 plus 30%	—$109,700
$176,800	—$311,900	$41,995.50 plus 35%	—$176,800
$311,900	. . . . . .	$89,280.50 plus 38.6%	—$311,900

TABLE 8—DAILY or MISCELLANEOUS Payroll Period

(a) SINGLE person (including head of household)—

If the amount of wages (after subtracting withholding allowances) divided by the number of days in the payroll period is: Not over $10.20 . . . . The amount of income tax to withhold per day is: $0

Over—	But not over—		of excess over—
$10.20	—$32.90	10%	—$10.20
$32.90	—$114.00	$2.27 plus 15%	—$32.90
$114.00	—$249.30	$14.44 plus 27%	—$114.00
$249.30	—$549.80	$50.97 plus 30%	—$249.30
$549.80	—$1,187.50	$141.12 plus 35%	—$549.80
$1,187.50	. . . . . .	$364.32 plus 38.6%	—$1,187.50

(b) MARRIED person—

If the amount of wages (after subtracting withholding allowances) divided by the number of days in the payroll period is: Not over $24.80 . . . . The amount of income tax to withhold per day is: $0

Over—	But not over—		of excess over—
$24.80	—$71.00	10%	—$24.80
$71.00	—$198.30	$4.62 plus 15%	—$71.00
$198.30	—$421.90	$23.72 plus 27%	—$198.30
$421.90	—$680.00	$84.09 plus 30%	—$421.90
$680.00	—$1,199.60	$161.52 plus 35%	—$680.00
$1,199.60	. . . . . .	$343.38 plus 38.6%	—$1,199.60

SINGLE Persons—**WEEKLY** Payroll Period

(For Wages Paid in 2002)

If the wages are–		And the number of withholding allowances claimed is—										
At least	But less than	0	1	2	3	4	5	6	7	8	9	10
		The amount of income tax to be withheld is—										
$0	**$55**	$0	$0	$0	$0	$0	$0	$0	$0	$0	$0	$0
55	**60**	1	0	0	0	0	0	0	0	0	0	0
60	**65**	1	0	0	0	0	0	0	0	0	0	0
65	**70**	2	0	0	0	0	0	0	0	0	0	0
70	**75**	2	0	0	0	0	0	0	0	0	0	0
75	**80**	3	0	0	0	0	0	0	0	0	0	0
80	**85**	3	0	0	0	0	0	0	0	0	0	0
85	**90**	4	0	0	0	0	0	0	0	0	0	0
90	**95**	4	0	0	0	0	0	0	0	0	0	0
95	**100**	5	0	0	0	0	0	0	0	0	0	0
100	**105**	5	0	0	0	0	0	0	0	0	0	0
105	**110**	6	0	0	0	0	0	0	0	0	0	0
110	**115**	6	0	0	0	0	0	0	0	0	0	0
115	**120**	7	1	0	0	0	0	0	0	0	0	0
120	**125**	7	1	0	0	0	0	0	0	0	0	0
125	**130**	8	2	0	0	0	0	0	0	0	0	0
130	**135**	8	2	0	0	0	0	0	0	0	0	0
135	**140**	9	3	0	0	0	0	0	0	0	0	0
140	**145**	9	3	0	0	0	0	0	0	0	0	0
145	**150**	10	4	0	0	0	0	0	0	0	0	0
150	**155**	10	4	0	0	0	0	0	0	0	0	0
155	**160**	11	5	0	0	0	0	0	0	0	0	0
160	**165**	11	5	0	0	0	0	0	0	0	0	0
165	**170**	12	6	0	0	0	0	0	0	0	0	0
170	**175**	13	6	1	0	0	0	0	0	0	0	0
175	**180**	13	7	1	0	0	0	0	0	0	0	0
180	**185**	14	7	2	0	0	0	0	0	0	0	0
185	**190**	15	8	2	0	0	0	0	0	0	0	0
190	**195**	16	8	3	0	0	0	0	0	0	0	0
195	**200**	16	9	3	0	0	0	0	0	0	0	0
200	**210**	17	10	4	0	0	0	0	0	0	0	0
210	**220**	19	11	5	0	0	0	0	0	0	0	0
220	**230**	20	12	6	0	0	0	0	0	0	0	0
230	**240**	22	13	7	1	0	0	0	0	0	0	0
240	**250**	23	15	8	2	0	0	0	0	0	0	0
250	**260**	25	16	9	3	0	0	0	0	0	0	0
260	**270**	26	18	10	4	0	0	0	0	0	0	0
270	**280**	28	19	11	5	0	0	0	0	0	0	0
280	**290**	29	21	12	6	0	0	0	0	0	0	0
290	**300**	31	22	14	7	1	0	0	0	0	0	0
300	**310**	32	24	15	8	2	0	0	0	0	0	0
310	**320**	34	25	17	9	3	0	0	0	0	0	0
320	**330**	35	27	18	10	4	0	0	0	0	0	0
330	**340**	37	28	20	11	5	0	0	0	0	0	0
340	**350**	38	30	21	12	6	1	0	0	0	0	0
350	**360**	40	31	23	14	7	2	0	0	0	0	0
360	**370**	41	33	24	15	8	3	0	0	0	0	0
370	**380**	43	34	26	17	9	4	0	0	0	0	0
380	**390**	44	36	27	18	10	5	0	0	0	0	0
390	**400**	46	37	29	20	11	6	0	0	0	0	0
400	**410**	47	39	30	21	13	7	1	0	0	0	0
410	**420**	49	40	32	23	14	8	2	0	0	0	0
420	**430**	50	42	33	24	16	9	3	0	0	0	0
430	**440**	52	43	35	26	17	10	4	0	0	0	0
440	**450**	53	45	36	27	19	11	5	0	0	0	0
450	**460**	55	46	38	29	20	12	6	0	0	0	0
460	**470**	56	48	39	30	22	13	7	1	0	0	0
470	**480**	58	49	41	32	23	15	8	2	0	0	0
480	**490**	59	51	42	33	25	16	9	3	0	0	0
490	**500**	61	52	44	35	26	18	10	4	0	0	0
500	**510**	62	54	45	36	28	19	11	5	0	0	0
510	**520**	64	55	47	38	29	21	12	6	0	0	0
520	**530**	65	57	48	39	31	22	14	7	1	0	0
530	**540**	67	58	50	41	32	24	15	8	2	0	0
540	**550**	68	60	51	42	34	25	17	9	3	0	0
550	**560**	70	61	53	44	35	27	18	10	4	0	0
560	**570**	71	63	54	45	37	28	20	11	5	0	0
570	**580**	74	64	56	47	38	30	21	12	6	0	0
580	**590**	76	66	57	48	40	31	23	14	7	1	0
590	**600**	79	67	59	50	41	33	24	15	8	2	0

Page 36

SINGLE Persons—WEEKLY Payroll Period

(For Wages Paid in 2002)

If the wages are–		And the number of withholding allowances claimed is—										
At least	But less than	0	1	2	3	4	5	6	7	8	9	10
		The amount of income tax to be withheld is—										
$600	**$610**	$82	$69	$60	$51	$43	$34	$26	$17	$9	$3	$0
610	**620**	84	70	62	53	44	36	27	18	10	4	0
620	**630**	87	72	63	54	46	37	29	20	11	5	0
630	**640**	90	74	65	56	47	39	30	21	13	6	1
640	**650**	92	77	66	57	49	40	32	23	14	7	2
650	**660**	95	80	68	59	50	42	33	24	16	8	3
660	**670**	98	82	69	60	52	43	35	26	17	9	4
670	**680**	101	85	71	62	53	45	36	27	19	10	5
680	**690**	103	88	72	63	55	46	38	29	20	12	6
690	**700**	106	90	75	65	56	48	39	30	22	13	7
700	**710**	109	93	77	66	58	49	41	32	23	15	8
710	**720**	111	96	80	68	59	51	42	33	25	16	9
720	**730**	114	98	83	69	61	52	44	35	26	18	10
730	**740**	117	101	86	71	62	54	45	36	28	19	11
740	**750**	119	104	88	73	64	55	47	38	29	21	12
750	**760**	122	107	91	75	65	57	48	39	31	22	13
760	**770**	125	109	94	78	67	58	50	41	32	24	15
770	**780**	128	112	96	81	68	60	51	42	34	25	16
780	**790**	130	115	99	83	70	61	53	44	35	27	18
790	**800**	133	117	102	86	71	63	54	45	37	28	19
800	**810**	136	120	104	89	73	64	56	47	38	30	21
810	**820**	138	123	107	92	76	66	57	48	40	31	22
820	**830**	141	125	110	94	79	67	59	50	41	33	24
830	**840**	144	128	113	97	81	69	60	51	43	34	25
840	**850**	146	131	115	100	84	70	62	53	44	36	27
850	**860**	149	134	118	102	87	72	63	54	46	37	28
860	**870**	152	136	121	105	90	74	65	56	47	39	30
870	**880**	155	139	123	108	92	77	66	57	49	40	31
880	**890**	157	142	126	110	95	79	68	59	50	42	33
890	**900**	160	144	129	113	98	82	69	60	52	43	34
900	**910**	163	147	131	116	100	85	71	62	53	45	36
910	**920**	165	150	134	119	103	87	72	63	55	46	37
920	**930**	168	152	137	121	106	90	75	65	56	48	39
930	**940**	171	155	140	124	108	93	77	66	58	49	40
940	**950**	173	158	142	127	111	96	80	68	59	51	42
950	**960**	176	161	145	129	114	98	83	69	61	52	43
960	**970**	179	163	148	132	117	101	85	71	62	54	45
970	**980**	182	166	150	135	119	104	88	72	64	55	46
980	**990**	184	169	153	137	122	106	91	75	65	57	48
990	**1,000**	187	171	156	140	125	109	93	78	67	58	49
1,000	**1,010**	190	174	158	143	127	112	96	81	68	60	51
1,010	**1,020**	192	177	161	146	130	114	99	83	70	61	52
1,020	**1,030**	195	179	164	148	133	117	102	86	71	63	54
1,030	**1,040**	198	182	167	151	135	120	104	89	73	64	55
1,040	**1,050**	200	185	169	154	138	123	107	91	76	66	57
1,050	**1,060**	203	188	172	156	141	125	110	94	78	67	58
1,060	**1,070**	206	190	175	159	144	128	112	97	81	69	60
1,070	**1,080**	209	193	177	162	146	131	115	99	84	70	61
1,080	**1,090**	211	196	180	164	149	133	118	102	87	72	63
1,090	**1,100**	214	198	183	167	152	136	120	105	89	74	64
1,100	**1,110**	217	201	185	170	154	139	123	108	92	76	66
1,110	**1,120**	219	204	188	173	157	141	126	110	95	79	67
1,120	**1,130**	222	206	191	175	160	144	129	113	97	82	69
1,130	**1,140**	225	209	194	178	162	147	131	116	100	85	70
1,140	**1,150**	227	212	196	181	165	150	134	118	103	87	72
1,150	**1,160**	230	215	199	183	168	152	137	121	105	90	74
1,160	**1,170**	233	217	202	186	171	155	139	124	108	93	77
1,170	**1,180**	236	220	204	189	173	158	142	126	111	95	80
1,180	**1,190**	238	223	207	191	176	160	145	129	114	98	82
1,190	**1,200**	241	225	210	194	179	163	147	132	116	101	85
1,200	**1,210**	244	228	212	197	181	166	150	135	119	103	88
1,210	**1,220**	246	231	215	200	184	168	153	137	122	106	91
1,220	**1,230**	249	233	218	202	187	171	156	140	124	109	93
1,230	**1,240**	252	236	221	205	189	174	158	143	127	112	96
1,240	**1,250**	254	239	223	208	192	177	161	145	130	114	99

$1,250 and over Use Table 1(a) for a **SINGLE person** on page 34. Also see the instructions on page 32.

MARRIED Persons—**WEEKLY** Payroll Period

(For Wages Paid in 2002)

If the wages are–		And the number of withholding allowances claimed is—										
At least	But less than	0	1	2	3	4	5	6	7	8	9	10
		The amount of income tax to be withheld is—										
$0	**$130**	$0	$0	$0	$0	$0	$0	$0	$0	$0	$0	$0
130	**135**	1	0	0	0	0	0	0	0	0	0	0
135	**140**	1	0	0	0	0	0	0	0	0	0	0
140	**145**	2	0	0	0	0	0	0	0	0	0	0
145	**150**	2	0	0	0	0	0	0	0	0	0	0
150	**155**	3	0	0	0	0	0	0	0	0	0	0
155	**160**	3	0	0	0	0	0	0	0	0	0	0
160	**165**	4	0	0	0	0	0	0	0	0	0	0
165	**170**	4	0	0	0	0	0	0	0	0	0	0
170	**175**	5	0	0	0	0	0	0	0	0	0	0
175	**180**	5	0	0	0	0	0	0	0	0	0	0
180	**185**	6	0	0	0	0	0	0	0	0	0	0
185	**190**	6	1	0	0	0	0	0	0	0	0	0
190	**195**	7	1	0	0	0	0	0	0	0	0	0
195	**200**	7	2	0	0	0	0	0	0	0	0	0
200	**210**	8	2	0	0	0	0	0	0	0	0	0
210	**220**	9	3	0	0	0	0	0	0	0	0	0
220	**230**	10	4	0	0	0	0	0	0	0	0	0
230	**240**	11	5	0	0	0	0	0	0	0	0	0
240	**250**	12	6	1	0	0	0	0	0	0	0	0
250	**260**	13	7	2	0	0	0	0	0	0	0	0
260	**270**	14	8	3	0	0	0	0	0	0	0	0
270	**280**	15	9	4	0	0	0	0	0	0	0	0
280	**290**	16	10	5	0	0	0	0	0	0	0	0
290	**300**	17	11	6	0	0	0	0	0	0	0	0
300	**310**	18	12	7	1	0	0	0	0	0	0	0
310	**320**	19	13	8	2	0	0	0	0	0	0	0
320	**330**	20	14	9	3	0	0	0	0	0	0	0
330	**340**	21	15	10	4	0	0	0	0	0	0	0
340	**350**	22	16	11	5	0	0	0	0	0	0	0
350	**360**	23	17	12	6	0	0	0	0	0	0	0
360	**370**	25	18	13	7	1	0	0	0	0	0	0
370	**380**	26	19	14	8	2	0	0	0	0	0	0
380	**390**	28	20	15	9	3	0	0	0	0	0	0
390	**400**	29	21	16	10	4	0	0	0	0	0	0
400	**410**	31	22	17	11	5	0	0	0	0	0	0
410	**420**	32	23	18	12	6	0	0	0	0	0	0
420	**430**	34	25	19	13	7	1	0	0	0	0	0
430	**440**	35	26	20	14	8	2	0	0	0	0	0
440	**450**	37	28	21	15	9	3	0	0	0	0	0
450	**460**	38	29	22	16	10	4	0	0	0	0	0
460	**470**	40	31	23	17	11	5	0	0	0	0	0
470	**480**	41	32	24	18	12	6	0	0	0	0	0
480	**490**	43	34	25	19	13	7	1	0	0	0	0
490	**500**	44	35	27	20	14	8	2	0	0	0	0
500	**510**	46	37	28	21	15	9	3	0	0	0	0
510	**520**	47	38	30	22	16	10	4	0	0	0	0
520	**530**	49	40	31	23	17	11	5	0	0	0	0
530	**540**	50	41	33	24	18	12	6	1	0	0	0
540	**550**	52	43	34	26	19	13	7	2	0	0	0
550	**560**	53	44	36	27	20	14	8	3	0	0	0
560	**570**	55	46	37	29	21	15	9	4	0	0	0
570	**580**	56	47	39	30	22	16	10	5	0	0	0
580	**590**	58	49	40	32	23	17	11	6	0	0	0
590	**600**	59	50	42	33	24	18	12	7	1	0	0
600	**610**	61	52	43	35	26	19	13	8	2	0	0
610	**620**	62	53	45	36	27	20	14	9	3	0	0
620	**630**	64	55	46	38	29	21	15	10	4	0	0
630	**640**	65	56	48	39	30	22	16	11	5	0	0
640	**650**	67	58	49	41	32	23	17	12	6	0	0
650	**660**	68	59	51	42	33	25	18	13	7	1	0
660	**670**	70	61	52	44	35	26	19	14	8	2	0
670	**680**	71	62	54	45	36	28	20	15	9	3	0
680	**690**	73	64	55	47	38	29	21	16	10	4	0
690	**700**	74	65	57	48	39	31	22	17	11	5	0
700	**710**	76	67	58	50	41	32	24	18	12	6	0
710	**720**	77	68	60	51	42	34	25	19	13	7	1
720	**730**	79	70	61	53	44	35	27	20	14	8	2
730	**740**	80	71	63	54	45	37	28	21	15	9	3
740	**750**	82	73	64	56	47	38	30	22	16	10	4

Page 38

MARRIED Persons—**WEEKLY** Payroll Period

(For Wages Paid in 2002)

If the wages are–		And the number of withholding allowances claimed is—										
At least	But less than	0	1	2	3	4	5	6	7	8	9	10
		The amount of income tax to be withheld is—										
$750	**$760**	$83	$74	$66	$57	$48	$40	$31	$23	$17	$11	$5
760	**770**	85	76	67	59	50	41	33	24	18	12	6
770	**780**	86	77	69	60	51	43	34	26	19	13	7
780	**790**	88	79	70	62	53	44	36	27	20	14	8
790	**800**	89	80	72	63	54	46	37	29	21	15	9
800	**810**	91	82	73	65	56	47	39	30	22	16	10
810	**820**	92	83	75	66	57	49	40	32	23	17	11
820	**830**	94	85	76	68	59	50	42	33	24	18	12
830	**840**	95	86	78	69	60	52	43	35	26	19	13
840	**850**	97	88	79	71	62	53	45	36	27	20	14
850	**860**	98	89	81	72	63	55	46	38	29	21	15
860	**870**	100	91	82	74	65	56	48	39	30	22	16
870	**880**	101	92	84	75	66	58	49	41	32	23	17
880	**890**	103	94	85	77	68	59	51	42	33	25	18
890	**900**	104	95	87	78	69	61	52	44	35	26	19
900	**910**	106	97	88	80	71	62	54	45	36	28	20
910	**920**	107	98	90	81	72	64	55	47	38	29	21
920	**930**	109	100	91	83	74	65	57	48	39	31	22
930	**940**	110	101	93	84	75	67	58	50	41	32	24
940	**950**	112	103	94	86	77	68	60	51	42	34	25
950	**960**	113	104	96	87	78	70	61	53	44	35	27
960	**970**	115	106	97	89	80	71	63	54	45	37	28
970	**980**	116	107	99	90	81	73	64	56	47	38	30
980	**990**	118	109	100	92	83	74	66	57	48	40	31
990	**1,000**	120	110	102	93	84	76	67	59	50	41	33
1,000	**1,010**	122	112	103	95	86	77	69	60	51	43	34
1,010	**1,020**	125	113	105	96	87	79	70	62	53	44	36
1,020	**1,030**	128	115	106	98	89	80	72	63	54	46	37
1,030	**1,040**	130	116	108	99	90	82	73	65	56	47	39
1,040	**1,050**	133	118	109	101	92	83	75	66	57	49	40
1,050	**1,060**	136	120	111	102	93	85	76	68	59	50	42
1,060	**1,070**	138	123	112	104	95	86	78	69	60	52	43
1,070	**1,080**	141	126	114	105	96	88	79	71	62	53	45
1,080	**1,090**	144	128	115	107	98	89	81	72	63	55	46
1,090	**1,100**	147	131	117	108	99	91	82	74	65	56	48
1,100	**1,110**	149	134	118	110	101	92	84	75	66	58	49
1,110	**1,120**	152	136	121	111	102	94	85	77	68	59	51
1,120	**1,130**	155	139	123	113	104	95	87	78	69	61	52
1,130	**1,140**	157	142	126	114	105	97	88	80	71	62	54
1,140	**1,150**	160	144	129	116	107	98	90	81	72	64	55
1,150	**1,160**	163	147	132	117	108	100	91	83	74	65	57
1,160	**1,170**	165	150	134	119	110	101	93	84	75	67	58
1,170	**1,180**	168	153	137	121	111	103	94	86	77	68	60
1,180	**1,190**	171	155	140	124	113	104	96	87	78	70	61
1,190	**1,200**	174	158	142	127	114	106	97	89	80	71	63
1,200	**1,210**	176	161	145	130	116	107	99	90	81	73	64
1,210	**1,220**	179	163	148	132	117	109	100	92	83	74	66
1,220	**1,230**	182	166	150	135	119	110	102	93	84	76	67
1,230	**1,240**	184	169	153	138	122	112	103	95	86	77	69
1,240	**1,250**	187	171	156	140	125	113	105	96	87	79	70
1,250	**1,260**	190	174	159	143	127	115	106	98	89	80	72
1,260	**1,270**	192	177	161	146	130	116	108	99	90	82	73
1,270	**1,280**	195	180	164	148	133	118	109	101	92	83	75
1,280	**1,290**	198	182	167	151	136	120	111	102	93	85	76
1,290	**1,300**	201	185	169	154	138	123	112	104	95	86	78
1,300	**1,310**	203	188	172	157	141	125	114	105	96	88	79
1,310	**1,320**	206	190	175	159	144	128	115	107	98	89	81
1,320	**1,330**	209	193	177	162	146	131	117	108	99	91	82
1,330	**1,340**	211	196	180	165	149	133	118	110	101	92	84
1,340	**1,350**	214	198	183	167	152	136	121	111	102	94	85
1,350	**1,360**	217	201	186	170	154	139	123	113	104	95	87
1,360	**1,370**	219	204	188	173	157	142	126	114	105	97	88
1,370	**1,380**	222	207	191	175	160	144	129	116	107	98	90
1,380	**1,390**	225	209	194	178	163	147	131	117	108	100	91
1,390	**1,400**	228	212	196	181	165	150	134	119	110	101	93
$1,400 and over		Use Table 1(b) for a **MARRIED person** on page 34. Also see the instructions on page 32.										

SINGLE Persons—**BIWEEKLY** Payroll Period

(For Wages Paid in 2002)

If the wages are–		And the number of withholding allowances claimed is—										
At least	But less than	0	1	2	3	4	5	6	7	8	9	10
		The amount of income tax to be withheld is—										
$0	**$105**	$0	$0	$0	$0	$0	$0	$0	$0	$0	$0	$0
105	**110**	1	0	0	0	0	0	0	0	0	0	0
110	**115**	1	0	0	0	0	0	0	0	0	0	0
115	**120**	2	0	0	0	0	0	0	0	0	0	0
120	**125**	2	0	0	0	0	0	0	0	0	0	0
125	**130**	3	0	0	0	0	0	0	0	0	0	0
130	**135**	3	0	0	0	0	0	0	0	0	0	0
135	**140**	4	0	0	0	0	0	0	0	0	0	0
140	**145**	4	0	0	0	0	0	0	0	0	0	0
145	**150**	5	0	0	0	0	0	0	0	0	0	0
150	**155**	5	0	0	0	0	0	0	0	0	0	0
155	**160**	6	0	0	0	0	0	0	0	0	0	0
160	**165**	6	0	0	0	0	0	0	0	0	0	0
165	**170**	7	0	0	0	0	0	0	0	0	0	0
170	**175**	7	0	0	0	0	0	0	0	0	0	0
175	**180**	8	0	0	0	0	0	0	0	0	0	0
180	**185**	8	0	0	0	0	0	0	0	0	0	0
185	**190**	9	0	0	0	0	0	0	0	0	0	0
190	**195**	9	0	0	0	0	0	0	0	0	0	0
195	**200**	10	0	0	0	0	0	0	0	0	0	0
200	**205**	10	0	0	0	0	0	0	0	0	0	0
205	**210**	11	0	0	0	0	0	0	0	0	0	0
210	**215**	11	0	0	0	0	0	0	0	0	0	0
215	**220**	12	0	0	0	0	0	0	0	0	0	0
220	**225**	12	1	0	0	0	0	0	0	0	0	0
225	**230**	13	1	0	0	0	0	0	0	0	0	0
230	**235**	13	2	0	0	0	0	0	0	0	0	0
235	**240**	14	2	0	0	0	0	0	0	0	0	0
240	**245**	14	3	0	0	0	0	0	0	0	0	0
245	**250**	15	3	0	0	0	0	0	0	0	0	0
250	**260**	15	4	0	0	0	0	0	0	0	0	0
260	**270**	16	5	0	0	0	0	0	0	0	0	0
270	**280**	17	6	0	0	0	0	0	0	0	0	0
280	**290**	18	7	0	0	0	0	0	0	0	0	0
290	**300**	19	8	0	0	0	0	0	0	0	0	0
300	**310**	20	9	0	0	0	0	0	0	0	0	0
310	**320**	21	10	0	0	0	0	0	0	0	0	0
320	**330**	22	11	0	0	0	0	0	0	0	0	0
330	**340**	24	12	0	0	0	0	0	0	0	0	0
340	**350**	25	13	1	0	0	0	0	0	0	0	0
350	**360**	27	14	2	0	0	0	0	0	0	0	0
360	**370**	28	15	3	0	0	0	0	0	0	0	0
370	**380**	30	16	4	0	0	0	0	0	0	0	0
380	**390**	31	17	5	0	0	0	0	0	0	0	0
390	**400**	33	18	6	0	0	0	0	0	0	0	0
400	**410**	34	19	7	0	0	0	0	0	0	0	0
410	**420**	36	20	8	0	0	0	0	0	0	0	0
420	**430**	37	21	9	0	0	0	0	0	0	0	0
430	**440**	39	22	10	0	0	0	0	0	0	0	0
440	**450**	40	23	11	0	0	0	0	0	0	0	0
450	**460**	42	24	12	1	0	0	0	0	0	0	0
460	**470**	43	26	13	2	0	0	0	0	0	0	0
470	**480**	45	27	14	3	0	0	0	0	0	0	0
480	**490**	46	29	15	4	0	0	0	0	0	0	0
490	**500**	48	30	16	5	0	0	0	0	0	0	0
500	**520**	50	33	18	6	0	0	0	0	0	0	0
520	**540**	53	36	20	8	0	0	0	0	0	0	0
540	**560**	56	39	22	10	0	0	0	0	0	0	0
560	**580**	59	42	24	12	1	0	0	0	0	0	0
580	**600**	62	45	27	14	3	0	0	0	0	0	0
600	**620**	65	48	30	16	5	0	0	0	0	0	0
620	**640**	68	51	33	18	7	0	0	0	0	0	0
640	**660**	71	54	36	20	9	0	0	0	0	0	0
660	**680**	74	57	39	22	11	0	0	0	0	0	0
680	**700**	77	60	42	25	13	1	0	0	0	0	0
700	**720**	80	63	45	28	15	3	0	0	0	0	0
720	**740**	83	66	48	31	17	5	0	0	0	0	0
740	**760**	86	69	51	34	19	7	0	0	0	0	0
760	**780**	89	72	54	37	21	9	0	0	0	0	0
780	**800**	92	75	57	40	23	11	0	0	0	0	0

SINGLE Persons—BIWEEKLY Payroll Period

(For Wages Paid in 2002)

If the wages are–		And the number of withholding allowances claimed is—										
At least	But less than	0	1	2	3	4	5	6	7	8	9	10
		The amount of income tax to be withheld is—										
$800	**$820**	$95	$78	$60	$43	$26	$13	$2	$0	$0	$0	$0
820	**840**	98	81	63	46	29	15	4	0	0	0	0
840	**860**	101	84	66	49	32	17	6	0	0	0	0
860	**880**	104	87	69	52	35	19	8	0	0	0	0
880	**900**	107	90	72	55	38	21	10	0	0	0	0
900	**920**	110	93	75	58	41	23	12	0	0	0	0
920	**940**	113	96	78	61	44	26	14	2	0	0	0
940	**960**	116	99	81	64	47	29	16	4	0	0	0
960	**980**	119	102	84	67	50	32	18	6	0	0	0
980	**1,000**	122	105	87	70	53	35	20	8	0	0	0
1,000	**1,020**	125	108	90	73	56	38	22	10	0	0	0
1,020	**1,040**	128	111	93	76	59	41	24	12	1	0	0
1,040	**1,060**	131	114	96	79	62	44	27	14	3	0	0
1,060	**1,080**	134	117	99	82	65	47	30	16	5	0	0
1,080	**1,100**	137	120	102	85	68	50	33	18	7	0	0
1,100	**1,120**	140	123	105	88	71	53	36	20	9	0	0
1,120	**1,140**	143	126	108	91	74	56	39	22	11	0	0
1,140	**1,160**	147	129	111	94	77	59	42	25	13	1	0
1,160	**1,180**	152	132	114	97	80	62	45	28	15	3	0
1,180	**1,200**	158	135	117	100	83	65	48	31	17	5	0
1,200	**1,220**	163	138	120	103	86	68	51	34	19	7	0
1,220	**1,240**	169	141	123	106	89	71	54	37	21	9	0
1,240	**1,260**	174	144	126	109	92	74	57	40	23	11	0
1,260	**1,280**	179	148	129	112	95	77	60	43	25	13	1
1,280	**1,300**	185	154	132	115	98	80	63	46	28	15	3
1,300	**1,320**	190	159	135	118	101	83	66	49	31	17	5
1,320	**1,340**	196	164	138	121	104	86	69	52	34	19	7
1,340	**1,360**	201	170	141	124	107	89	72	55	37	21	9
1,360	**1,380**	206	175	144	127	110	92	75	58	40	23	11
1,380	**1,400**	212	181	150	130	113	95	78	61	43	26	13
1,400	**1,420**	217	186	155	133	116	98	81	64	46	29	15
1,420	**1,440**	223	191	160	136	119	101	84	67	49	32	17
1,440	**1,460**	228	197	166	139	122	104	87	70	52	35	19
1,460	**1,480**	233	202	171	142	125	107	90	73	55	38	21
1,480	**1,500**	239	208	177	145	128	110	93	76	58	41	24
1,500	**1,520**	244	213	182	151	131	113	96	79	61	44	27
1,520	**1,540**	250	218	187	156	134	116	99	82	64	47	30
1,540	**1,560**	255	224	193	162	137	119	102	85	67	50	33
1,560	**1,580**	260	229	198	167	140	122	105	88	70	53	36
1,580	**1,600**	266	235	204	172	143	125	108	91	73	56	39
1,600	**1,620**	271	240	209	178	147	128	111	94	76	59	42
1,620	**1,640**	277	245	214	183	152	131	114	97	79	62	45
1,640	**1,660**	282	251	220	189	157	134	117	100	82	65	48
1,660	**1,680**	287	256	225	194	163	137	120	103	85	68	51
1,680	**1,700**	293	262	231	199	168	140	123	106	88	71	54
1,700	**1,720**	298	267	236	205	174	143	126	109	91	74	57
1,720	**1,740**	304	272	241	210	179	148	129	112	94	77	60
1,740	**1,760**	309	278	247	216	184	153	132	115	97	80	63
1,760	**1,780**	314	283	252	221	190	159	135	118	100	83	66
1,780	**1,800**	320	289	258	226	195	164	138	121	103	86	69
1,800	**1,820**	325	294	263	232	201	169	141	124	106	89	72
1,820	**1,840**	331	299	268	237	206	175	144	127	109	92	75
1,840	**1,860**	336	305	274	243	211	180	149	130	112	95	78
1,860	**1,880**	341	310	279	248	217	186	154	133	115	98	81
1,880	**1,900**	347	316	285	253	222	191	160	136	118	101	84
1,900	**1,920**	352	321	290	259	228	196	165	139	121	104	87
1,920	**1,940**	358	326	295	264	233	202	171	142	124	107	90
1,940	**1,960**	363	332	301	270	238	207	176	145	127	110	93
1,960	**1,980**	368	337	306	275	244	213	181	150	130	113	96
1,980	**2,000**	374	343	312	280	249	218	187	156	133	116	99
2,000	**2,020**	379	348	317	286	255	223	192	161	136	119	102
2,020	**2,040**	385	353	322	291	260	229	198	167	139	122	105
2,040	**2,060**	390	359	328	297	265	234	203	172	142	125	108
2,060	**2,080**	395	364	333	302	271	240	208	177	146	128	111
2,080	**2,100**	401	370	339	307	276	245	214	183	152	131	114

$2,100 and over Use Table 2(a) for a **SINGLE person** on page 34. Also see the instructions on page 32.

MARRIED Persons—**BIWEEKLY** Payroll Period

(For Wages Paid in 2002)

If the wages are–		And the number of withholding allowances claimed is—										
At least	But less than	0	1	2	3	4	5	6	7	8	9	10
		The amount of income tax to be withheld is—										
$0	**$250**	$0	$0	$0	$0	$0	$0	$0	$0	$0	$0	$0
250	**260**	1	0	0	0	0	0	0	0	0	0	0
260	**270**	2	0	0	0	0	0	0	0	0	0	0
270	**280**	3	0	0	0	0	0	0	0	0	0	0
280	**290**	4	0	0	0	0	0	0	0	0	0	0
290	**300**	5	0	0	0	0	0	0	0	0	0	0
300	**310**	6	0	0	0	0	0	0	0	0	0	0
310	**320**	7	0	0	0	0	0	0	0	0	0	0
320	**330**	8	0	0	0	0	0	0	0	0	0	0
330	**340**	9	0	0	0	0	0	0	0	0	0	0
340	**350**	10	0	0	0	0	0	0	0	0	0	0
350	**360**	11	0	0	0	0	0	0	0	0	0	0
360	**370**	12	0	0	0	0	0	0	0	0	0	0
370	**380**	13	1	0	0	0	0	0	0	0	0	0
380	**390**	14	2	0	0	0	0	0	0	0	0	0
390	**400**	15	3	0	0	0	0	0	0	0	0	0
400	**410**	16	4	0	0	0	0	0	0	0	0	0
410	**420**	17	5	0	0	0	0	0	0	0	0	0
420	**430**	18	6	0	0	0	0	0	0	0	0	0
430	**440**	19	7	0	0	0	0	0	0	0	0	0
440	**450**	20	8	0	0	0	0	0	0	0	0	0
450	**460**	21	9	0	0	0	0	0	0	0	0	0
460	**470**	22	10	0	0	0	0	0	0	0	0	0
470	**480**	23	11	0	0	0	0	0	0	0	0	0
480	**490**	24	12	1	0	0	0	0	0	0	0	0
490	**500**	25	13	2	0	0	0	0	0	0	0	0
500	**520**	26	15	3	0	0	0	0	0	0	0	0
520	**540**	28	17	5	0	0	0	0	0	0	0	0
540	**560**	30	19	7	0	0	0	0	0	0	0	0
560	**580**	32	21	9	0	0	0	0	0	0	0	0
580	**600**	34	23	11	0	0	0	0	0	0	0	0
600	**620**	36	25	13	2	0	0	0	0	0	0	0
620	**640**	38	27	15	4	0	0	0	0	0	0	0
640	**660**	40	29	17	6	0	0	0	0	0	0	0
660	**680**	42	31	19	8	0	0	0	0	0	0	0
680	**700**	44	33	21	10	0	0	0	0	0	0	0
700	**720**	46	35	23	12	0	0	0	0	0	0	0
720	**740**	49	37	25	14	2	0	0	0	0	0	0
740	**760**	52	39	27	16	4	0	0	0	0	0	0
760	**780**	55	41	29	18	6	0	0	0	0	0	0
780	**800**	58	43	31	20	8	0	0	0	0	0	0
800	**820**	61	45	33	22	10	0	0	0	0	0	0
820	**840**	64	47	35	24	12	1	0	0	0	0	0
840	**860**	67	50	37	26	14	3	0	0	0	0	0
860	**880**	70	53	39	28	16	5	0	0	0	0	0
880	**900**	73	56	41	30	18	7	0	0	0	0	0
900	**920**	76	59	43	32	20	9	0	0	0	0	0
920	**940**	79	62	45	34	22	11	0	0	0	0	0
940	**960**	82	65	48	36	24	13	1	0	0	0	0
960	**980**	85	68	51	38	26	15	3	0	0	0	0
980	**1,000**	88	71	54	40	28	17	5	0	0	0	0
1,000	**1,020**	91	74	57	42	30	19	7	0	0	0	0
1,020	**1,040**	94	77	60	44	32	21	9	0	0	0	0
1,040	**1,060**	97	80	63	46	34	23	11	0	0	0	0
1,060	**1,080**	100	83	66	48	36	25	13	1	0	0	0
1,080	**1,100**	103	86	69	51	38	27	15	3	0	0	0
1,100	**1,120**	106	89	72	54	40	29	17	5	0	0	0
1,120	**1,140**	109	92	75	57	42	31	19	7	0	0	0
1,140	**1,160**	112	95	78	60	44	33	21	9	0	0	0
1,160	**1,180**	115	98	81	63	46	35	23	11	0	0	0
1,180	**1,200**	118	101	84	66	49	37	25	13	2	0	0
1,200	**1,220**	121	104	87	69	52	39	27	15	4	0	0
1,220	**1,240**	124	107	90	72	55	41	29	17	6	0	0
1,240	**1,260**	127	110	93	75	58	43	31	19	8	0	0
1,260	**1,280**	130	113	96	78	61	45	33	21	10	0	0
1,280	**1,300**	133	116	99	81	64	47	35	23	12	0	0
1,300	**1,320**	136	119	102	84	67	50	37	25	14	2	0
1,320	**1,340**	139	122	105	87	70	53	39	27	16	4	0
1,340	**1,360**	142	125	108	90	73	56	41	29	18	6	0
1,360	**1,380**	145	128	111	93	76	59	43	31	20	8	0

MARRIED Persons—BIWEEKLY Payroll Period

(For Wages Paid in 2002)

If the wages are–		And the number of withholding allowances claimed is—										
At least	But less than	0	1	2	3	4	5	6	7	8	9	10
		The amount of income tax to be withheld is—										
$1,380	$1,400	$148	$131	$114	$96	$79	$62	$45	$33	$22	$10	$0
1,400	1,420	151	134	117	99	82	65	47	35	24	12	1
1,420	1,440	154	137	120	102	85	68	50	37	26	14	3
1,440	1,460	157	140	123	105	88	71	53	39	28	16	5
1,460	1,480	160	143	126	108	91	74	56	41	30	18	7
1,480	1,500	163	146	129	111	94	77	59	43	32	20	9
1,500	1,520	166	149	132	114	97	80	62	45	34	22	11
1,520	1,540	169	152	135	117	100	83	65	48	36	24	13
1,540	1,560	172	155	138	120	103	86	68	51	38	26	15
1,560	1,580	175	158	141	123	106	89	71	54	40	28	17
1,580	1,600	178	161	144	126	109	92	74	57	42	30	19
1,600	1,620	181	164	147	129	112	95	77	60	44	32	21
1,620	1,640	184	167	150	132	115	98	80	63	46	34	23
1,640	1,660	187	170	153	135	118	101	83	66	49	36	25
1,660	1,680	190	173	156	138	121	104	86	69	52	38	27
1,680	1,700	193	176	159	141	124	107	89	72	55	40	29
1,700	1,720	196	179	162	144	127	110	92	75	58	42	31
1,720	1,740	199	182	165	147	130	113	95	78	61	44	33
1,740	1,760	202	185	168	150	133	116	98	81	64	46	35
1,760	1,780	205	188	171	153	136	119	101	84	67	49	37
1,780	1,800	208	191	174	156	139	122	104	87	70	52	39
1,800	1,820	211	194	177	159	142	125	107	90	73	55	41
1,820	1,840	214	197	180	162	145	128	110	93	76	58	43
1,840	1,860	217	200	183	165	148	131	113	96	79	61	45
1,860	1,880	220	203	186	168	151	134	116	99	82	64	47
1,880	1,900	223	206	189	171	154	137	119	102	85	67	50
1,900	1,920	226	209	192	174	157	140	122	105	88	70	53
1,920	1,940	229	212	195	177	160	143	125	108	91	73	56
1,940	1,960	232	215	198	180	163	146	128	111	94	76	59
1,960	1,980	235	218	201	183	166	149	131	114	97	79	62
1,980	2,000	239	221	204	186	169	152	134	117	100	82	65
2,000	2,020	244	224	207	189	172	155	137	120	103	85	68
2,020	2,040	250	227	210	192	175	158	140	123	106	88	71
2,040	2,060	255	230	213	195	178	161	143	126	109	91	74
2,060	2,080	261	233	216	198	181	164	146	129	112	94	77
2,080	2,100	266	236	219	201	184	167	149	132	115	97	80
2,100	2,120	271	240	222	204	187	170	152	135	118	100	83
2,120	2,140	277	246	225	207	190	173	155	138	121	103	86
2,140	2,160	282	251	228	210	193	176	158	141	124	106	89
2,160	2,180	288	257	231	213	196	179	161	144	127	109	92
2,180	2,200	293	262	234	216	199	182	164	147	130	112	95
2,200	2,220	298	267	237	219	202	185	167	150	133	115	98
2,220	2,240	304	273	242	222	205	188	170	153	136	118	101
2,240	2,260	309	278	247	225	208	191	173	156	139	121	104
2,260	2,280	315	284	252	228	211	194	176	159	142	124	107
2,280	2,300	320	289	258	231	214	197	179	162	145	127	110
2,300	2,320	325	294	263	234	217	200	182	165	148	130	113
2,320	2,340	331	300	269	237	220	203	185	168	151	133	116
2,340	2,360	336	305	274	243	223	206	188	171	154	136	119
2,360	2,380	342	311	279	248	226	209	191	174	157	139	122
2,380	2,400	347	316	285	254	229	212	194	177	160	142	125
2,400	2,420	352	321	290	259	232	215	197	180	163	145	128
2,420	2,440	358	327	296	264	235	218	200	183	166	148	131
2,440	2,460	363	332	301	270	239	221	203	186	169	151	134
2,460	2,480	369	338	306	275	244	224	206	189	172	154	137
2,480	2,500	374	343	312	281	249	227	209	192	175	157	140
2,500	2,520	379	348	317	286	255	230	212	195	178	160	143
2,520	2,540	385	354	323	291	260	233	215	198	181	163	146
2,540	2,560	390	359	328	297	266	236	218	201	184	166	149
2,560	2,580	396	365	333	302	271	240	221	204	187	169	152
2,580	2,600	401	370	339	308	276	245	224	207	190	172	155
2,600	2,620	406	375	344	313	282	251	227	210	193	175	158
2,620	2,640	412	381	350	318	287	256	230	213	196	178	161
2,640	2,660	417	386	355	324	293	262	233	216	199	181	164
2,660	2,680	423	392	360	329	298	267	236	219	202	184	167
$2,680 and over		Use Table 2(b) for a **MARRIED person** on page 34. Also see the instructions on page 32.										

Page 43

SINGLE Persons—**SEMIMONTHLY** Payroll Period

(For Wages Paid in 2002)

If the wages are–		And the number of withholding allowances claimed is—										
At least	But less than	0	1	2	3	4	5	6	7	8	9	10
		The amount of income tax to be withheld is—										
$0	**$115**	$0	$0	$0	$0	$0	$0	$0	$0	$0	$0	$0
115	**120**	1	0	0	0	0	0	0	0	0	0	0
120	**125**	1	0	0	0	0	0	0	0	0	0	0
125	**130**	2	0	0	0	0	0	0	0	0	0	0
130	**135**	2	0	0	0	0	0	0	0	0	0	0
135	**140**	3	0	0	0	0	0	0	0	0	0	0
140	**145**	3	0	0	0	0	0	0	0	0	0	0
145	**150**	4	0	0	0	0	0	0	0	0	0	0
150	**155**	4	0	0	0	0	0	0	0	0	0	0
155	**160**	5	0	0	0	0	0	0	0	0	0	0
160	**165**	5	0	0	0	0	0	0	0	0	0	0
165	**170**	6	0	0	0	0	0	0	0	0	0	0
170	**175**	6	0	0	0	0	0	0	0	0	0	0
175	**180**	7	0	0	0	0	0	0	0	0	0	0
180	**185**	7	0	0	0	0	0	0	0	0	0	0
185	**190**	8	0	0	0	0	0	0	0	0	0	0
190	**195**	8	0	0	0	0	0	0	0	0	0	0
195	**200**	9	0	0	0	0	0	0	0	0	0	0
200	**205**	9	0	0	0	0	0	0	0	0	0	0
205	**210**	10	0	0	0	0	0	0	0	0	0	0
210	**215**	10	0	0	0	0	0	0	0	0	0	0
215	**220**	11	0	0	0	0	0	0	0	0	0	0
220	**225**	11	0	0	0	0	0	0	0	0	0	0
225	**230**	12	0	0	0	0	0	0	0	0	0	0
230	**235**	12	0	0	0	0	0	0	0	0	0	0
235	**240**	13	0	0	0	0	0	0	0	0	0	0
240	**245**	13	1	0	0	0	0	0	0	0	0	0
245	**250**	14	1	0	0	0	0	0	0	0	0	0
250	**260**	14	2	0	0	0	0	0	0	0	0	0
260	**270**	15	3	0	0	0	0	0	0	0	0	0
270	**280**	16	4	0	0	0	0	0	0	0	0	0
280	**290**	17	5	0	0	0	0	0	0	0	0	0
290	**300**	18	6	0	0	0	0	0	0	0	0	0
300	**310**	19	7	0	0	0	0	0	0	0	0	0
310	**320**	20	8	0	0	0	0	0	0	0	0	0
320	**330**	21	9	0	0	0	0	0	0	0	0	0
330	**340**	22	10	0	0	0	0	0	0	0	0	0
340	**350**	23	11	0	0	0	0	0	0	0	0	0
350	**360**	24	12	0	0	0	0	0	0	0	0	0
360	**370**	26	13	0	0	0	0	0	0	0	0	0
370	**380**	27	14	1	0	0	0	0	0	0	0	0
380	**390**	29	15	2	0	0	0	0	0	0	0	0
390	**400**	30	16	3	0	0	0	0	0	0	0	0
400	**410**	32	17	4	0	0	0	0	0	0	0	0
410	**420**	33	18	5	0	0	0	0	0	0	0	0
420	**430**	35	19	6	0	0	0	0	0	0	0	0
430	**440**	36	20	7	0	0	0	0	0	0	0	0
440	**450**	38	21	8	0	0	0	0	0	0	0	0
450	**460**	39	22	9	0	0	0	0	0	0	0	0
460	**470**	41	23	10	0	0	0	0	0	0	0	0
470	**480**	42	24	11	0	0	0	0	0	0	0	0
480	**490**	44	25	12	0	0	0	0	0	0	0	0
490	**500**	45	27	13	1	0	0	0	0	0	0	0
500	**520**	48	29	15	2	0	0	0	0	0	0	0
520	**540**	51	32	17	4	0	0	0	0	0	0	0
540	**560**	54	35	19	6	0	0	0	0	0	0	0
560	**580**	57	38	21	8	0	0	0	0	0	0	0
580	**600**	60	41	23	10	0	0	0	0	0	0	0
600	**620**	63	44	25	12	0	0	0	0	0	0	0
620	**640**	66	47	28	14	2	0	0	0	0	0	0
640	**660**	69	50	31	16	4	0	0	0	0	0	0
660	**680**	72	53	34	18	6	0	0	0	0	0	0
680	**700**	75	56	37	20	8	0	0	0	0	0	0
700	**720**	78	59	40	22	10	0	0	0	0	0	0
720	**740**	81	62	43	24	12	0	0	0	0	0	0
740	**760**	84	65	46	27	14	1	0	0	0	0	0
760	**780**	87	68	49	30	16	3	0	0	0	0	0
780	**800**	90	71	52	33	18	5	0	0	0	0	0
800	**820**	93	74	55	36	20	7	0	0	0	0	0
820	**840**	96	77	58	39	22	9	0	0	0	0	0

SINGLE Persons—SEMIMONTHLY Payroll Period

(For Wages Paid in 2002)

If the wages are–		And the number of withholding allowances claimed is—										
At least	But less than	0	1	2	3	4	5	6	7	8	9	10
		The amount of income tax to be withheld is—										
$840	**$860**	$99	$80	$61	$42	$24	$11	$0	$0	$0	$0	$0
860	**880**	102	83	64	45	27	13	1	0	0	0	0
880	**900**	105	86	67	48	30	15	3	0	0	0	0
900	**920**	108	89	70	51	33	17	5	0	0	0	0
920	**940**	111	92	73	54	36	19	7	0	0	0	0
940	**960**	114	95	76	57	39	21	9	0	0	0	0
960	**980**	117	98	79	60	42	23	11	0	0	0	0
980	**1,000**	120	101	82	63	45	26	13	0	0	0	0
1,000	**1,020**	123	104	85	66	48	29	15	2	0	0	0
1,020	**1,040**	126	107	88	69	51	32	17	4	0	0	0
1,040	**1,060**	129	110	91	72	54	35	19	6	0	0	0
1,060	**1,080**	132	113	94	75	57	38	21	8	0	0	0
1,080	**1,100**	135	116	97	78	60	41	23	10	0	0	0
1,100	**1,120**	138	119	100	81	63	44	25	12	0	0	0
1,120	**1,140**	141	122	103	84	66	47	28	14	2	0	0
1,140	**1,160**	144	125	106	87	69	50	31	16	4	0	0
1,160	**1,180**	147	128	109	90	72	53	34	18	6	0	0
1,180	**1,200**	150	131	112	93	75	56	37	20	8	0	0
1,200	**1,220**	153	134	115	96	78	59	40	22	10	0	0
1,220	**1,240**	156	137	118	99	81	62	43	24	12	0	0
1,240	**1,260**	160	140	121	102	84	65	46	27	14	1	0
1,260	**1,280**	166	143	124	105	87	68	49	30	16	3	0
1,280	**1,300**	171	146	127	108	90	71	52	33	18	5	0
1,300	**1,320**	177	149	130	111	93	74	55	36	20	7	0
1,320	**1,340**	182	152	133	114	96	77	58	39	22	9	0
1,340	**1,360**	187	155	136	117	99	80	61	42	24	11	0
1,360	**1,380**	193	159	139	120	102	83	64	45	27	13	1
1,380	**1,400**	198	164	142	123	105	86	67	48	30	15	3
1,400	**1,420**	204	170	145	126	108	89	70	51	33	17	5
1,420	**1,440**	209	175	148	129	111	92	73	54	36	19	7
1,440	**1,460**	214	181	151	132	114	95	76	57	39	21	9
1,460	**1,480**	220	186	154	135	117	98	79	60	42	23	11
1,480	**1,500**	225	191	158	138	120	101	82	63	45	26	13
1,500	**1,520**	231	197	163	141	123	104	85	66	48	29	15
1,520	**1,540**	236	202	168	144	126	107	88	69	51	32	17
1,540	**1,560**	241	208	174	147	129	110	91	72	54	35	19
1,560	**1,580**	247	213	179	150	132	113	94	75	57	38	21
1,580	**1,600**	252	218	185	153	135	116	97	78	60	41	23
1,600	**1,620**	258	224	190	156	138	119	100	81	63	44	25
1,620	**1,640**	263	229	195	162	141	122	103	84	66	47	28
1,640	**1,660**	268	235	201	167	144	125	106	87	69	50	31
1,660	**1,680**	274	240	206	173	147	128	109	90	72	53	34
1,680	**1,700**	279	245	212	178	150	131	112	93	75	56	37
1,700	**1,720**	285	251	217	183	153	134	115	96	78	59	40
1,720	**1,740**	290	256	222	189	156	137	118	99	81	62	43
1,740	**1,760**	295	262	228	194	160	140	121	102	84	65	46
1,760	**1,780**	301	267	233	200	166	143	124	105	87	68	49
1,780	**1,800**	306	272	239	205	171	146	127	108	90	71	52
1,800	**1,820**	312	278	244	210	177	149	130	111	93	74	55
1,820	**1,840**	317	283	249	216	182	152	133	114	96	77	58
1,840	**1,860**	322	289	255	221	187	155	136	117	99	80	61
1,860	**1,880**	328	294	260	227	193	159	139	120	102	83	64
1,880	**1,900**	333	299	266	232	198	164	142	123	105	86	67
1,900	**1,920**	339	305	271	237	204	170	145	126	108	89	70
1,920	**1,940**	344	310	276	243	209	175	148	129	111	92	73
1,940	**1,960**	349	316	282	248	214	181	151	132	114	95	76
1,960	**1,980**	355	321	287	254	220	186	154	135	117	98	79
1,980	**2,000**	360	326	293	259	225	191	158	138	120	101	82
2,000	**2,020**	366	332	298	264	231	197	163	141	123	104	85
2,020	**2,040**	371	337	303	270	236	202	168	144	126	107	88
2,040	**2,060**	376	343	309	275	241	208	174	147	129	110	91
2,060	**2,080**	382	348	314	281	247	213	179	150	132	113	94
2,080	**2,100**	387	353	320	286	252	218	185	153	135	116	97
2,100	**2,120**	393	359	325	291	258	224	190	156	138	119	100
2,120	**2,140**	398	364	330	297	263	229	195	162	141	122	103

$2,140 and over Use Table 3(a) for a **SINGLE person** on page 34. Also see the instructions on page 32.

Page 45

MARRIED Persons—**SEMIMONTHLY** Payroll Period

(For Wages Paid in 2002)

If the wages are–		And the number of withholding allowances claimed is—										
At least	But less than	0	1	2	3	4	5	6	7	8	9	10
		The amount of income tax to be withheld is—										
$0	**$270**	$0	$0	$0	$0	$0	$0	$0	$0	$0	$0	$0
270	**280**	1	0	0	0	0	0	0	0	0	0	0
280	**290**	2	0	0	0	0	0	0	0	0	0	0
290	**300**	3	0	0	0	0	0	0	0	0	0	0
300	**310**	4	0	0	0	0	0	0	0	0	0	0
310	**320**	5	0	0	0	0	0	0	0	0	0	0
320	**330**	6	0	0	0	0	0	0	0	0	0	0
330	**340**	7	0	0	0	0	0	0	0	0	0	0
340	**350**	8	0	0	0	0	0	0	0	0	0	0
350	**360**	9	0	0	0	0	0	0	0	0	0	0
360	**370**	10	0	0	0	0	0	0	0	0	0	0
370	**380**	11	0	0	0	0	0	0	0	0	0	0
380	**390**	12	0	0	0	0	0	0	0	0	0	0
390	**400**	13	0	0	0	0	0	0	0	0	0	0
400	**410**	14	1	0	0	0	0	0	0	0	0	0
410	**420**	15	2	0	0	0	0	0	0	0	0	0
420	**430**	16	3	0	0	0	0	0	0	0	0	0
430	**440**	17	4	0	0	0	0	0	0	0	0	0
440	**450**	18	5	0	0	0	0	0	0	0	0	0
450	**460**	19	6	0	0	0	0	0	0	0	0	0
460	**470**	20	7	0	0	0	0	0	0	0	0	0
470	**480**	21	8	0	0	0	0	0	0	0	0	0
480	**490**	22	9	0	0	0	0	0	0	0	0	0
490	**500**	23	10	0	0	0	0	0	0	0	0	0
500	**520**	24	12	0	0	0	0	0	0	0	0	0
520	**540**	26	14	1	0	0	0	0	0	0	0	0
540	**560**	28	16	3	0	0	0	0	0	0	0	0
560	**580**	30	18	5	0	0	0	0	0	0	0	0
580	**600**	32	20	7	0	0	0	0	0	0	0	0
600	**620**	34	22	9	0	0	0	0	0	0	0	0
620	**640**	36	24	11	0	0	0	0	0	0	0	0
640	**660**	38	26	13	1	0	0	0	0	0	0	0
660	**680**	40	28	15	3	0	0	0	0	0	0	0
680	**700**	42	30	17	5	0	0	0	0	0	0	0
700	**720**	44	32	19	7	0	0	0	0	0	0	0
720	**740**	46	34	21	9	0	0	0	0	0	0	0
740	**760**	48	36	23	11	0	0	0	0	0	0	0
760	**780**	50	38	25	13	0	0	0	0	0	0	0
780	**800**	53	40	27	15	2	0	0	0	0	0	0
800	**820**	56	42	29	17	4	0	0	0	0	0	0
820	**840**	59	44	31	19	6	0	0	0	0	0	0
840	**860**	62	46	33	21	8	0	0	0	0	0	0
860	**880**	65	48	35	23	10	0	0	0	0	0	0
880	**900**	68	50	37	25	12	0	0	0	0	0	0
900	**920**	71	52	39	27	14	2	0	0	0	0	0
920	**940**	74	55	41	29	16	4	0	0	0	0	0
940	**960**	77	58	43	31	18	6	0	0	0	0	0
960	**980**	80	61	45	33	20	8	0	0	0	0	0
980	**1,000**	83	64	47	35	22	10	0	0	0	0	0
1,000	**1,020**	86	67	49	37	24	12	0	0	0	0	0
1,020	**1,040**	89	70	52	39	26	14	1	0	0	0	0
1,040	**1,060**	92	73	55	41	28	16	3	0	0	0	0
1,060	**1,080**	95	76	58	43	30	18	5	0	0	0	0
1,080	**1,100**	98	79	61	45	32	20	7	0	0	0	0
1,100	**1,120**	101	82	64	47	34	22	9	0	0	0	0
1,120	**1,140**	104	85	67	49	36	24	11	0	0	0	0
1,140	**1,160**	107	88	70	51	38	26	13	1	0	0	0
1,160	**1,180**	110	91	73	54	40	28	15	3	0	0	0
1,180	**1,200**	113	94	76	57	42	30	17	5	0	0	0
1,200	**1,220**	116	97	79	60	44	32	19	7	0	0	0
1,220	**1,240**	119	100	82	63	46	34	21	9	0	0	0
1,240	**1,260**	122	103	85	66	48	36	23	11	0	0	0
1,260	**1,280**	125	106	88	69	50	38	25	13	0	0	0
1,280	**1,300**	128	109	91	72	53	40	27	15	2	0	0
1,300	**1,320**	131	112	94	75	56	42	29	17	4	0	0
1,320	**1,340**	134	115	97	78	59	44	31	19	6	0	0
1,340	**1,360**	137	118	100	81	62	46	33	21	8	0	0
1,360	**1,380**	140	121	103	84	65	48	35	23	10	0	0
1,380	**1,400**	143	124	106	87	68	50	37	25	12	0	0
1,400	**1,420**	146	127	109	90	71	52	39	27	14	2	0

Page 46

MARRIED Persons—**SEMIMONTHLY** Payroll Period

(For Wages Paid in 2002)

If the wages are–		And the number of withholding allowances claimed is—										
At least	But less than	0	1	2	3	4	5	6	7	8	9	10
		The amount of income tax to be withheld is—										
$1,420	**$1,440**	$149	$130	$112	$93	$74	$55	$41	$29	$16	$4	$0
1,440	**1,460**	152	133	115	96	77	58	43	31	18	6	0
1,460	**1,480**	155	136	118	99	80	61	45	33	20	8	0
1,480	**1,500**	158	139	121	102	83	64	47	35	22	10	0
1,500	**1,520**	161	142	124	105	86	67	49	37	24	12	0
1,520	**1,540**	164	145	127	108	89	70	52	39	26	14	1
1,540	**1,560**	167	148	130	111	92	73	55	41	28	16	3
1,560	**1,580**	170	151	133	114	95	76	58	43	30	18	5
1,580	**1,600**	173	154	136	117	98	79	61	45	32	20	7
1,600	**1,620**	176	157	139	120	101	82	64	47	34	22	9
1,620	**1,640**	179	160	142	123	104	85	67	49	36	24	11
1,640	**1,660**	182	163	145	126	107	88	70	51	38	26	13
1,660	**1,680**	185	166	148	129	110	91	73	54	40	28	15
1,680	**1,700**	188	169	151	132	113	94	76	57	42	30	17
1,700	**1,720**	191	172	154	135	116	97	79	60	44	32	19
1,720	**1,740**	194	175	157	138	119	100	82	63	46	34	21
1,740	**1,760**	197	178	160	141	122	103	85	66	48	36	23
1,760	**1,780**	200	181	163	144	125	106	88	69	50	38	25
1,780	**1,800**	203	184	166	147	128	109	91	72	53	40	27
1,800	**1,820**	206	187	169	150	131	112	94	75	56	42	29
1,820	**1,840**	209	190	172	153	134	115	97	78	59	44	31
1,840	**1,860**	212	193	175	156	137	118	100	81	62	46	33
1,860	**1,880**	215	196	178	159	140	121	103	84	65	48	35
1,880	**1,900**	218	199	181	162	143	124	106	87	68	50	37
1,900	**1,920**	221	202	184	165	146	127	109	90	71	52	39
1,920	**1,940**	224	205	187	168	149	130	112	93	74	55	41
1,940	**1,960**	227	208	190	171	152	133	115	96	77	58	43
1,960	**1,980**	230	211	193	174	155	136	118	99	80	61	45
1,980	**2,000**	233	214	196	177	158	139	121	102	83	64	47
2,000	**2,020**	236	217	199	180	161	142	124	105	86	67	49
2,020	**2,040**	239	220	202	183	164	145	127	108	89	70	52
2,040	**2,060**	242	223	205	186	167	148	130	111	92	73	55
2,060	**2,080**	245	226	208	189	170	151	133	114	95	76	58
2,080	**2,100**	248	229	211	192	173	154	136	117	98	79	61
2,100	**2,120**	251	232	214	195	176	157	139	120	101	82	64
2,120	**2,140**	254	235	217	198	179	160	142	123	104	85	67
2,140	**2,160**	257	238	220	201	182	163	145	126	107	88	70
2,160	**2,180**	263	241	223	204	185	166	148	129	110	91	73
2,180	**2,200**	268	244	226	207	188	169	151	132	113	94	76
2,200	**2,220**	274	247	229	210	191	172	154	135	116	97	79
2,220	**2,240**	279	250	232	213	194	175	157	138	119	100	82
2,240	**2,260**	284	253	235	216	197	178	160	141	122	103	85
2,260	**2,280**	290	256	238	219	200	181	163	144	125	106	88
2,280	**2,300**	295	261	241	222	203	184	166	147	128	109	91
2,300	**2,320**	301	267	244	225	206	187	169	150	131	112	94
2,320	**2,340**	306	272	247	228	209	190	172	153	134	115	97
2,340	**2,360**	311	278	250	231	212	193	175	156	137	118	100
2,360	**2,380**	317	283	253	234	215	196	178	159	140	121	103
2,380	**2,400**	322	288	256	237	218	199	181	162	143	124	106
2,400	**2,420**	328	294	260	240	221	202	184	165	146	127	109
2,420	**2,440**	333	299	266	243	224	205	187	168	149	130	112
2,440	**2,460**	338	305	271	246	227	208	190	171	152	133	115
2,460	**2,480**	344	310	276	249	230	211	193	174	155	136	118
2,480	**2,500**	349	315	282	252	233	214	196	177	158	139	121
2,500	**2,520**	355	321	287	255	236	217	199	180	161	142	124
2,520	**2,540**	360	326	293	259	239	220	202	183	164	145	127
2,540	**2,560**	365	332	298	264	242	223	205	186	167	148	130
2,560	**2,580**	371	337	303	270	245	226	208	189	170	151	133
2,580	**2,600**	376	342	309	275	248	229	211	192	173	154	136
2,600	**2,620**	382	348	314	280	251	232	214	195	176	157	139
2,620	**2,640**	387	353	320	286	254	235	217	198	179	160	142
2,640	**2,660**	392	359	325	291	257	238	220	201	182	163	145
2,660	**2,680**	398	364	330	297	263	241	223	204	185	166	148
2,680	**2,700**	403	369	336	302	268	244	226	207	188	169	151
2,700	**2,720**	409	375	341	307	274	247	229	210	191	172	154
$2,720 and over		Use Table 3(b) for a **MARRIED person** on page 34. Also see the instructions on page 32.										

SINGLE Persons—**MONTHLY** Payroll Period

(For Wages Paid in 2002)

If the wages are–		And the number of withholding allowances claimed is—										
At least	But less than	0	1	2	3	4	5	6	7	8	9	10
		The amount of income tax to be withheld is—										
$0	$230	$0	$0	$0	$0	$0	$0	$0	$0	$0	$0	$0
230	240	1	0	0	0	0	0	0	0	0	0	0
240	250	2	0	0	0	0	0	0	0	0	0	0
250	260	3	0	0	0	0	0	0	0	0	0	0
260	270	4	0	0	0	0	0	0	0	0	0	0
270	280	5	0	0	0	0	0	0	0	0	0	0
280	290	6	0	0	0	0	0	0	0	0	0	0
290	300	7	0	0	0	0	0	0	0	0	0	0
300	320	9	0	0	0	0	0	0	0	0	0	0
320	340	11	0	0	0	0	0	0	0	0	0	0
340	360	13	0	0	0	0	0	0	0	0	0	0
360	380	15	0	0	0	0	0	0	0	0	0	0
380	400	17	0	0	0	0	0	0	0	0	0	0
400	420	19	0	0	0	0	0	0	0	0	0	0
420	440	21	0	0	0	0	0	0	0	0	0	0
440	460	23	0	0	0	0	0	0	0	0	0	0
460	480	25	0	0	0	0	0	0	0	0	0	0
480	500	27	2	0	0	0	0	0	0	0	0	0
500	520	29	4	0	0	0	0	0	0	0	0	0
520	540	31	6	0	0	0	0	0	0	0	0	0
540	560	33	8	0	0	0	0	0	0	0	0	0
560	580	35	10	0	0	0	0	0	0	0	0	0
580	600	37	12	0	0	0	0	0	0	0	0	0
600	640	40	15	0	0	0	0	0	0	0	0	0
640	680	44	19	0	0	0	0	0	0	0	0	0
680	720	48	23	0	0	0	0	0	0	0	0	0
720	760	53	27	2	0	0	0	0	0	0	0	0
760	800	59	31	6	0	0	0	0	0	0	0	0
800	840	65	35	10	0	0	0	0	0	0	0	0
840	880	71	39	14	0	0	0	0	0	0	0	0
880	920	77	43	18	0	0	0	0	0	0	0	0
920	960	83	47	22	0	0	0	0	0	0	0	0
960	1,000	89	52	26	1	0	0	0	0	0	0	0
1,000	1,040	95	58	30	5	0	0	0	0	0	0	0
1,040	1,080	101	64	34	9	0	0	0	0	0	0	0
1,080	1,120	107	70	38	13	0	0	0	0	0	0	0
1,120	1,160	113	76	42	17	0	0	0	0	0	0	0
1,160	1,200	119	82	46	21	0	0	0	0	0	0	0
1,200	1,240	125	88	50	25	0	0	0	0	0	0	0
1,240	1,280	131	94	56	29	4	0	0	0	0	0	0
1,280	1,320	137	100	62	33	8	0	0	0	0	0	0
1,320	1,360	143	106	68	37	12	0	0	0	0	0	0
1,360	1,400	149	112	74	41	16	0	0	0	0	0	0
1,400	1,440	155	118	80	45	20	0	0	0	0	0	0
1,440	1,480	161	124	86	49	24	0	0	0	0	0	0
1,480	1,520	167	130	92	55	28	3	0	0	0	0	0
1,520	1,560	173	136	98	61	32	7	0	0	0	0	0
1,560	1,600	179	142	104	67	36	11	0	0	0	0	0
1,600	1,640	185	148	110	73	40	15	0	0	0	0	0
1,640	1,680	191	154	116	79	44	19	0	0	0	0	0
1,680	1,720	197	160	122	85	48	23	0	0	0	0	0
1,720	1,760	203	166	128	91	53	27	2	0	0	0	0
1,760	1,800	209	172	134	97	59	31	6	0	0	0	0
1,800	1,840	215	178	140	103	65	35	10	0	0	0	0
1,840	1,880	221	184	146	109	71	39	14	0	0	0	0
1,880	1,920	227	190	152	115	77	43	18	0	0	0	0
1,920	1,960	233	196	158	121	83	47	22	0	0	0	0
1,960	2,000	239	202	164	127	89	52	26	1	0	0	0
2,000	2,040	245	208	170	133	95	58	30	5	0	0	0
2,040	2,080	251	214	176	139	101	64	34	9	0	0	0
2,080	2,120	257	220	182	145	107	70	38	13	0	0	0
2,120	2,160	263	226	188	151	113	76	42	17	0	0	0
2,160	2,200	269	232	194	157	119	82	46	21	0	0	0
2,200	2,240	275	238	200	163	125	88	50	25	0	0	0
2,240	2,280	281	244	206	169	131	94	56	29	4	0	0
2,280	2,320	287	250	212	175	137	100	62	33	8	0	0
2,320	2,360	293	256	218	181	143	106	68	37	12	0	0
2,360	2,400	299	262	224	187	149	112	74	41	16	0	0
2,400	2,440	305	268	230	193	155	118	80	45	20	0	0
2,440	2,480	311	274	236	199	161	124	86	49	24	0	0

Page 48

SINGLE Persons—MONTHLY Payroll Period

(For Wages Paid in 2002)

If the wages are–		And the number of withholding allowances claimed is—										
At least	But less than	0	1	2	3	4	5	6	7	8	9	10
		The amount of income tax to be withheld is—										
$2,480	**$2,520**	$321	$280	$242	$205	$167	$130	$92	$55	$28	$3	$0
2,520	**2,560**	332	286	248	211	173	136	98	61	32	7	0
2,560	**2,600**	342	292	254	217	179	142	104	67	36	11	0
2,600	**2,640**	353	298	260	223	185	148	110	73	40	15	0
2,640	**2,680**	364	304	266	229	191	154	116	79	44	19	0
2,680	**2,720**	375	310	272	235	197	160	122	85	48	23	0
2,720	**2,760**	386	318	278	241	203	166	128	91	53	27	2
2,760	**2,800**	396	329	284	247	209	172	134	97	59	31	6
2,800	**2,840**	407	340	290	253	215	178	140	103	65	35	10
2,840	**2,880**	418	350	296	259	221	184	146	109	71	39	14
2,880	**2,920**	429	361	302	265	227	190	152	115	77	43	18
2,920	**2,960**	440	372	308	271	233	196	158	121	83	47	22
2,960	**3,000**	450	383	315	277	239	202	164	127	89	52	26
3,000	**3,040**	461	394	326	283	245	208	170	133	95	58	30
3,040	**3,080**	472	404	337	289	251	214	176	139	101	64	34
3,080	**3,120**	483	415	348	295	257	220	182	145	107	70	38
3,120	**3,160**	494	426	359	301	263	226	188	151	113	76	42
3,160	**3,200**	504	437	369	307	269	232	194	157	119	82	46
3,200	**3,240**	515	448	380	313	275	238	200	163	125	88	50
3,240	**3,280**	526	458	391	323	281	244	206	169	131	94	56
3,280	**3,320**	537	469	402	334	287	250	212	175	137	100	62
3,320	**3,360**	548	480	413	345	293	256	218	181	143	106	68
3,360	**3,400**	558	491	423	356	299	262	224	187	149	112	74
3,400	**3,440**	569	502	434	367	305	268	230	193	155	118	80
3,440	**3,480**	580	512	445	377	311	274	236	199	161	124	86
3,480	**3,520**	591	523	456	388	321	280	242	205	167	130	92
3,520	**3,560**	602	534	467	399	332	286	248	211	173	136	98
3,560	**3,600**	612	545	477	410	342	292	254	217	179	142	104
3,600	**3,640**	623	556	488	421	353	298	260	223	185	148	110
3,640	**3,680**	634	566	499	431	364	304	266	229	191	154	116
3,680	**3,720**	645	577	510	442	375	310	272	235	197	160	122
3,720	**3,760**	656	588	521	453	386	318	278	241	203	166	128
3,760	**3,800**	666	599	531	464	396	329	284	247	209	172	134
3,800	**3,840**	677	610	542	475	407	340	290	253	215	178	140
3,840	**3,880**	688	620	553	485	418	350	296	259	221	184	146
3,880	**3,920**	699	631	564	496	429	361	302	265	227	190	152
3,920	**3,960**	710	642	575	507	440	372	308	271	233	196	158
3,960	**4,000**	720	653	585	518	450	383	315	277	239	202	164
4,000	**4,040**	731	664	596	529	461	394	326	283	245	208	170
4,040	**4,080**	742	674	607	539	472	404	337	289	251	214	176
4,080	**4,120**	753	685	618	550	483	415	348	295	257	220	182
4,120	**4,160**	764	696	629	561	494	426	359	301	263	226	188
4,160	**4,200**	774	707	639	572	504	437	369	307	269	232	194
4,200	**4,240**	785	718	650	583	515	448	380	313	275	238	200
4,240	**4,280**	796	728	661	593	526	458	391	323	281	244	206
4,280	**4,320**	807	739	672	604	537	469	402	334	287	250	212
4,320	**4,360**	818	750	683	615	548	480	413	345	293	256	218
4,360	**4,400**	828	761	693	626	558	491	423	356	299	262	224
4,400	**4,440**	839	772	704	637	569	502	434	367	305	268	230
4,440	**4,480**	850	782	715	647	580	512	445	377	311	274	236
4,480	**4,520**	861	793	726	658	591	523	456	388	321	280	242
4,520	**4,560**	872	804	737	669	602	534	467	399	332	286	248
4,560	**4,600**	882	815	747	680	612	545	477	410	342	292	254
4,600	**4,640**	893	826	758	691	623	556	488	421	353	298	260
4,640	**4,680**	904	836	769	701	634	566	499	431	364	304	266
4,680	**4,720**	915	847	780	712	645	577	510	442	375	310	272
4,720	**4,760**	926	858	791	723	656	588	521	453	386	318	278
4,760	**4,800**	936	869	801	734	666	599	531	464	396	329	284
4,800	**4,840**	947	880	812	745	677	610	542	475	407	340	290
4,840	**4,880**	958	890	823	755	688	620	553	485	418	350	296
4,880	**4,920**	969	901	834	766	699	631	564	496	429	361	302
4,920	**4,960**	980	912	845	777	710	642	575	507	440	372	308
4,960	**5,000**	990	923	855	788	720	653	585	518	450	383	315
5,000	**5,040**	1,001	934	866	799	731	664	596	529	461	394	326
5,040	**5,080**	1,012	944	877	809	742	674	607	539	472	404	337

$5,080 and over Use Table 4(a) for a **SINGLE person** on page 34. Also see the instructions on page 32.

MARRIED Persons—**MONTHLY** Payroll Period

(For Wages Paid in 2002)

If the wages are–		And the number of withholding allowances claimed is—										
At least	But less than	0	1	2	3	4	5	6	7	8	9	10
		The amount of income tax to be withheld is—										
$0	$540	$0	$0	$0	$0	$0	$0	$0	$0	$0	$0	$0
540	560	1	0	0	0	0	0	0	0	0	0	0
560	580	3	0	0	0	0	0	0	0	0	0	0
580	600	5	0	0	0	0	0	0	0	0	0	0
600	640	8	0	0	0	0	0	0	0	0	0	0
640	680	12	0	0	0	0	0	0	0	0	0	0
680	720	16	0	0	0	0	0	0	0	0	0	0
720	760	20	0	0	0	0	0	0	0	0	0	0
760	800	24	0	0	0	0	0	0	0	0	0	0
800	840	28	3	0	0	0	0	0	0	0	0	0
840	880	32	7	0	0	0	0	0	0	0	0	0
880	920	36	11	0	0	0	0	0	0	0	0	0
920	960	40	15	0	0	0	0	0	0	0	0	0
960	1,000	44	19	0	0	0	0	0	0	0	0	0
1,000	1,040	48	23	0	0	0	0	0	0	0	0	0
1,040	1,080	52	27	2	0	0	0	0	0	0	0	0
1,080	1,120	56	31	6	0	0	0	0	0	0	0	0
1,120	1,160	60	35	10	0	0	0	0	0	0	0	0
1,160	1,200	64	39	14	0	0	0	0	0	0	0	0
1,200	1,240	68	43	18	0	0	0	0	0	0	0	0
1,240	1,280	72	47	22	0	0	0	0	0	0	0	0
1,280	1,320	76	51	26	1	0	0	0	0	0	0	0
1,320	1,360	80	55	30	5	0	0	0	0	0	0	0
1,360	1,400	84	59	34	9	0	0	0	0	0	0	0
1,400	1,440	88	63	38	13	0	0	0	0	0	0	0
1,440	1,480	92	67	42	17	0	0	0	0	0	0	0
1,480	1,520	96	71	46	21	0	0	0	0	0	0	0
1,520	1,560	100	75	50	25	0	0	0	0	0	0	0
1,560	1,600	106	79	54	29	4	0	0	0	0	0	0
1,600	1,640	112	83	58	33	8	0	0	0	0	0	0
1,640	1,680	118	87	62	37	12	0	0	0	0	0	0
1,680	1,720	124	91	66	41	16	0	0	0	0	0	0
1,720	1,760	130	95	70	45	20	0	0	0	0	0	0
1,760	1,800	136	99	74	49	24	0	0	0	0	0	0
1,800	1,840	142	105	78	53	28	3	0	0	0	0	0
1,840	1,880	148	111	82	57	32	7	0	0	0	0	0
1,880	1,920	154	117	86	61	36	11	0	0	0	0	0
1,920	1,960	160	123	90	65	40	15	0	0	0	0	0
1,960	2,000	166	129	94	69	44	19	0	0	0	0	0
2,000	2,040	172	135	98	73	48	23	0	0	0	0	0
2,040	2,080	178	141	103	77	52	27	2	0	0	0	0
2,080	2,120	184	147	109	81	56	31	6	0	0	0	0
2,120	2,160	190	153	115	85	60	35	10	0	0	0	0
2,160	2,200	196	159	121	89	64	39	14	0	0	0	0
2,200	2,240	202	165	127	93	68	43	18	0	0	0	0
2,240	2,280	208	171	133	97	72	47	22	0	0	0	0
2,280	2,320	214	177	139	102	76	51	26	1	0	0	0
2,320	2,360	220	183	145	108	80	55	30	5	0	0	0
2,360	2,400	226	189	151	114	84	59	34	9	0	0	0
2,400	2,440	232	195	157	120	88	63	38	13	0	0	0
2,440	2,480	238	201	163	126	92	67	42	17	0	0	0
2,480	2,520	244	207	169	132	96	71	46	21	0	0	0
2,520	2,560	250	213	175	138	100	75	50	25	0	0	0
2,560	2,600	256	219	181	144	106	79	54	29	4	0	0
2,600	2,640	262	225	187	150	112	83	58	33	8	0	0
2,640	2,680	268	231	193	156	118	87	62	37	12	0	0
2,680	2,720	274	237	199	162	124	91	66	41	16	0	0
2,720	2,760	280	243	205	168	130	95	70	45	20	0	0
2,760	2,800	286	249	211	174	136	99	74	49	24	0	0
2,800	2,840	292	255	217	180	142	105	78	53	28	3	0
2,840	2,880	298	261	223	186	148	111	82	57	32	7	0
2,880	2,920	304	267	229	192	154	117	86	61	36	11	0
2,920	2,960	310	273	235	198	160	123	90	65	40	15	0
2,960	3,000	316	279	241	204	166	129	94	69	44	19	0
3,000	3,040	322	285	247	210	172	135	98	73	48	23	0
3,040	3,080	328	291	253	216	178	141	103	77	52	27	2
3,080	3,120	334	297	259	222	184	147	109	81	56	31	6
3,120	3,160	340	303	265	228	190	153	115	85	60	35	10
3,160	3,200	346	309	271	234	196	159	121	89	64	39	14
3,200	3,240	352	315	277	240	202	165	127	93	68	43	18

Page 50

MARRIED Persons—MONTHLY Payroll Period

(For Wages Paid in 2002)

If the wages are–		And the number of withholding allowances claimed is—										
At least	But less than	0	1	2	3	4	5	6	7	8	9	10
		The amount of income tax to be withheld is—										
$3,240	$3,280	$358	$321	$283	$246	$208	$171	$133	$97	$72	$47	$22
3,280	3,320	364	327	289	252	214	177	139	102	76	51	26
3,320	3,360	370	333	295	258	220	183	145	108	80	55	30
3,360	3,400	376	339	301	264	226	189	151	114	84	59	34
3,400	3,440	382	345	307	270	232	195	157	120	88	63	38
3,440	3,480	388	351	313	276	238	201	163	126	92	67	42
3,480	3,520	394	357	319	282	244	207	169	132	96	71	46
3,520	3,560	400	363	325	288	250	213	175	138	100	75	50
3,560	3,600	406	369	331	294	256	219	181	144	106	79	54
3,600	3,640	412	375	337	300	262	225	187	150	112	83	58
3,640	3,680	418	381	343	306	268	231	193	156	118	87	62
3,680	3,720	424	387	349	312	274	237	199	162	124	91	66
3,720	3,760	430	393	355	318	280	243	205	168	130	95	70
3,760	3,800	436	399	361	324	286	249	211	174	136	99	74
3,800	3,840	442	405	367	330	292	255	217	180	142	105	78
3,840	3,880	448	411	373	336	298	261	223	186	148	111	82
3,880	3,920	454	417	379	342	304	267	229	192	154	117	86
3,920	3,960	460	423	385	348	310	273	235	198	160	123	90
3,960	4,000	466	429	391	354	316	279	241	204	166	129	94
4,000	4,040	472	435	397	360	322	285	247	210	172	135	98
4,040	4,080	478	441	403	366	328	291	253	216	178	141	103
4,080	4,120	484	447	409	372	334	297	259	222	184	147	109
4,120	4,160	490	453	415	378	340	303	265	228	190	153	115
4,160	4,200	496	459	421	384	346	309	271	234	196	159	121
4,200	4,240	502	465	427	390	352	315	277	240	202	165	127
4,240	4,280	508	471	433	396	358	321	283	246	208	171	133
4,280	4,320	515	477	439	402	364	327	289	252	214	177	139
4,320	4,360	526	483	445	408	370	333	295	258	220	183	145
4,360	4,400	536	489	451	414	376	339	301	264	226	189	151
4,400	4,440	547	495	457	420	382	345	307	270	232	195	157
4,440	4,480	558	501	463	426	388	351	313	276	238	201	163
4,480	4,520	569	507	469	432	394	357	319	282	244	207	169
4,520	4,560	580	513	475	438	400	363	325	288	250	213	175
4,560	4,600	590	523	481	444	406	369	331	294	256	219	181
4,600	4,640	601	534	487	450	412	375	337	300	262	225	187
4,640	4,680	612	545	493	456	418	381	343	306	268	231	193
4,680	4,720	623	555	499	462	424	387	349	312	274	237	199
4,720	4,760	634	566	505	468	430	393	355	318	280	243	205
4,760	4,800	644	577	511	474	436	399	361	324	286	249	211
4,800	4,840	655	588	520	480	442	405	367	330	292	255	217
4,840	4,880	666	599	531	486	448	411	373	336	298	261	223
4,880	4,920	677	609	542	492	454	417	379	342	304	267	229
4,920	4,960	688	620	553	498	460	423	385	348	310	273	235
4,960	5,000	698	631	563	504	466	429	391	354	316	279	241
5,000	5,040	709	642	574	510	472	435	397	360	322	285	247
5,040	5,080	720	653	585	518	478	441	403	366	328	291	253
5,080	5,120	731	663	596	528	484	447	409	372	334	297	259
5,120	5,160	742	674	607	539	490	453	415	378	340	303	265
5,160	5,200	752	685	617	550	496	459	421	384	346	309	271
5,200	5,240	763	696	628	561	502	465	427	390	352	315	277
5,240	5,280	774	707	639	572	508	471	433	396	358	321	283
5,280	5,320	785	717	650	582	515	477	439	402	364	327	289
5,320	5,360	796	728	661	593	526	483	445	408	370	333	295
5,360	5,400	806	739	671	604	536	489	451	414	376	339	301
5,400	5,440	817	750	682	615	547	495	457	420	382	345	307
5,440	5,480	828	761	693	626	558	501	463	426	388	351	313
5,480	5,520	839	771	704	636	569	507	469	432	394	357	319
5,520	5,560	850	782	715	647	580	513	475	438	400	363	325
5,560	5,600	860	793	725	658	590	523	481	444	406	369	331
5,600	5,640	871	804	736	669	601	534	487	450	412	375	337
5,640	5,680	882	815	747	680	612	545	493	456	418	381	343
5,680	5,720	893	825	758	690	623	555	499	462	424	387	349
5,720	5,760	904	836	769	701	634	566	505	468	430	393	355
5,760	5,800	914	847	779	712	644	577	511	474	436	399	361
5,800	5,840	925	858	790	723	655	588	520	480	442	405	367

$5,840 and over Use Table 4(b) for a **MARRIED person** on page 34. Also see the instructions on page 32.

SINGLE Persons—**DAILY OR MISCELLANEOUS** Payroll Period

(For Wages Paid in 2002)

If the wages are–		And the number of withholding allowances claimed is—										
At least	But less than	0	1	2	3	4	5	6	7	8	9	10
		The amount of income tax to be withheld is—										
$0	**$15**	$0	$0	$0	$0	$0	$0	$0	$0	$0	$0	$0
15	**18**	1	0	0	0	0	0	0	0	0	0	0
18	**21**	1	0	0	0	0	0	0	0	0	0	0
21	**24**	1	0	0	0	0	0	0	0	0	0	0
24	**27**	2	0	0	0	0	0	0	0	0	0	0
27	**30**	2	1	0	0	0	0	0	0	0	0	0
30	**33**	2	1	0	0	0	0	0	0	0	0	0
33	**36**	3	1	0	0	0	0	0	0	0	0	0
36	**39**	3	2	0	0	0	0	0	0	0	0	0
39	**42**	3	2	1	0	0	0	0	0	0	0	0
42	**45**	4	2	1	0	0	0	0	0	0	0	0
45	**48**	4	3	1	0	0	0	0	0	0	0	0
48	**51**	5	3	2	0	0	0	0	0	0	0	0
51	**54**	5	3	2	1	0	0	0	0	0	0	0
54	**57**	6	4	2	1	0	0	0	0	0	0	0
57	**60**	6	4	3	1	0	0	0	0	0	0	0
60	**63**	7	5	3	2	1	0	0	0	0	0	0
63	**66**	7	5	4	2	1	0	0	0	0	0	0
66	**69**	7	6	4	2	1	0	0	0	0	0	0
69	**72**	8	6	4	3	1	0	0	0	0	0	0
72	**75**	8	7	5	3	2	1	0	0	0	0	0
75	**78**	9	7	5	4	2	1	0	0	0	0	0
78	**81**	9	8	6	4	2	1	0	0	0	0	0
81	**84**	10	8	6	5	3	1	0	0	0	0	0
84	**87**	10	8	7	5	3	2	1	0	0	0	0
87	**90**	11	9	7	5	4	2	1	0	0	0	0
90	**93**	11	9	8	6	4	2	1	0	0	0	0
93	**96**	12	10	8	6	5	3	2	0	0	0	0
96	**99**	12	10	9	7	5	3	2	1	0	0	0
99	**102**	12	11	9	7	5	4	2	1	0	0	0
102	**105**	13	11	9	8	6	4	2	1	0	0	0
105	**108**	13	12	10	8	6	5	3	2	0	0	0
108	**111**	14	12	10	9	7	5	3	2	1	0	0
111	**114**	14	12	11	9	7	6	4	2	1	0	0
114	**117**	15	13	11	9	8	6	4	3	1	0	0
117	**120**	16	13	12	10	8	6	5	3	2	0	0
120	**123**	16	14	12	10	9	7	5	3	2	1	0
123	**126**	17	14	13	11	9	7	6	4	2	1	0
126	**129**	18	15	13	11	10	8	6	4	3	1	0
129	**132**	19	16	13	12	10	8	7	5	3	2	0
132	**135**	20	17	14	12	10	9	7	5	4	2	1
135	**138**	21	17	14	13	11	9	7	6	4	2	1
138	**141**	21	18	15	13	11	10	8	6	4	3	1
141	**144**	22	19	16	14	12	10	8	7	5	3	2
144	**147**	23	20	17	14	12	11	9	7	5	4	2
147	**150**	24	21	18	14	13	11	9	7	6	4	2
150	**153**	25	21	18	15	13	11	10	8	6	4	3
153	**156**	25	22	19	16	14	12	10	8	7	5	3
156	**159**	26	23	20	17	14	12	11	9	7	5	4
159	**162**	27	24	21	18	15	13	11	9	8	6	4
162	**165**	28	25	22	18	15	13	11	10	8	6	5
165	**168**	29	25	22	19	16	14	12	10	8	7	5
168	**171**	29	26	23	20	17	14	12	11	9	7	5
171	**174**	30	27	24	21	18	15	13	11	9	8	6
174	**177**	31	28	25	22	19	15	13	12	10	8	6
177	**180**	32	29	26	23	19	16	14	12	10	9	7
180	**183**	33	30	26	23	20	17	14	12	11	9	7
183	**186**	33	30	27	24	21	18	15	13	11	9	8
186	**189**	34	31	28	25	22	19	16	13	12	10	8
189	**192**	35	32	29	26	23	20	16	14	12	10	9
192	**195**	36	33	30	27	23	20	17	14	13	11	9
195	**198**	37	34	30	27	24	21	18	15	13	11	10
198	**201**	38	34	31	28	25	22	19	16	13	12	10
201	**204**	38	35	32	29	26	23	20	17	14	12	10
204	**207**	39	36	33	30	27	24	20	17	14	13	11
207	**210**	40	37	34	31	27	24	21	18	15	13	11
210	**213**	41	38	35	31	28	25	22	19	16	13	12
213	**216**	42	38	35	32	29	26	23	20	17	14	12
216	**219**	42	39	36	33	30	27	24	21	17	14	13
219	**222**	43	40	37	34	31	28	24	21	18	15	13

Page 52

SINGLE Persons—**DAILY OR MISCELLANEOUS** Payroll Period

(For Wages Paid in 2002)

If the wages are–		And the number of withholding allowances claimed is—										
At least	But less than	0	1	2	3	4	5	6	7	8	9	10
		The amount of income tax to be withheld is—										
$222	**$225**	$44	$41	$38	$35	$32	$28	$25	$22	$19	$16	$14
225	**228**	45	42	39	35	32	29	26	23	20	17	14
228	**231**	46	43	39	36	33	30	27	24	21	18	14
231	**234**	46	43	40	37	34	31	28	25	22	18	15
234	**237**	47	44	41	38	35	32	29	25	22	19	16
237	**240**	48	45	42	39	36	32	29	26	23	20	17
240	**243**	49	46	43	40	36	33	30	27	24	21	18
243	**246**	50	47	43	40	37	34	31	28	25	22	19
246	**249**	50	47	44	41	38	35	32	29	26	22	19
249	**252**	51	48	45	42	39	36	33	29	26	23	20
252	**255**	52	49	46	43	40	37	33	30	27	24	21
255	**258**	53	50	47	44	40	37	34	31	28	25	22
258	**261**	54	51	47	44	41	38	35	32	29	26	23
261	**264**	55	51	48	45	42	39	36	33	30	26	23
264	**267**	56	52	49	46	43	40	37	34	30	27	24
267	**270**	57	53	50	47	44	41	37	34	31	28	25
270	**273**	58	54	51	48	44	41	38	35	32	29	26
273	**276**	59	55	52	48	45	42	39	36	33	30	27
276	**279**	59	56	52	49	46	43	40	37	34	31	27
279	**282**	60	57	53	50	47	44	41	38	34	31	28
282	**285**	61	58	54	51	48	45	42	38	35	32	29
285	**288**	62	59	55	52	49	45	42	39	36	33	30
288	**291**	63	60	56	53	49	46	43	40	37	34	31
291	**294**	64	60	57	54	50	47	44	41	38	35	31
294	**297**	65	61	58	54	51	48	45	42	39	35	32
297	**300**	66	62	59	55	52	49	46	42	39	36	33
300	**303**	67	63	60	56	53	49	46	43	40	37	34
303	**306**	68	64	61	57	54	50	47	44	41	38	35
306	**309**	68	65	61	58	55	51	48	45	42	39	36
309	**312**	69	66	62	59	55	52	49	46	43	39	36
312	**315**	70	67	63	60	56	53	50	46	43	40	37
315	**318**	71	68	64	61	57	54	50	47	44	41	38
318	**321**	72	69	65	62	58	55	51	48	45	42	39
321	**324**	73	69	66	63	59	56	52	49	46	43	40
324	**327**	74	70	67	63	60	57	53	50	47	43	40
327	**330**	75	71	68	64	61	57	54	51	47	44	41
330	**333**	76	72	69	65	62	58	55	51	48	45	42
333	**336**	77	73	70	66	63	59	56	52	49	46	43
336	**339**	77	74	70	67	64	60	57	53	50	47	44
339	**341**	78	75	71	68	64	61	57	54	51	47	44
341	**343**	79	75	72	68	65	61	58	55	51	48	45
343	**345**	79	76	72	69	66	62	59	55	52	48	45
345	**347**	80	77	73	70	66	63	59	56	52	49	46
347	**349**	81	77	74	70	67	63	60	56	53	50	46
349	**351**	81	78	74	71	67	64	60	57	53	50	47
351	**353**	82	78	75	71	68	64	61	58	54	51	48
353	**355**	82	79	75	72	69	65	62	58	55	51	48
355	**357**	83	80	76	73	69	66	62	59	55	52	49
357	**359**	84	80	77	73	70	66	63	59	56	52	49
359	**361**	84	81	77	74	70	67	63	60	56	53	50
361	**363**	85	81	78	74	71	67	64	61	57	54	50
363	**365**	85	82	78	75	72	68	65	61	58	54	51
365	**367**	86	83	79	76	72	69	65	62	58	55	51
367	**369**	87	83	80	76	73	69	66	62	59	55	52
369	**371**	87	84	80	77	73	70	66	63	59	56	53
371	**373**	88	84	81	77	74	70	67	64	60	57	53
373	**375**	88	85	81	78	75	71	68	64	61	57	54
375	**377**	89	86	82	79	75	72	68	65	61	58	54
377	**379**	90	86	83	79	76	72	69	65	62	58	55
379	**381**	90	87	83	80	76	73	69	66	62	59	56
381	**383**	91	87	84	80	77	73	70	67	63	60	56
383	**385**	91	88	84	81	78	74	71	67	64	60	57
385	**387**	92	89	85	82	78	75	71	68	64	61	57
387	**389**	93	89	86	82	79	75	72	68	65	61	58
389	**391**	93	90	86	83	79	76	72	69	65	62	59
$391 and over		Use Table 8(a) for a **SINGLE person** on page 35. Also see the instructions on page 32.										

Page 53

MARRIED Persons—**DAILY OR MISCELLANEOUS** Payroll Period

(For Wages Paid in 2002)

If the wages are–		And the number of withholding allowances claimed is—										
At least	But less than	0	1	2	3	4	5	6	7	8	9	10
		The amount of income tax to be withheld is—										
$0	$30	$0	$0	$0	$0	$0	$0	$0	$0	$0	$0	$0
30	33	1	0	0	0	0	0	0	0	0	0	0
33	36	1	0	0	0	0	0	0	0	0	0	0
36	39	1	0	0	0	0	0	0	0	0	0	0
39	42	2	0	0	0	0	0	0	0	0	0	0
42	45	2	1	0	0	0	0	0	0	0	0	0
45	48	2	1	0	0	0	0	0	0	0	0	0
48	51	2	1	0	0	0	0	0	0	0	0	0
51	54	3	2	0	0	0	0	0	0	0	0	0
54	57	3	2	1	0	0	0	0	0	0	0	0
57	60	3	2	1	0	0	0	0	0	0	0	0
60	63	4	3	1	0	0	0	0	0	0	0	0
63	66	4	3	2	1	0	0	0	0	0	0	0
66	69	4	3	2	1	0	0	0	0	0	0	0
69	72	5	3	2	1	0	0	0	0	0	0	0
72	75	5	4	3	1	0	0	0	0	0	0	0
75	78	5	4	3	2	1	0	0	0	0	0	0
78	81	6	4	3	2	1	0	0	0	0	0	0
81	84	6	5	3	2	1	0	0	0	0	0	0
84	87	7	5	4	3	1	0	0	0	0	0	0
87	90	7	6	4	3	2	1	0	0	0	0	0
90	93	8	6	4	3	2	1	0	0	0	0	0
93	96	8	6	5	4	2	1	0	0	0	0	0
96	99	9	7	5	4	3	2	0	0	0	0	0
99	102	9	7	6	4	3	2	1	0	0	0	0
102	105	9	8	6	4	3	2	1	0	0	0	0
105	108	10	8	6	5	4	2	1	0	0	0	0
108	111	10	9	7	5	4	3	2	0	0	0	0
111	114	11	9	7	6	4	3	2	1	0	0	0
114	117	11	10	8	6	4	3	2	1	0	0	0
117	120	12	10	8	7	5	4	2	1	0	0	0
120	123	12	10	9	7	5	4	3	2	0	0	0
123	126	13	11	9	7	6	4	3	2	1	0	0
126	129	13	11	10	8	6	5	3	2	1	0	0
129	132	14	12	10	8	7	5	4	2	1	0	0
132	135	14	12	11	9	7	5	4	3	2	0	0
135	138	14	13	11	9	8	6	4	3	2	1	0
138	141	15	13	11	10	8	6	5	3	2	1	0
141	144	15	14	12	10	8	7	5	4	3	1	0
144	147	16	14	12	11	9	7	5	4	3	2	1
147	150	16	15	13	11	9	8	6	4	3	2	1
150	153	17	15	13	12	10	8	6	5	3	2	1
153	156	17	15	14	12	10	8	7	5	4	3	1
156	159	18	16	14	12	11	9	7	5	4	3	2
159	162	18	16	15	13	11	9	8	6	4	3	2
162	165	18	17	15	13	12	10	8	6	5	3	2
165	168	19	17	15	14	12	10	9	7	5	4	3
168	171	19	18	16	14	12	11	9	7	6	4	3
171	174	20	18	16	15	13	11	9	8	6	4	3
174	177	20	19	17	15	13	12	10	8	6	5	4
177	180	21	19	17	16	14	12	10	9	7	5	4
180	183	21	19	18	16	14	13	11	9	7	6	4
183	186	22	20	18	16	15	13	11	10	8	6	4
186	189	22	20	19	17	15	13	12	10	8	7	5
189	192	23	21	19	17	16	14	12	10	9	7	5
192	195	23	21	20	18	16	14	13	11	9	7	6
195	198	23	22	20	18	17	15	13	11	10	8	6
198	201	24	22	20	19	17	15	14	12	10	8	7
201	204	25	23	21	19	17	16	14	12	11	9	7
204	207	26	23	21	20	18	16	14	13	11	9	7
207	210	26	24	22	20	18	17	15	13	11	10	8
210	213	27	24	22	21	19	17	15	14	12	10	8
213	216	28	25	23	21	19	17	16	14	12	11	9
216	219	29	26	23	21	20	18	16	14	13	11	9
219	222	30	27	24	22	20	18	17	15	13	11	10
222	225	31	27	24	22	21	19	17	15	14	12	10
225	228	31	28	25	23	21	19	18	16	14	12	11
228	231	32	29	26	23	21	20	18	16	15	13	11
231	234	33	30	27	24	22	20	18	17	15	13	12
234	237	34	31	28	24	22	21	19	17	15	14	12

Page 54

MARRIED Persons—**DAILY OR MISCELLANEOUS** Payroll Period

(For Wages Paid in 2002)

If the wages are–		And the number of withholding allowances claimed is—										
At least	But less than	0	1	2	3	4	5	6	7	8	9	10
		The amount of income tax to be withheld is—										
$237	**$240**	$35	$31	$28	$25	$23	$21	$19	$18	$16	$14	$12
240	**243**	35	32	29	26	23	22	20	18	16	15	13
243	**246**	36	33	30	27	24	22	20	19	17	15	13
246	**249**	37	34	31	28	25	22	21	19	17	16	14
249	**252**	38	35	32	28	25	23	21	19	18	16	14
252	**255**	39	36	32	29	26	23	22	20	18	16	15
255	**258**	39	36	33	30	27	24	22	20	19	17	15
258	**261**	40	37	34	31	28	25	23	21	19	17	16
261	**264**	41	38	35	32	29	25	23	21	20	18	16
264	**267**	42	39	36	33	29	26	23	22	20	18	16
267	**270**	43	40	36	33	30	27	24	22	20	19	17
270	**273**	43	40	37	34	31	28	25	23	21	19	17
273	**276**	44	41	38	35	32	29	26	23	21	20	18
276	**279**	45	42	39	36	33	30	26	23	22	20	18
279	**282**	46	43	40	37	33	30	27	24	22	20	19
282	**285**	47	44	40	37	34	31	28	25	23	21	19
285	**288**	48	44	41	38	35	32	29	26	23	21	20
288	**291**	48	45	42	39	36	33	30	27	24	22	20
291	**294**	49	46	43	40	37	34	30	27	24	22	21
294	**297**	50	47	44	41	38	34	31	28	25	23	21
297	**300**	51	48	45	41	38	35	32	29	26	23	21
300	**303**	52	48	45	42	39	36	33	30	27	24	22
303	**306**	52	49	46	43	40	37	34	31	27	24	22
306	**309**	53	50	47	44	41	38	35	31	28	25	23
309	**312**	54	51	48	45	42	38	35	32	29	26	23
312	**315**	55	52	49	45	42	39	36	33	30	27	24
315	**318**	56	53	49	46	43	40	37	34	31	28	24
318	**321**	56	53	50	47	44	41	38	35	32	28	25
321	**324**	57	54	51	48	45	42	39	35	32	29	26
324	**327**	58	55	52	49	46	42	39	36	33	30	27
327	**330**	59	56	53	50	46	43	40	37	34	31	28
330	**333**	60	57	53	50	47	44	41	38	35	32	29
333	**336**	60	57	54	51	48	45	42	39	36	32	29
336	**339**	61	58	55	52	49	46	43	39	36	33	30
339	**341**	62	59	56	53	50	46	43	40	37	34	31
341	**343**	63	59	56	53	50	47	44	41	38	34	31
343	**345**	63	60	57	54	51	47	44	41	38	35	32
345	**347**	64	60	57	54	51	48	45	42	39	36	32
347	**349**	64	61	58	55	52	49	45	42	39	36	33
349	**351**	65	62	58	55	52	49	46	43	40	37	34
351	**353**	65	62	59	56	53	50	47	43	40	37	34
353	**355**	66	63	60	56	53	50	47	44	41	38	35
355	**357**	66	63	60	57	54	51	48	44	41	38	35
357	**359**	67	64	61	57	54	51	48	45	42	39	36
359	**361**	67	64	61	58	55	52	49	46	42	39	36
361	**363**	68	65	62	59	55	52	49	46	43	40	37
363	**365**	68	65	62	59	56	53	50	47	44	40	37
365	**367**	69	66	63	60	57	53	50	47	44	41	38
367	**369**	70	66	63	60	57	54	51	48	45	42	38
369	**371**	70	67	64	61	58	55	51	48	45	42	39
371	**373**	71	68	64	61	58	55	52	49	46	43	39
373	**375**	71	68	65	62	59	56	52	49	46	43	40
375	**377**	72	69	65	62	59	56	53	50	47	44	41
377	**379**	72	69	66	63	60	57	54	50	47	44	41
379	**381**	73	70	67	63	60	57	54	51	48	45	42
381	**383**	73	70	67	64	61	58	55	52	48	45	42
383	**385**	74	71	68	65	61	58	55	52	49	46	43
385	**387**	74	71	68	65	62	59	56	53	49	46	43
387	**389**	75	72	69	66	62	59	56	53	50	47	44
389	**391**	75	72	69	66	63	60	57	54	51	47	44
391	**393**	76	73	70	67	64	60	57	54	51	48	45
393	**395**	77	73	70	67	64	61	58	55	52	49	45
395	**397**	77	74	71	68	65	62	58	55	52	49	46
397	**399**	78	75	71	68	65	62	59	56	53	50	46
399	**401**	78	75	72	69	66	63	59	56	53	50	47
$401 and over		Use Table 8(b) for a **MARRIED person** on page 35. Also see the instructions on page 32.										

Tables for Percentage Method of Advance EIC Payments

(For Wages Paid in 2002)

Table 1. WEEKLY Payroll Period

(a) SINGLE or HEAD OF HOUSEHOLD

If the amount of wages (before deducting withholding allowances) is: Over—	But not over—	The amount of payment to be made is:
$0	$141	20.40% of wages
$141	$260	$29
$260		$29 less 9.588% of wages in excess of $260

(b) MARRIED Without Spouse Filing Certificate

If the amount of wages (before deducting withholding allowances) is: Over—	But not over—	The amount of payment to be made is:
$0	$141	20.40% of wages
$141	$279	$29
$279		$29 less 9.588% of wages in excess of $279

(c) MARRIED With Both Spouses Filing Certificate

If the amount of wages (before deducting withholding allowances) is: Over—	But not over—	The amount of payment to be made is:
$0	$70	20.40% of wages
$70	$139	$14
$139		$14 less 9.588% of wages in excess of $139

Table 2. BIWEEKLY Payroll Period

(a) SINGLE or HEAD OF HOUSEHOLD

If the amount of wages (before deducting withholding allowances) is: Over—	But not over—	The amount of payment to be made is:
$0	$283	20.40% of wages
$283	$520	$58
$520		$58 less 9.588% of wages in excess of $520

(b) MARRIED Without Spouse Filing Certificate

If the amount of wages (before deducting withholding allowances) is: Over—	But not over—	The amount of payment to be made is:
$0	$283	20.40% of wages
$283	$558	$58
$558		$58 less 9.588% of wages in excess of $558

(c) MARRIED With Both Spouses Filing Certificate

If the amount of wages (before deducting withholding allowances) is: Over—	But not over—	The amount of payment to be made is:
$0	$141	20.40% of wages
$141	$279	$29
$279		$29 less 9.588% of wages in excess of $279

Table 3. SEMIMONTHLY Payroll Period

(a) SINGLE or HEAD OF HOUSEHOLD

If the amount of wages (before deducting withholding allowances) is: Over—	But not over—	The amount of payment to be made is:
$0	$307	20.40% of wages
$307	$563	$63
$563		$63 less 9.588% of wages in excess of $563

(b) MARRIED Without Spouse Filing Certificate

If the amount of wages (before deducting withholding allowances) is: Over—	But not over—	The amount of payment to be made is:
$0	$307	20.40% of wages
$307	$605	$63
$605		$63 less 9.588% of wages in excess of $605

(c) MARRIED With Both Spouses Filing Certificate

If the amount of wages (before deducting withholding allowances) is: Over—	But not over—	The amount of payment to be made is:
$0	$153	20.40% of wages
$153	$302	$31
$302		$31 less 9.588% of wages in excess of $302

Table 4. MONTHLY Payroll Period

(a) SINGLE or HEAD OF HOUSEHOLD

If the amount of wages (before deducting withholding allowances) is: Over—	But not over—	The amount of payment to be made is:
$0	$614	20.40% of wages
$614	$1,126	$125
$1,126		$125 less 9.588% of wages in excess of $1,126

(b) MARRIED Without Spouse Filing Certificate

If the amount of wages (before deducting withholding allowances) is: Over—	But not over—	The amount of payment to be made is:
$0	$614	20.40% of wages
$614	$1,210	$125
$1,210		$125 less 9.588% of wages in excess of $1,210

(c) MARRIED With Both Spouses Filing Certificate

If the amount of wages (before deducting withholding allowances) is: Over—	But not over—	The amount of payment to be made is:
$0	$307	20.40% of wages
$307	$605	$63
$605		$63 less 9.588% of wages in excess of $605

Page 56

Tables for Percentage Method of Advance EIC Payments (Continued)

(For Wages Paid in 2002)

Table 5. QUARTERLY Payroll Period

(a) SINGLE or HEAD OF HOUSEHOLD

If the amount of wages (before deducting withholding allowances) is: Over—	But not over—	The amount of payment to be made is:
$0	$1,842	20.40% of wages
$1,842	$3,380	$376
$3,380		$376 less 9.588% of wages in excess of $3,380

(b) MARRIED Without Spouse Filing Certificate

If the amount of wages (before deducting withholding allowances) is: Over—	But not over—	The amount of payment to be made is:
$0	$1,842	20.40% of wages
$1,842	$3,630	$376
$3,630		$376 less 9.588% of wages in excess of $3,630

(c) MARRIED With Both Spouses Filing Certificate

If the amount of wages (before deducting withholding allowances) is: Over—	But not over—	The amount of payment to be made is:
$0	$921	20.40% of wages
$921	$1,815	$188
$1,815		$188 less 9.588% of wages in excess of $1,815

Table 6. SEMIANNUAL Payroll Period

(a) SINGLE or HEAD OF HOUSEHOLD

If the amount of wages (before deducting withholding allowances) is: Over—	But not over—	The amount of payment to be made is:
$0	$3,685	20.40% of wages
$3,685	$6,760	$752
$6,760		$752 less 9.588% of wages in excess of $6,760

(b) MARRIED Without Spouse Filing Certificate

If the amount of wages (before deducting withholding allowances) is: Over—	But not over—	The amount of payment to be made is:
$0	$3,685	20.40% of wages
$3,685	$7,260	$752
$7,260		$752 less 9.588% of wages in excess of $7,260

(c) MARRIED With Both Spouses Filing Certificate

If the amount of wages (before deducting withholding allowances) is: Over—	But not over—	The amount of payment to be made is:
$0	$1,842	20.40% of wages
$1,842	$3,630	$376
$3,630		$376 less 9.588% of wages in excess of $3,630

Table 7. ANNUAL Payroll Period

(a) SINGLE or HEAD OF HOUSEHOLD

If the amount of wages (before deducting withholding allowances) is: Over—	But not over—	The amount of payment to be made is:
$0	$7,370	20.40% of wages
$7,370	$13,520	$1,503
$13,520		$1,503 less 9.588% of wages in excess of $13,520

(b) MARRIED Without Spouse Filing Certificate

If the amount of wages (before deducting withholding allowances) is: Over—	But not over—	The amount of payment to be made is:
$0	$7,370	20.40% of wages
$7,370	$14,520	$1,503
$14,520		$1,503 less 9.588% of wages in excess of $14,520

(c) MARRIED With Both Spouses Filing Certificate

If the amount of wages (before deducting withholding allowances) is: Over—	But not over—	The amount of payment to be made is:
$0	$3,685	20.40% of wages
$3,685	$7,260	$752
$7,260		$752 less 9.588% of wages in excess of $7,260

Table 8. DAILY or MISCELLANEOUS Payroll Period

(a) SINGLE or HEAD OF HOUSEHOLD

If the wages divided by the number of days in such period (before deducting withholding allowances) are: Over—	But not over—	The amount of payment to be made is the following amount multiplied by the number of days in such period:
$0	$28	20.40% of wages
$28	$52	$6
$52		$6 less 9.588% of wages in excess of $52

(b) MARRIED Without Spouse Filing Certificate

If the wages divided by the number of days in such period (before deducting withholding allowances) are: Over—	But not over—	The amount of payment to be made is the following amount multiplied by the number of days in such period:
$0	$28	20.40% of wages
$28	$55	$6
$55		$6 less 9.588% of wages in excess of $55

(c) MARRIED With Both Spouses Filing Certificate

If the wages divided by the number of days in such period (before deducting withholding allowances) are: Over—	But not over—	The amount of payment to be made is the following amount multiplied by the number of days in such period:
$0	$14	20.40% of wages
$14	$27	$3
$27		$3 less 9.588% of wages in excess of $27

Tables for Wage Bracket Method of Advance EIC Payments (For Wages Paid in 2002)

WEEKLY Payroll Period

SINGLE or HEAD OF HOUSEHOLD

Wages— At least	But less than	Payment to be made	Wages— At least	But less than	Payment to be made	Wages— At least	But less than	Payment to be made	Wages— At least	But less than	Payment to be made	Wages— At least	But less than	Payment to be made	Wages— At least	But less than	Payment to be made
$0	$5	$0	$50	$55	$10	$100	$105	$20	$270	$280	$27	$370	$380	$17	$470	$480	$8
5	10	1	55	60	11	105	110	21	280	290	26	380	390	16	480	490	7
10	15	2	60	65	12	110	115	22	290	300	25	390	400	15	490	500	6
15	20	3	65	70	13	115	120	23	300	310	24	400	410	15	500	510	5
20	25	4	70	75	14	120	125	24	310	320	23	410	420	14	510	520	4
25	30	5	75	80	15	125	130	26	320	330	22	420	430	13	520	530	3
30	35	6	80	85	16	130	135	27	330	340	21	430	440	12	530	540	2
35	40	7	85	90	17	135	140	28	340	350	20	440	450	11	540	550	1
40	45	8	90	95	18	140	260	29	350	360	19	450	460	10	550	- - -	0
45	50	9	95	100	19	260	270	28	360	370	18	460	470	9			

MARRIED Without Spouse Filing Certificate

Wages— At least	But less than	Payment to be made	Wages— At least	But less than	Payment to be made	Wages— At least	But less than	Payment to be made	Wages— At least	But less than	Payment to be made	Wages— At least	But less than	Payment to be made	Wages— At least	But less than	Payment to be made
$0	$5	$0	$50	$55	$10	$100	$105	$20	$285	$295	$27	$385	$395	$18	$485	$495	$8
5	10	1	55	60	11	105	110	21	295	305	26	395	405	17	495	505	7
10	15	2	60	65	12	110	115	22	305	315	25	405	415	16	505	515	6
15	20	3	65	70	13	115	120	23	315	325	25	415	425	15	515	525	5
20	25	4	70	75	14	120	125	24	325	335	24	425	435	14	525	535	4
25	30	5	75	80	15	125	130	26	335	345	23	435	445	13	535	545	3
30	35	6	80	85	16	130	135	27	345	355	22	445	455	12	545	555	2
35	40	7	85	90	17	135	140	28	355	365	21	455	465	11	555	565	2
40	45	8	90	95	18	140	275	29	365	375	20	465	475	10	565	575	1
45	50	9	95	100	19	275	285	28	375	385	19	475	485	9	575	- - -	0

MARRIED With Both Spouses Filing Certificate

Wages— At least	But less than	Payment to be made	Wages— At least	But less than	Payment to be made	Wages— At least	But less than	Payment to be made	Wages— At least	But less than	Payment to be made	Wages— At least	But less than	Payment to be made	Wages— At least	But less than	Payment to be made
$0	$5	$0	$30	$35	$6	$60	$65	$12	$165	$175	$11	$225	$235	$5	$285	- - -	$0
5	10	1	35	40	7	65	70	13	175	185	10	235	245	4			
10	15	2	40	45	8	70	135	14	185	195	9	245	255	3			
15	20	3	45	50	9	135	145	14	195	205	8	255	265	2			
20	25	4	50	55	10	145	155	13	205	215	7	265	275	1			
25	30	5	55	60	11	155	165	12	215	225	6	275	285	1			

BIWEEKLY Payroll Period

SINGLE or HEAD OF HOUSEHOLD

Wages— At least	But less than	Payment to be made	Wages— At least	But less than	Payment to be made	Wages— At least	But less than	Payment to be made	Wages— At least	But less than	Payment to be made	Wages— At least	But less than	Payment to be made	Wages— At least	But less than	Payment to be made
$0	$5	$0	$100	$105	$20	$200	$205	$41	$550	$560	$54	$750	$760	$35	$950	$960	$16
5	10	1	105	110	21	205	210	42	560	570	53	760	770	34	960	970	15
10	15	2	110	115	22	210	215	43	570	580	52	770	780	33	970	980	14
15	20	3	115	120	23	215	220	44	580	590	51	780	790	32	980	990	13
20	25	4	120	125	24	220	225	45	590	600	50	790	800	31	990	1,000	12
25	30	5	125	130	26	225	230	46	600	610	49	800	810	30	1,000	1,010	11
30	35	6	130	135	27	230	235	47	610	620	48	810	820	29	1,010	1,020	10
35	40	7	135	140	28	235	240	48	620	630	47	820	830	28	1,020	1,030	9
40	45	8	140	145	29	240	245	49	630	640	46	830	840	27	1,030	1,040	8
45	50	9	145	150	30	245	250	50	640	650	45	840	850	26	1,040	1,050	7
50	55	10	150	155	31	250	255	51	650	660	44	850	860	25	1,050	1,060	6
55	60	11	155	160	32	255	260	52	660	670	43	860	870	24	1,060	1,070	5
60	65	12	160	165	33	260	265	53	670	680	42	870	880	23	1,070	1,080	4
65	70	13	165	170	34	265	270	54	680	690	42	880	890	22	1,080	1,090	3
70	75	14	170	175	35	270	275	55	690	700	41	890	900	21	1,090	1,100	2
75	80	15	175	180	36	275	280	56	700	710	40	900	910	20	1,100	1,110	1
80	85	16	180	185	37	280	520	57	710	720	39	910	920	19	1,110	- - -	0
85	90	17	185	190	38	520	530	57	720	730	38	920	930	18			
90	95	18	190	195	39	530	540	56	730	740	37	930	940	18			
95	100	19	195	200	40	540	550	55	740	750	36	940	950	17			

MARRIED Without Spouse Filing Certificate

Wages— At least	But less than	Payment to be made	Wages— At least	But less than	Payment to be made	Wages— At least	But less than	Payment to be made	Wages— At least	But less than	Payment to be made	Wages— At least	But less than	Payment to be made	Wages— At least	But less than	Payment to be made
$0	$5	$0	$20	$25	$4	$40	$45	$8	$60	$65	$12	$80	$85	$16	$100	$105	$20
5	10	1	25	30	5	45	50	9	65	70	13	85	90	17	105	110	21
10	15	2	30	35	6	50	55	10	70	75	14	90	95	18	110	115	22
15	20	3	35	40	7	55	60	11	75	80	15	95	100	19			

(continued on next page)

BIWEEKLY Payroll Period

MARRIED Without Spouse Filing Certificate

Wages— At least	Wages— But less than	Payment to be made	Wages— At least	Wages— But less than	Payment to be made	Wages— At least	Wages— But less than	Payment to be made	Wages— At least	Wages— But less than	Payment to be made	Wages— At least	Wages— But less than	Payment to be made	Wages— At least	Wages— But less than	Payment to be made
$115	$120	$23	$195	$200	$40	$275	$280	$56	$695	$705	$44	$855	$865	$28	$1,015	$1,025	$13
120	125	24	200	205	41	280	555	57	705	715	43	865	875	27	1,025	1,035	12
125	130	26	205	210	42	555	565	57	715	725	42	875	885	26	1,035	1,045	11
130	135	27	210	215	43	565	575	56	725	735	41	885	895	26	1,045	1,055	10
135	140	28	215	220	44	575	585	55	735	745	40	895	905	25	1,055	1,065	9
140	145	29	220	225	45	585	595	54	745	755	39	905	915	24	1,065	1,075	8
145	150	30	225	230	46	595	605	53	755	765	38	915	925	23	1,075	1,085	7
150	155	31	230	235	47	605	615	52	765	775	37	925	935	22	1,085	1,095	6
155	160	32	235	240	48	615	625	51	775	785	36	935	945	21	1,095	1,105	5
160	165	33	240	245	49	625	635	50	785	795	35	945	955	20	1,105	1,115	4
165	170	34	245	250	50	635	645	50	795	805	34	955	965	19	1,115	1,125	3
170	175	35	250	255	51	645	655	49	805	815	33	965	975	18	1,125	1,135	3
175	180	36	255	260	52	655	665	48	815	825	32	975	985	17	1,135	1,145	2
180	185	37	260	265	53	665	675	47	825	835	31	985	995	16	1,145	1,155	1
185	190	38	265	270	54	675	685	46	835	845	30	995	1,005	15	1,155	- - -	0
190	195	39	270	275	55	685	695	45	845	855	29	1,005	1,015	14			

MARRIED With Both Spouses Filing Certificate

Wages— At least	Wages— But less than	Payment to be made	Wages— At least	Wages— But less than	Payment to be made	Wages— At least	Wages— But less than	Payment to be made	Wages— At least	Wages— But less than	Payment to be made	Wages— At least	Wages— But less than	Payment to be made	Wages— At least	Wages— But less than	Payment to be made
$0	$5	$0	$50	$55	$10	$100	$105	$20	$285	$295	$27	$385	$395	$18	$485	$495	$8
5	10	1	55	60	11	105	110	21	295	305	26	395	405	17	495	505	7
10	15	2	60	65	12	110	115	22	305	315	25	405	415	16	505	515	6
15	20	3	65	70	13	115	120	23	315	325	25	415	425	15	515	525	5
20	25	4	70	75	14	120	125	24	325	335	24	425	435	14	525	535	4
25	30	5	75	80	15	125	130	26	335	345	23	435	445	13	535	545	3
30	35	6	80	85	16	130	135	27	345	355	22	445	455	12	545	555	2
35	40	7	85	90	17	135	140	28	355	365	21	455	465	11	555	565	2
40	45	8	90	95	18	140	275	29	365	375	20	465	475	10	565	575	1
45	50	9	95	100	19	275	285	28	375	385	19	475	485	9	575	- - -	0

SEMIMONTHLY Payroll Period

SINGLE or HEAD OF HOUSEHOLD

Wages— At least	Wages— But less than	Payment to be made	Wages— At least	Wages— But less than	Payment to be made	Wages— At least	Wages— But less than	Payment to be made	Wages— At least	Wages— But less than	Payment to be made	Wages— At least	Wages— But less than	Payment to be made	Wages— At least	Wages— But less than	Payment to be made
$0	$5	$0	$110	$115	$22	$220	$225	$45	$600	$610	$58	$820	$830	$37	$1,040	$1,050	$16
5	10	1	115	120	23	225	230	46	610	620	57	830	840	36	1,050	1,060	15
10	15	2	120	125	24	230	235	47	620	630	56	840	850	35	1,060	1,070	14
15	20	3	125	130	26	235	240	48	630	640	55	850	860	34	1,070	1,080	13
20	25	4	130	135	27	240	245	49	640	650	54	860	870	33	1,080	1,090	12
25	30	5	135	140	28	245	250	50	650	660	53	870	880	32	1,090	1,100	11
30	35	6	140	145	29	250	255	51	660	670	52	880	890	31	1,100	1,110	10
35	40	7	145	150	30	255	260	52	670	680	51	890	900	30	1,110	1,120	9
40	45	8	150	155	31	260	265	53	680	690	50	900	910	29	1,120	1,130	8
45	50	9	155	160	32	265	270	54	690	700	50	910	920	28	1,130	1,140	7
50	55	10	160	165	33	270	275	55	700	710	49	920	930	27	1,140	1,150	6
55	60	11	165	170	34	275	280	56	710	720	48	930	940	27	1,150	1,160	5
60	65	12	170	175	35	280	285	57	720	730	47	940	950	26	1,160	1,170	4
65	70	13	175	180	36	285	290	58	730	740	46	950	960	25	1,170	1,180	3
70	75	14	180	185	37	290	295	59	740	750	45	960	970	24	1,180	1,190	3
75	80	15	185	190	38	295	300	60	750	760	44	970	980	23	1,190	1,200	2
80	85	16	190	195	39	300	305	61	760	770	43	980	990	22	1,200	1,210	1
85	90	17	195	200	40	305	560	62	770	780	42	990	1,000	21	1,210	- - -	0
90	95	18	200	205	41	560	570	62	780	790	41	1,000	1,010	20			
95	100	19	205	210	42	570	580	61	790	800	40	1,010	1,020	19			
100	105	20	210	215	43	580	590	60	800	810	39	1,020	1,030	18			
105	110	21	215	220	44	590	600	59	810	820	38	1,030	1,040	17			

MARRIED Without Spouse Filing Certificate

Wages— At least	Wages— But less than	Payment to be made	Wages— At least	Wages— But less than	Payment to be made	Wages— At least	Wages— But less than	Payment to be made	Wages— At least	Wages— But less than	Payment to be made	Wages— At least	Wages— But less than	Payment to be made	Wages— At least	Wages— But less than	Payment to be made
$0	$5	$0	$50	$55	$10	$100	$105	$20	$150	$155	$31	$200	$205	$41	$250	$255	$51
5	10	1	55	60	11	105	110	21	155	160	32	205	210	42	255	260	52
10	15	2	60	65	12	110	115	22	160	165	33	210	215	43	260	265	53
15	20	3	65	70	13	115	120	23	165	170	34	215	220	44	265	270	54
20	25	4	70	75	14	120	125	24	170	175	35	220	225	45	270	275	55
25	30	5	75	80	15	125	130	26	175	180	36	225	230	46	275	280	56
30	35	6	80	85	16	130	135	27	180	185	37	230	235	47	280	285	57
35	40	7	85	90	17	135	140	28	185	190	38	235	240	48	285	290	58
40	45	8	90	95	18	140	145	29	190	195	39	240	245	49	290	295	59
45	50	9	95	100	19	145	150	30	195	200	40	245	250	50			

(continued on next page)

SEMIMONTHLY Payroll Period

MARRIED Without Spouse Filing Certificate

Wages— At least	Wages— But less than	Payment to be made	Wages— At least	Wages— But less than	Payment to be made	Wages— At least	Wages— But less than	Payment to be made	Wages— At least	Wages— But less than	Payment to be made	Wages— At least	Wages— But less than	Payment to be made	Wages— At least	Wages— But less than	Payment to be made
$295	$300	$60	$695	$705	$53	$815	$825	$42	$935	$945	$30	$1,055	$1,065	$19	$1,175	$1,185	$7
300	305	61	705	715	52	825	835	41	945	955	29	1,065	1,075	18	1,185	1,195	6
305	605	62	715	725	51	835	845	40	955	965	28	1,075	1,085	17	1,195	1,205	5
605	615	62	725	735	50	845	855	39	965	975	27	1,085	1,095	16	1,205	1,215	4
615	625	61	735	745	49	855	865	38	975	985	26	1,095	1,105	15	1,215	1,225	3
625	635	60	745	755	48	865	875	37	985	995	25	1,105	1,115	14	1,225	1,235	2
635	645	59	755	765	47	875	885	36	995	1,005	24	1,115	1,125	13	1,235	1,245	1
645	655	58	765	775	46	885	895	35	1,005	1,015	23	1,125	1,135	12	1,245	- - -	0
655	665	57	775	785	45	895	905	34	1,015	1,025	22	1,135	1,145	11			
665	675	56	785	795	44	905	915	33	1,025	1,035	21	1,145	1,155	10			
675	685	55	795	805	43	915	925	32	1,035	1,045	20	1,155	1,165	9			
685	695	54	805	815	42	925	935	31	1,045	1,055	19	1,165	1,175	8			

MARRIED With Both Spouses Filing Certificate

Wages— At least	Wages— But less than	Payment to be made	Wages— At least	Wages— But less than	Payment to be made	Wages— At least	Wages— But less than	Payment to be made	Wages— At least	Wages— But less than	Payment to be made	Wages— At least	Wages— But less than	Payment to be made	Wages— At least	Wages— But less than	Payment to be made
$0	$5	$0	$55	$60	$11	$110	$115	$22	$320	$330	$29	$430	$440	$18	$540	$550	$8
5	10	1	60	65	12	115	120	23	330	340	28	440	450	17	550	560	7
10	15	2	65	70	13	120	125	24	340	350	27	450	460	16	560	570	6
15	20	3	70	75	14	125	130	26	350	360	26	460	470	15	570	580	5
20	25	4	75	80	15	130	135	27	360	370	25	470	480	14	580	590	4
25	30	5	80	85	16	135	140	28	370	380	24	480	490	13	590	600	3
30	35	6	85	90	17	140	145	29	380	390	23	490	500	12	600	610	2
35	40	7	90	95	18	145	150	30	390	400	22	500	510	11	610	620	1
40	45	8	95	100	19	150	300	31	400	410	21	510	520	10	620	- - -	0
45	50	9	100	105	20	300	310	31	410	420	20	520	530	10			
50	55	10	105	110	21	310	320	30	420	430	19	530	540	9			

MONTHLY Payroll Period

SINGLE or HEAD OF HOUSEHOLD

Wages— At least	Wages— But less than	Payment to be made	Wages— At least	Wages— But less than	Payment to be made	Wages— At least	Wages— But less than	Payment to be made	Wages— At least	Wages— But less than	Payment to be made	Wages— At least	Wages— But less than	Payment to be made	Wages— At least	Wages— But less than	Payment to be made
$0	$5	$0	$215	$220	$44	$430	$435	$88	$1,185	$1,195	$119	$1,615	$1,625	$77	$2,045	$2,055	$36
5	10	1	220	225	45	435	440	89	1,195	1,205	118	1,625	1,635	77	2,055	2,065	35
10	15	2	225	230	46	440	445	90	1,205	1,215	117	1,635	1,645	76	2,065	2,075	34
15	20	3	230	235	47	445	450	91	1,215	1,225	116	1,645	1,655	75	2,075	2,085	33
20	25	4	235	240	48	450	455	92	1,225	1,235	115	1,655	1,665	74	2,085	2,095	32
25	30	5	240	245	49	455	460	93	1,235	1,245	114	1,665	1,675	73	2,095	2,105	31
30	35	6	245	250	50	460	465	94	1,245	1,255	113	1,675	1,685	72	2,105	2,115	30
35	40	7	250	255	51	465	470	95	1,255	1,265	112	1,685	1,695	71	2,115	2,125	30
40	45	8	255	260	52	470	475	96	1,265	1,275	111	1,695	1,705	70	2,125	2,135	29
45	50	9	260	265	53	475	480	97	1,275	1,285	110	1,705	1,715	69	2,135	2,145	28
50	55	10	265	270	54	480	485	98	1,285	1,295	109	1,715	1,725	68	2,145	2,155	27
55	60	11	270	275	55	485	490	99	1,295	1,305	108	1,725	1,735	67	2,155	2,165	26
60	65	12	275	280	56	490	495	100	1,305	1,315	107	1,735	1,745	66	2,165	2,175	25
65	70	13	280	285	57	495	500	101	1,315	1,325	106	1,745	1,755	65	2,175	2,185	24
70	75	14	285	290	58	500	505	102	1,325	1,335	105	1,755	1,765	64	2,185	2,195	23
75	80	15	290	295	59	505	510	103	1,335	1,345	104	1,765	1,775	63	2,195	2,205	22
80	85	16	295	300	60	510	515	104	1,345	1,355	103	1,775	1,785	62	2,205	2,215	21
85	90	17	300	305	61	515	520	105	1,355	1,365	102	1,785	1,795	61	2,215	2,225	20
90	95	18	305	310	62	520	525	106	1,365	1,375	101	1,795	1,805	60	2,225	2,235	19
95	100	19	310	315	63	525	530	107	1,375	1,385	100	1,805	1,815	59	2,235	2,245	18
100	105	20	315	320	64	530	535	108	1,385	1,395	100	1,815	1,825	58	2,245	2,255	17
105	110	21	320	325	65	535	540	109	1,395	1,405	99	1,825	1,835	57	2,255	2,265	16
110	115	22	325	330	66	540	545	110	1,405	1,415	98	1,835	1,845	56	2,265	2,275	15
115	120	23	330	335	67	545	550	111	1,415	1,425	97	1,845	1,855	55	2,275	2,285	14
120	125	24	335	340	68	550	555	112	1,425	1,435	96	1,855	1,865	54	2,285	2,295	13
125	130	26	340	345	69	555	560	113	1,435	1,445	95	1,865	1,875	53	2,295	2,305	12
130	135	27	345	350	70	560	565	114	1,445	1,455	94	1,875	1,885	53	2,305	2,315	11
135	140	28	350	355	71	565	570	115	1,455	1,465	93	1,885	1,895	52	2,315	2,325	10
140	145	29	355	360	72	570	575	116	1,465	1,475	92	1,895	1,905	51	2,325	2,335	9
145	150	30	360	365	73	575	580	117	1,475	1,485	91	1,905	1,915	50	2,335	2,345	8
150	155	31	365	370	74	580	585	118	1,485	1,495	90	1,915	1,925	49	2,345	2,355	7
155	160	32	370	375	75	585	590	119	1,495	1,505	89	1,925	1,935	48	2,355	2,365	7
160	165	33	375	380	77	590	595	120	1,505	1,515	88	1,935	1,945	47	2,365	2,375	6
165	170	34	380	385	78	595	600	121	1,515	1,525	87	1,945	1,955	46	2,375	2,385	5
170	175	35	385	390	79	600	605	122	1,525	1,535	86	1,955	1,965	45	2,385	2,395	4
175	180	36	390	395	80	605	610	123	1,535	1,545	85	1,965	1,975	44	2,395	2,405	3
180	185	37	395	400	81	610	1,125	124	1,545	1,555	84	1,975	1,985	43	2,405	2,415	2
185	190	38	400	405	82	1,125	1,135	124	1,555	1,565	83	1,985	1,995	42	2,415	2,425	1 0
190	195	39	405	410	83	1,135	1,145	123	1,565	1,575	82	1,995	2,005	41	2,425	- - -	
195	200	40	410	415	84	1,145	1,155	123	1,575	1,585	81	2,005	2,015	40			
200	205	41	415	420	85	1,155	1,165	122	1,585	1,595	80	2,015	2,025	39			
205	210	42	420	425	86	1,165	1,175	121	1,595	1,605	79	2,025	2,035	38			
210	215	43	425	430	87	1,175	1,185	120	1,605	1,615	78	2,035	2,045	37			

Page 60

MONTHLY Payroll Period

MARRIED Without Spouse Filing Certificate

Wages— At least	But less than	Payment to be made	Wages— At least	But less than	Payment to be made	Wages— At least	But less than	Payment to be made	Wages— At least	But less than	Payment to be made	Wages— At least	But less than	Payment to be made	Wages— At least	But less than	Payment to be made
$0	$5	$0	$215	$220	$44	$430	$435	$88	$1,270	$1,280	$119	$1,700	$1,710	$77	$2,130	$2,140	$36
5	10	1	220	225	45	435	440	89	1,280	1,290	118	1,710	1,720	76	2,140	2,150	35
10	15	2	225	230	46	440	445	90	1,290	1,300	117	1,720	1,730	75	2,150	2,160	34
15	20	3	230	235	47	445	450	91	1,300	1,310	116	1,730	1,740	74	2,160	2,170	33
20	25	4	235	240	48	450	455	92	1,310	1,320	115	1,740	1,750	73	2,170	2,180	32
25	30	5	240	245	49	455	460	93	1,320	1,330	114	1,750	1,760	73	2,180	2,190	31
30	35	6	245	250	50	460	465	94	1,330	1,340	113	1,760	1,770	72	2,190	2,200	30
35	40	7	250	255	51	465	470	95	1,340	1,350	112	1,770	1,780	71	2,200	2,210	29
40	45	8	255	260	52	470	475	96	1,350	1,360	111	1,780	1,790	70	2,210	2,220	28
45	50	9	260	265	53	475	480	97	1,360	1,370	110	1,790	1,800	69	2,220	2,230	27
50	55	10	265	270	54	480	485	98	1,370	1,380	109	1,800	1,810	68	2,230	2,240	26
55	60	11	270	275	55	485	490	99	1,380	1,390	108	1,810	1,820	67	2,240	2,250	26
60	65	12	275	280	56	490	495	100	1,390	1,400	107	1,820	1,830	66	2,250	2,260	25
65	70	13	280	285	57	495	500	101	1,400	1,410	106	1,830	1,840	65	2,260	2,270	24
70	75	14	285	290	58	500	505	102	1,410	1,420	105	1,840	1,850	64	2,270	2,280	23
75	80	15	290	295	59	505	510	103	1,420	1,430	104	1,850	1,860	63	2,280	2,290	22
80	85	16	295	300	60	510	515	104	1,430	1,440	103	1,860	1,870	62	2,290	2,300	21
85	90	17	300	305	61	515	520	105	1,440	1,450	102	1,870	1,880	61	2,300	2,310	20
90	95	18	305	310	62	520	525	106	1,450	1,460	101	1,880	1,890	60	2,310	2,320	19
95	100	19	310	315	63	525	530	107	1,460	1,470	100	1,890	1,900	59	2,320	2,330	18
100	105	20	315	320	64	530	535	108	1,470	1,480	99	1,900	1,910	58	2,330	2,340	17
105	110	21	320	325	65	535	540	109	1,480	1,490	98	1,910	1,920	57	2,340	2,350	16
110	115	22	325	330	66	540	545	110	1,490	1,500	97	1,920	1,930	56	2,350	2,360	15
115	120	23	330	335	67	545	550	111	1,500	1,510	96	1,930	1,940	55	2,360	2,370	14
120	125	24	335	340	68	550	555	112	1,510	1,520	96	1,940	1,950	54	2,370	2,380	13
125	130	26	340	345	69	555	560	113	1,520	1,530	95	1,950	1,960	53	2,380	2,390	12
130	135	27	345	350	70	560	565	114	1,530	1,540	94	1,960	1,970	52	2,390	2,400	11
135	140	28	350	355	71	565	570	115	1,540	1,550	93	1,970	1,980	51	2,400	2,410	10
140	145	29	355	360	72	570	575	116	1,550	1,560	92	1,980	1,990	50	2,410	2,420	9
145	150	30	360	365	73	575	580	117	1,560	1,570	91	1,990	2,000	50	2,420	2,430	8
150	155	31	365	370	74	580	585	118	1,570	1,580	90	2,000	2,010	49	2,430	2,440	7
155	160	32	370	375	75	585	590	119	1,580	1,590	89	2,010	2,020	48	2,440	2,450	6
160	165	33	375	380	77	590	595	120	1,590	1,600	88	2,020	2,030	47	2,450	2,460	5
165	170	34	380	385	78	595	600	121	1,600	1,610	87	2,030	2,040	46	2,460	2,470	4
170	175	35	385	390	79	600	605	122	1,610	1,620	86	2,040	2,050	45	2,470	2,480	3
175	180	36	390	395	80	605	610	123	1,620	1,630	85	2,050	2,060	44	2,480	2,490	3
180	185	37	395	400	81	610	1,210	124	1,630	1,640	84	2,060	2,070	43	2,490	2,500	2
185	190	38	400	405	82	1,210	1,220	124	1,640	1,650	83	2,070	2,080	42	2,500	2,510	1
190	195	39	405	410	83	1,220	1,230	123	1,650	1,660	82	2,080	2,090	41	2,510	- - -	0
195	200	40	410	415	84	1,230	1,240	122	1,660	1,670	81	2,090	2,100	40			
200	205	41	415	420	85	1,240	1,250	121	1,670	1,680	80	2,100	2,110	39			
205	210	42	420	425	86	1,250	1,260	120	1,680	1,690	79	2,110	2,120	38			
210	215	43	425	430	87	1,260	1,270	119	1,690	1,700	78	2,120	2,130	37			

MARRIED With Both Spouses Filing Certificate

Wages— At least	But less than	Payment to be made	Wages— At least	But less than	Payment to be made	Wages— At least	But less than	Payment to be made	Wages— At least	But less than	Payment to be made	Wages— At least	But less than	Payment to be made	Wages— At least	But less than	Payment to be made
$0	$5	$0	$110	$115	$22	$220	$225	$45	$645	$655	$58	$865	$875	$37	$1,085	$1,095	$16
5	10	1	115	120	23	225	230	46	655	665	57	875	885	36	1,095	1,105	15
10	15	2	120	125	24	230	235	47	665	675	56	885	895	35	1,105	1,115	14
15	20	3	125	130	26	235	240	48	675	685	55	895	905	34	1,115	1,125	13
20	25	4	130	135	27	240	245	49	685	695	54	905	915	33	1,125	1,135	12
25	30	5	135	140	28	245	250	50	695	705	53	915	925	32	1,135	1,145	11
30	35	6	140	145	29	250	255	51	705	715	52	925	935	31	1,145	1,155	10
35	40	7	145	150	30	255	260	52	715	725	51	935	945	30	1,155	1,165	9
40	45	8	150	155	31	260	265	53	725	735	50	945	955	29	1,165	1,175	8
45	50	9	155	160	32	265	270	54	735	745	49	955	965	28	1,175	1,185	7
50	55	10	160	165	33	270	275	55	745	755	48	965	975	27	1,185	1,195	6
55	60	11	165	170	34	275	280	56	755	765	47	975	985	26	1,195	1,205	5
60	65	12	170	175	35	280	285	57	765	775	46	985	995	25	1,205	1,215	4
65	70	13	175	180	36	285	290	58	775	785	45	995	1,005	24	1,215	1,225	3
70	75	14	180	185	37	290	295	59	785	795	44	1,005	1,015	23	1,225	1,235	2
75	80	15	185	190	38	295	300	60	795	805	43	1,015	1,025	22	1,235	1,245	1
80	85	16	190	195	39	300	305	61	805	815	43	1,025	1,035	21	1,245	- - -	0
85	90	17	195	200	40	305	605	62	815	825	42	1,035	1,045	20			
90	95	18	200	205	41	605	615	62	825	835	41	1,045	1,055	20			
95	100	19	205	210	42	615	625	61	835	845	40	1,055	1,065	19			
100	105	20	210	215	43	625	635	60	845	855	39	1,065	1,075	18			
105	110	21	215	220	44	635	645	59	855	865	38	1,075	1,085	17			

DAILY Payroll Period

SINGLE or HEAD OF HOUSEHOLD						MARRIED Without Spouse Filing Certificate						MARRIED With Both Spouses Filing Certificate					
Wages— At least	But less than	Payment to be made	Wages— At least	But less than	Payment to be made	Wages— At least	But less than	Payment to be made	Wages— At least	But less than	Payment to be made	Wages— At least	But less than	Payment to be made	Wages— At least	But less than	Payment to be made
$0	$5	$0	$50	$60	$5	$0	$5	$0	$55	$65	$5	$0	$5	$0	$25	$35	$2
5	10	1	60	70	4	5	10	1	65	75	4	5	10	1	35	45	1
10	15	2	70	80	3	10	15	2	75	85	3	10	25	2	45	- - -	0
15	20	3	80	90	2	15	20	3	85	95	2						
20	25	4	90	100	1	20	25	4	95	105	1						
25	50	5	100	- - -	0	25	55	5	105	- - -	0						

Page 61

Index

■

Form **7018-A**
(Rev. November 2001)
Department of the Treasury
Internal Revenue Service

Employer's Order Blank for 2002 Forms

Visit IRS's Web Site @ www.irs.gov

OMB No. 1545-1059

Instructions. Enter the quantity next to the form you are ordering. **Please order the number of forms needed, not the number of sheets. Note:** None of the items on the order blank are available from the IRS in continuous feed version. All forms on this order blank that require multiple copies are carbonized so that you will not have to insert carbons. You will automatically receive one instruction with any form on this order blank. **Type or print** your name and complete mail delivery address in the space provided below. An accurate mail delivery address is necessary to ensure delivery of your order.

USE THIS PORTION FOR 2002 FORMS ONLY

Quantity	Item	Title	Quantity	Item	Title
	W-2	Wage and Tax Statement		**1099 G**	Certain Government Payments
	W-2 C	Corrected Wage and Tax Statement		**1099 INT**	Interest Income
	W-3	Transmittal of Wage and Tax Statements		**1099 LTC**	Long-Term Care and Accelerated Death Benefits
	W-3 C	Transmittal of Corrected Wage and Tax Statements		**1099 MISC**	Miscellaneous Income
	W-4	Employee's Withholding Allowance Certificate		**1099 MSA**	Distributions From an Archer MSA or Medicare+Choice MSA
	W-4 P	Withholding Certificate for Pension or Annuity Payments		**1099 OID**	Original Issue Discount
	W-4 S	Request for Federal Income Tax Withholding From Sick Pay		**1099 PATR**	Taxable Distributions Received From Cooperatives
	W-5	Earned Income Credit Advance Payment Certificate		**1099 Q**	Qualified Tuition Program Payments (Under Section 529)
	1096	Annual Summary and Transmittal of U.S. Information Returns		**1099 R**	Distributions From Pensions, Annuities, Retirement or Profit-Sharing Plans, IRAs, Insurance Contracts, etc.
	1098	Mortgage Interest Statement		**1099 S**	Proceeds From Real Estate Transactions
	1098 E	Student Loan Interest Statement		**5498**	IRA and Coverdell ESA Contribution Information
	1098 T	Tuition Payments Statement		**5498 MSA**	Archer MSA or Medicare+Choice MSA Information
	1099 A	Acquisition or Abandonment of Secured Property		**Pub 15 A**	Employer's Supplemental Tax Guide
	1099 B	Proceeds From Broker and Barter Exchange Transactions		**Pub 15 B**	Employer's Tax Guide to Fringe Benefits
	1099 C	Cancellation of Debt		**Pub 1494**	Table for Figuring Amount Exempt From Levy On Wages, Salary, and Other Income (Forms 668-W(c) and 668-W(c)(DO))
	1099 DIV	Dividends and Distributions			

Print or Type Only			
	Attention:		Daytime Telephone Number: ( )
	Company Name:		
	Postal Mailing Address:	Ste/Room	
	City:	State:	Zip Code:
	Foreign Country:	International Postal Code:	

Where To Send Your Order

Send your order to the Internal Revenue Service address for the Area Distribution Center closest to your state.

Central Area Distribution Center
P.O. Box 8908
Bloomington, IL 61702-8908

Eastern Area Distribution Center
P.O. Box 85075
Richmond, VA 23261-5075

Western Area Distribution Center
Rancho Cordova, CA 95743-0001

Paperwork Reduction Act Notice. We ask for the information on this form to carry out the Internal Revenue laws of the United States. Your response is voluntary.

You are not required to provide the information requested on a form that is subject to the Paperwork Reduction Act unless the form displays a valid OMB control number. Books or records relating to a form or its instructions must be retained as long as their contents may become material in the administration of any Internal Revenue law. Generally, tax returns and return information are confidential, as required by Code section 6103.

The time needed to complete this form will vary depending on the individual circumstances. The estimated average time is 3 minutes. If you have comments concerning the accuracy of this time estimate or suggestions for making this form simpler, we would be happy to hear from you. You can write to the Tax Forms Committee, Western Area Distribution Center, Rancho Cordova, CA 95743-0001.

DO NOT send your order Form 7018-A to the Tax Forms Committee. Instead, send your Forms order to the IRS Area Distribution Center closest to your state.

Cat. No. 43709Q

Page 63

Quick and Easy Access to Tax Help and Forms

Personal Computer

You can access the IRS Web Site 24 hours a day, 7 days a week, at **www.irs.gov** to:

- Download forms, instructions, and publications
- See answers to frequently asked tax questions
- Search publications on-line by topic or keyword
- Figure your withholding allowances using our W-4 calculator
- Send us comments or request help by e-mail
- Sign up to receive local and national tax news by e-mail

You can also reach us using File Transfer Protocol at ftp.irs.gov

Fax

You can get over 100 of the most requested forms and instructions 24 hours a day, 7 days a week, by fax. Just call **703-368-9694** from the telephone connected to the fax machine.

For help with transmission problems, call the FedWorld Help Desk at **703-487-4608.**

Long-distance charges may apply.

Mail

You can order forms, instructions, and publications by completing the order blank on page 63. You should receive your order within 10 days after we receive your request.

Phone

You can order forms and publications and receive automated information 24 hours a day, 7 days a week, by phone.

Forms and Publications

Call **1-800-TAX-FORM** (1-800-829-3676) to order current year forms, instructions, and publications, and prior year forms and instructions. You should receive your order within 10 days.

TeleTax Topics

Call **1-800-829-4477** to listen to pre-recorded messages covering about 150 tax topics. See page 5 for a list of the topics.

Walk-In

You can pick up some of the most requested forms, instructions, and publications at many IRS offices, post offices, and libraries. Some IRS offices, libraries, city and county government offices, credit unions, grocery stores, office supply stores, and copy centers have an extensive collection of products available to photocopy or print from a CD-ROM.

CD-ROM

Order **Pub. 1796,** Federal Tax Products on CD-ROM, and get:

- Current year forms, instructions, and publications
- Prior year forms, instructions, and publications
- Frequently requested tax forms that may be filled in electronically, printed out for submission, and saved for recordkeeping
- The Internal Revenue Bulletin

Buy the CD-ROM on the Internet at **www.irs.gov/cdorders** from the National Technical Information Service (NTIS) for $21 (no handling fee) or call **1-877-CDFORMS** (1-877-233-6767) toll free to buy the CD-ROM for $21 (plus a $5 handling fee).

Internal Revenue Service
WADC–9999
Rancho Cordova, CA 95743-9999

Official Business
Penalty for Private Use $300

Deliver to Payroll Department

PRSRT STD
Postage and Fees Paid
Internal Revenue Service
Permit No. G-48

Page 64

Page 514, add after Appendix A:

APPENDIX A-1

IRS EMPLOYER'S SUPPLEMENTAL TAX GUIDE

[*Note: A full-sized version of this guide can be downloaded at <<http://www.irs.gov/forms_pubs/pubs.html>>.*]

Department of the Treasury
Internal Revenue Service

Publication 15-A
(Rev. January 2002)
Cat. No. 21453T

Employer's Supplemental Tax Guide

(Supplement to Circular E, Employer's Tax Guide (Publication 15))

Get forms and other information faster and easier by:

Computer • www.irs.gov or **FTP** • ftp.irs.gov

FAX • 703-368-9694 (from your fax machine)

www.irs.gov/elec_svs/efile-bus.html

Contents

Introduction

This publication supplements **Circular E,** Employer's Tax Guide. It contains specialized and detailed employment tax information supplementing the basic information provided in Circular E. **Publication 15-B,** Employer's Tax Guide to Fringe Benefits, contains information about the employment tax treatment of various types of noncash compensation. This publication contains:

- Alternative methods and tables for figuring income tax withholding.
- Combined income tax, employee social security tax, and employee Medicare tax withholding tables.
- Tables for withholding on distributions of Indian gaming profits to tribal members.

Telephone help. You can call the IRS with your tax questions. Check your telephone book for the local number or call 1-800-829-1040.

Help for people with disabilities. Telephone help is available using TTY/TDD equipment. You can call 1-800-829-4059 with your tax question or to order forms and publications. You may also use this number for problem resolution assistance.

Ordering publications and forms. See page 61 for information on how to obtain forms and publications.

Useful Items

You may want to see:

Publication

- ❑ **15** Circular E, Employer's Tax Guide
- ❑ **15-B** Employer's Tax Guide to Fringe Benefits
- ❑ **51** Agricultural Employer's Tax Guide
- ❑ **509** Tax Calendars for 2002
- ❑ **225** Farmer's Tax Guide
- ❑ **515** Withholding of Tax on Nonresident Aliens and Foreign Corporations
- ❑ **535** Business Expenses
- ❑ **553** Highlights of 2001 Tax Changes
- ❑ **583** Starting a Business and Keeping Records
- ❑ **1635** Understanding Your EIN

Items To Note

Furnishing Form W-2 to employees electronically. You may set up a system to furnish Forms W-2 to employees who choose to receive them in this format beginning with Forms W-2 due after December 31, 2000. Each employee participating must consent electronically, and you must notify the employees of all hardware and software requirements to receive them. You may not send Form W-2 electronically to any employee who does not consent or who has revoked consent previously provided.

To furnish Forms W-2 electronically, you must meet the following disclosure requirements and provide a clear and conspicuous statement of each of them to your employees.

1) The employee must be informed that he or she may receive a paper Form W-2 if consent is not given to receive it electronically. The consent statement must be made electronically in a way that demonstrates that the employee can access the Form W-2 in the electronic form that will be used to furnish the statement.
2) The employee must be informed how to obtain a paper copy and whether any fee will be charged for a paper copy.
3) The employee may withdraw consent in writing at any time on 30 days notice. The employer will confirm the withdrawal in writing, and inform the employee of the consequences of the withdrawal.
4) The employer will notify the employee of the scope and duration of the consent.
5) The employer will inform the employee that the form may be required to be attached to his or her tax returns, and that the employee may need to print the forms.

The employer must furnish the electronic statements by the due date of the paper forms. The employer must notify the employees that the Forms W-2 will be posted on a web site by January 31. This notice may be delivered by mail, electronic mail, or in person.

For more information, see Temporary Regulation 31.6051-1T.

Electronic deposit requirement. Certain employers are required to make deposits of employment taxes using the Electronic Federal Tax Payment System (EFTPS). If you are required to use EFTPS and fail to do so, you may be subject to a 10% penalty. See Circular E for more information.

If you are not required to use EFTPS, you may participate voluntarily. To enroll in or get more information about EFTPS, call 1-800-945-8400 or 1-800-555-4477 or visit the EFTPS Web Site at **www.eftps.gov.**

Electronic submission of Forms W-4, W-4P, W-4S, W-4V, and W-5. You may set up a system to electronically receive any or all of the following forms from an employee or payee:

- **Form W-4**, Employee's Withholding Allowance Certificate
- **Form W-4P,** Withholding Certificate for Pension or Annuity Payments
- **Form W-4S,** Request for Federal Income Tax Withholding From Sick Pay
- **Form W-4V,** Voluntary Withholding Request
- **Form W-5,** Employee's Advance Earned Income Credit Certificate

If you establish an electronic system to receive any of these forms, you do not need to process that form in a paper version.

For each form that you establish an electronic submission system for, you must meet the following requirements:

1) The electronic system must ensure that the information received by the payer is the information sent by the payee. The system must document all occasions of user access that result in a submission. In addition, the design and operation of the electronic system, including access procedures, must make it reasonably certain that the person accessing the system and submitting the form is the person identified on the form.
2) The electronic system must provide exactly the same information as the paper form.
3) The electronic submission must be signed with an electronic signature by the payee whose name is on the form. The electronic signature must be the final entry in the submission.
4) Upon request, you must furnish a hard copy of any completed electronic form to the IRS and a state-

ment that, to the best of the payer's knowledge, the electronic form was submitted by the named payee. The hard copy of the electronic form must provide exactly the same information as, but need not be a facsimile of, the paper form. For Forms W-4 and W-5, the signature must be under penalty of perjury, and must contain the same language that appears on the paper version of the form. The electronic system must inform the employee that he or she must make a declaration contained in the perjury statement and that the declaration is made by signing the Form W-4 or W-5.

5) You must meet all recordkeeping requirements that apply to the paper forms.

For more information, see:

- Form W-4—Regulations section 31.3402(f)(5)-1
- Form W-5—Announcement 99-3 (99-3 IRB 15)
- Forms W-4P, W-4S, and W-4V—Announcement 99-6 (99-4 IRB 24)

Photographs of Missing Children

The Internal Revenue Service is a proud partner with the National Center for Missing and Exploited Children. Photographs of missing children selected by the Center may appear in this booklet on pages that would otherwise be blank. You can help bring these children home by looking at the photographs and calling **1-800-THE-LOST** (1-800-843-5678) if you recognize a child.

1. Who Are Employees?

Before you can know how to treat payments you make for services, you must first know the business relationship that exists between you and the person performing the services. The person performing the services may be—

- An independent contractor.
- A common-law employee.
- A statutory employee.
- A statutory nonemployee.

This discussion explains these four categories. A later discussion, **Employee or Independent Contractor?** (section 2), points out the differences between an independent contractor and an employee and gives examples from various types of occupations. If an individual who works for you is not an employee under the common-law rules (see section 2), you generally do not have to withhold Federal income tax from that individual's pay. However, in some cases you may be required to withhold under backup withholding requirements on these payments. See Circular E for information on backup withholding.

Independent Contractors

People such as lawyers, contractors, subcontractors, public stenographers, and auctioneers who follow an independent trade, business, or profession in which they offer their services to the public, are generally not employees. However, whether such people are employees or independent contractors depends on the facts in each case. The general rule is that an individual is an independent contractor if you, the person for whom the services are performed, have the right to control or direct only the result of the work and not the means and methods of accomplishing the result.

Common-Law Employees

Under common-law rules, anyone who performs services for you is your employee if you can control what will be done and how it will be done. This is so even when you give the employee freedom of action. What matters is that you have the right to control the details of how the services are performed. For a discussion of facts that indicate whether an individual providing services is an independent contractor or employee, see **Employee or Independent Contractor?** (section 2).

If you have an employer-employee relationship, it makes no difference how it is labeled. The ***substance*** of the relationship, ***not the label,*** governs the worker's status. Nor does it matter whether the individual is employed full time or part time.

For employment tax purposes, no distinction is made between classes of employees. Superintendents, managers, and other supervisory personnel are all employees. An ***officer of a corporation*** is generally an employee; however, an officer who performs no services or only minor services, and neither receives nor is entitled to receive any pay, is not considered an employee. A ***director*** of a corporation is not an employee with respect to services performed as a director.

You generally have to withhold and pay income, social security, and Medicare taxes on wages you pay to common-law employees. However, the wages of certain employees may be exempt from one or more of these taxes. See **Employees of Exempt Organizations** (section 3) and **Religious Exemptions** (section 4).

Leased employees. Under certain circumstances, a corporation furnishing workers to various professional people and firms is the employer of those workers for employment tax purposes. For example, a professional service corporation may provide the services of secretaries, nurses, and other similarly trained workers to its subscribers.

The service corporation enters into contracts with the subscribers under which the subscribers specify the services to be provided and the fee to be paid to the service corporation for each individual furnished. The service corporation has the right to control and direct the worker's services for the subscriber, including the right to discharge or reassign the worker. The service corporation hires the workers, controls the payment of their wages, provides them with unemployment insurance and other benefits, and is the employer for employment tax purposes. For information on employee leasing as it relates to pension plan qualification requirements, see **Leased employees** in **Pub. 560,** Retirement Plans for Small Business (SEP, SIMPLE, and Keogh Plans).

Additional information. For more information about the treatment of special types of employment, the treatment of special types of payments, and similar subjects, get Circular E or Circular A (for agricultural employers).

Statutory Employees

If workers are independent contractors under the common law rules, such workers may nevertheless be treated as employees by statute ("statutory employees") for certain employment tax purposes if they fall within any one of the following four categories and meet the three conditions described under **Social security and Medicare taxes,** below.

1) A driver who distributes beverages (other than milk) or meat, vegetable, fruit, or bakery products; or who picks up and delivers laundry or dry cleaning, if the driver is your agent or is paid on commission.
2) A full-time life insurance sales agent whose principal business activity is selling life insurance or annuity contracts, or both, primarily for one life insurance company.
3) An individual who works at home on materials or goods that you supply and that must be returned to you or to a person you name, if you also furnish specifications for the work to be done.
4) A full-time traveling or city salesperson who works on your behalf and turns in orders to you from wholesalers, retailers, contractors, or operators of hotels, restaurants, or other similar establishments. The goods sold must be merchandise for resale or supplies for use in the buyer's business operation. The work performed for you must be the salesperson's principal business activity. See **Salesperson** in section 2.

Social security and Medicare taxes. Withhold social security and Medicare taxes from the wages of statutory employees if all three of the following conditions apply.

- The service contract states or implies that substantially all the services are to be performed personally by them.
- They do not have a substantial investment in the equipment and property used to perform the services (other than an investment in transportation facilities).
- The services are performed on a continuing basis for the same payer.

Federal unemployment (FUTA) tax. For FUTA tax, the term ***employee*** means the same as it does for social security and Medicare taxes, except that it does not include statutory employees in categories 2 and 3 above. Thus, any individual who is an employee under category 1 or 4 is also an employee for FUTA tax purposes and subject to FUTA tax.

Income tax. Do not withhold income tax from the wages of statutory employees.

Reporting payments to statutory employees. Furnish a Form W-2 to a statutory employee, and check "statutory employee" in box 13. Show your payments to the employee as other compensation in box 1. Also, show social security wages in box 3, social security tax withheld in box 4, Medicare wages in box 5, and Medicare tax withheld in box 6. The statutory employee can deduct his or her trade or business expenses from the payments shown on Form W-2. He or she reports earnings as a statutory employee on line 1 of Schedule C or C-EZ (Form 1040). (A statutory employee's business expenses are deductible on Schedule C or C-EZ (Form 1040) and are not subject to the reduction by 2% of his or her adjusted gross income that applies to common-law employees.)

Statutory Nonemployees

There are two categories of statutory nonemployees: ***direct sellers*** and ***licensed real estate agents.*** They are treated as self-employed for all Federal tax purposes, including income and employment taxes, if:

1) Substantially all payments for their services as direct sellers or real estate agents are directly related to sales or other output, rather than to the number of hours worked and
2) Their services are performed under a written contract providing that they will not be treated as employees for Federal tax purposes.

Direct sellers. Direct sellers include persons falling within any of the following three groups:

1) Persons engaged in selling (or soliciting the sale of) consumer products in the home or place of business other than in a permanent retail establishment.
2) Persons engaged in selling (or soliciting the sale of) consumer products to any buyer on a buy-sell basis, a deposit-commission basis, or any similar basis prescribed by regulations, for resale in the home or at a place of business other than in a permanent retail establishment.
3) Persons engaged in the trade or business of delivering or distributing newspapers or shopping news (including any services directly related to such delivery or distribution).

Direct selling includes activities of individuals who attempt to increase direct sales activities of their direct sellers and who earn income based on the productivity of their direct sellers. Such activities include providing motivation and encouragement; imparting skills, knowledge, or experience; and recruiting. For more information on direct sellers, see **Pub. 911,** Direct Sellers.

Licensed real estate agents. This category includes individuals engaged in appraisal activities for real estate sales if they earn income based on sales or other output.

Misclassification of Employees

Consequences of treating an employee as an independent contractor. If you classify an employee as an independent contractor and you have no reasonable basis for doing so, you may be held liable for employment taxes for that worker (the relief provisions, discussed below, will not apply). See Internal Revenue Code section 3509 for more information.

Relief provisions. If you have a reasonable basis for not treating a worker as an employee, you may be relieved from having to pay employment taxes for that worker. To get this relief, you must file all required Federal information returns on a basis consistent with your treatment of the worker. You (or your predecessor) must not have treated any worker holding a substantially similar position as an employee for any periods beginning after 1977.

Technical service specialists. This relief provision does not apply to a worker who provides services to another business (the client) as a technical service specialist under an arrangement between the business providing the worker, such as a technical services firm, and the client. A technical service specialist is an engineer, designer, drafter, computer programmer, systems analyst, or other similarly skilled worker engaged in a similar line of work.

This rule does not affect the determination of whether such workers are employees under the common-law rules. The common-law rules control whether the specialist is treated as an employee or an independent contractor. However, if you directly contract with a technical service specialist to provide services for your business rather than for another business, you may still be entitled to the relief provision. See **Employee or Independent Contractor?** below.

2. Employee or Independent Contractor?

An employer must generally withhold income taxes, withhold and pay social security and Medicare taxes, and pay unemployment tax on wages paid to an employee. An employer does not generally have to withhold or pay any taxes on payments to independent contractors.

Common-Law Rules

To determine whether an individual is an employee or an independent contractor under the common law, the relationship of the worker and the business must be examined. All evidence of control and independence must be considered. In any employee-independent contractor determination, all information that provides evidence of the degree of control and the degree of independence must be considered.

Facts that provide evidence of the degree of control and independence fall into three categories: behavioral control, financial control, and the type of relationship of the parties. These facts are discussed below.

Behavioral control. Facts that show whether the business has a right to direct and control how the worker does the task for which the worker is hired include the type and degree of—

Instructions the business gives the worker. An employee is generally subject to the business' instructions about when, where, and how to work. All of the following are examples of types of instructions about how to do work:

- When and where to do the work
- What tools or equipment to use
- What workers to hire or to assist with the work
- Where to purchase supplies and services
- What work must be performed by a specified individual
- What order or sequence to follow

The amount of instruction needed varies among different jobs. Even if no instructions are given, sufficient behavioral control may exist if the employer has the right to control how the work results are achieved. A business may lack the knowledge to instruct some highly specialized professionals; in other cases, the task may require little or no instruction. The key consideration is whether the business has retained the right to control the details of a worker's performance or instead has given up that right.

Training the business gives the worker. An employee may be trained to perform services in a particular manner. Independent contractors ordinarily use their own methods.

Financial control. Facts that show whether the business has a right to control the business aspects of the worker's job include:

The extent to which the worker has unreimbursed business expenses. Independent contractors are more likely to have unreimbursed expenses than are employees. Fixed ongoing costs that are incurred regardless of whether work is currently being performed are especially important. However, employees may also incur unreimbursed expenses in connection with the services they perform for their business.

The extent of the worker's investment. An independent contractor often has a significant investment in the facilities he or she uses in performing services for someone else. However, a significant investment is not necessary for independent contractor status.

The extent to which the worker makes services available to the relevant market. An independent contractor is generally free to seek out business opportunities.

Independent contractors often advertise, maintain a visible business location, and are available to work in the relevant market.

How the business pays the worker. An employee is generally guaranteed a regular wage amount for an hourly, weekly, or other period of time. This usually indicates that a worker is an employee, even when the wage or salary is supplemented by a commission. An independent contractor is usually paid by a flat fee for the job. However, it is common in some professions, such as law, to pay independent contractors hourly.

The extent to which the worker can realize a profit or loss. An independent contractor can make a profit or loss.

Type of relationship. Facts that show the parties' type of relationship include:

Written contracts describing the relationship the parties intended to create.

Whether the business provides the worker with employee-type benefits, such as insurance, a pension plan, vacation pay, or sick pay.

The permanency of the relationship. If you engage a worker with the expectation that the relationship will continue indefinitely, rather than for a specific project or period, this is generally considered evidence that your intent was to create an employer-employee relationship.

The extent to which services performed by the worker are a key aspect of the regular business of the company. If a worker provides services that are a key aspect of your regular business activity, it is more likely that you will have the right to direct and control his or her activities. For example, if a law firm hires an attorney, it is likely that it will present the attorney's work as its own and would have the right to control or direct that work. This would indicate an employer-employee relationship.

IRS help. If you want the IRS to determine whether a worker is an employee, file **Form SS-8,** Determination of Worker Status for Purposes of Federal Employment Taxes and Income Tax Withholding, with the IRS.

Industry Examples

The following examples may help you properly classify your workers.

Building and Construction Industry

Example 1. Jerry Jones has an agreement with Wilma White to supervise the remodeling of her house. She did not advance funds to help him carry on the work. She makes direct payments to the suppliers for all necessary materials. She carries liability and workers' compensation insurance covering Jerry and others he engaged to assist him. She pays them an hourly rate and exercises almost constant supervision over the work. Jerry is not free to transfer his assistants to other jobs. He may not work on other jobs while working for Wilma. He assumes no responsibility to complete the work and will incur no contractual liability if he fails to do so. He and his assistants perform personal services for hourly wages. They are employees of Wilma White.

Example 2. Milton Manning, an experienced tilesetter, orally agreed with a corporation to perform full-time services at construction sites. He uses his own tools and performs services in the order designated by the corporation and according to its specifications. The corporation supplies all materials, makes frequent inspections of his work, pays him on a piecework basis, and carries workers' compensation insurance on him. He does not have a place of business or hold himself out to perform similar services for others. Either party can end the services at any time. Milton Manning is an employee of the corporation.

Example 3. Wallace Black agreed with the Sawdust Co. to supply the construction labor for a group of houses. The company agreed to pay all construction costs. However, he supplies all the tools and equipment. He performs personal services as a carpenter and mechanic for an hourly wage. He also acts as superintendent and foreman and engages other individuals to assist him. The company has the right to select, approve, or discharge any helper. A company representative makes frequent inspections of the construction site. When a house is finished, Wallace is paid a certain percentage of its costs. He is not responsible for faults, defects of construction, or wasteful operation. At the end of each week, he presents the company with a statement of the amount he has spent, including the payroll. The company gives him a check for that amount from which he pays the assistants, although he is not personally liable for their wages. Wallace Black and his assistants are employees of the Sawdust Co.

Example 4. Bill Plum contracted with Elm Corporation to complete the roofing on a housing complex. A signed contract established a flat amount for the services rendered by Bill Plum. Bill is a licensed roofer and carries workers' compensation and liability insurance under the business name, Plum Roofing. He hires his own roofers who are treated as employees for Federal employment tax purposes. If there is a problem with the roofing work, Plum Roofing is responsible for paying for any repairs. Bill Plum, doing business as Plum Roofing, is an independent contractor.

Example 5. Vera Elm, an electrician, submitted a job estimate to a housing complex for electrical work at $16 per hour for 400 hours. She is to receive $1,280 every 2 weeks for the next 10 weeks. This is not considered payment by the hour. Even if she works more or less than 400 hours to complete the work, Vera Elm will receive $6,400. She also performs additional electrical installations under contracts with other companies, which she obtained through advertisements. Vera is an independent contractor.

Trucking Industry

Example. Rose Trucking contracts to deliver material for Forest Inc. at $140 per ton. Rose Trucking is not paid for any articles that are not delivered. At times, Jan Rose, who operates as Rose Trucking, may also lease another truck and engage a driver to complete the contract. All operating expenses, including insurance coverage, are paid by Jan Rose. All equipment is owned or rented by Jan, and she is responsible for all maintenance. None of the drivers are provided by Forest Inc. Jan Rose, operating as Rose Trucking, is an independent contractor.

Computer Industry

Example. Steve Smith, a computer programmer, is laid off when Megabyte Inc. downsizes. Megabyte agrees to pay Steve a flat amount to complete a one-time project to create a certain product. It is not clear how long it will take to complete the project, and Steve is not guaranteed any minimum payment for the hours spent on the program. Megabyte provides Steve with no instructions beyond the specifications for the product itself. Steve and Megabyte have a written contract, which provides that Steve is considered to be an independent contractor, is required to pay Federal and state taxes, and receives no benefits from Megabyte. Megabyte will file a Form 1099-MISC. Steve does the work on a new high-end computer which cost him $7,000. Steve works at home and is not expected or allowed to attend meetings of the software development group. Steve is an independent contractor.

Automobile Industry

Example 1. Donna Lee is a salesperson employed on a full-time basis by Bob Blue, an auto dealer. She works 6 days a week and is on duty in Bob's showroom on certain assigned days and times. She appraises trade-ins, but her appraisals are subject to the sales manager's approval. Lists of prospective customers belong to the dealer. She has to develop leads and report results to the sales manager. Because of her experience, she requires only minimal assistance in closing and financing sales and in other phases of her work. She is paid a commission and is eligible for prizes and bonuses offered by Bob. Bob also pays the cost of health insurance and group-term life insurance for Donna. Donna is an employee of Bob Blue.

Example 2. Sam Sparks performs auto repair services in the repair department of an auto sales company. He works regular hours and is paid on a percentage basis. He has no investment in the repair department. The sales company supplies all facilities, repair parts, and supplies; issues instructions on the amounts to be charged, parts to be used, and the time for completion of each job; and checks all estimates and repair orders. Sam is an employee of the sales company.

Example 3. An auto sales agency furnishes space for Helen Bach to perform auto repair services. She provides her own tools, equipment, and supplies. She seeks out business from insurance adjusters and other individuals and does all the body and paint work that comes to the agency. She hires and discharges her own helpers, determines her own and her helpers' working hours, quotes prices for repair work, makes all necessary adjustments, assumes all losses from uncollectible accounts, and receives, as compensation for her services, a large percentage of the gross collections from the auto repair shop. Helen is an independent contractor and the helpers are her employees.

Attorney

Example. Donna Yuma is a sole practitioner who rents office space and pays for the following items: telephone, computer, on-line legal research linkup, fax machine, and photocopier. Donna buys office supplies and pays bar dues and membership dues for three other professional organizations. Donna has a part-time receptionist who also does the bookkeeping. She pays the receptionist, withholds and pays Federal and state employment taxes, and files a Form W-2 each year. For the past 2 years, Donna has had only three clients, corporations with which there have been long-standing relationships. Donna charges the corporations an hourly rate for her services, sending monthly bills detailing the work performed for the prior month. The bills include charges for long distance calls, on-line research time, fax charges, photocopies, postage, and travel, costs for which the corporations have agreed to reimburse her. Donna is an independent contractor.

Taxicab Driver

Example. Tom Spruce rents a cab from Taft Cab Co. for $150 per day. He pays the costs of maintaining and operating the cab. Tom Spruce keeps all fares he receives from customers. Although he receives the benefit of Taft's two-way radio communication equipment, dispatcher, and advertising, these items benefit both Taft and Tom Spruce. Tom Spruce is an independent contractor.

Salesperson

To determine whether salespersons are employees under the usual common-law rules, you must evaluate each individual case. If a salesperson who works for you does not meet the tests for a common-law employee, discussed earlier, you do not have to withhold income tax from his or her pay (see **Statutory Employees** earlier). However, even if a salesperson is not an employee under the usual common-law rules, his or her pay may still be subject to social security, Medicare, and FUTA taxes. To determine whether a salesperson is an employee for social security, Medicare, and FUTA tax purposes, the salesperson must meet **all** eight elements of the statutory employee test. A salesperson is a statutory employee for social security, Medicare, and FUTA tax purposes if he or she:

1) Works full time for one person or company except, possibly, for sideline sales activities on behalf of some other person,
2) Sells on behalf of, and turns his or her orders over to, the person or company for which he or she works,
3) Sells to wholesalers, retailers, contractors, or operators of hotels, restaurants, or similar establishments,
4) Sells merchandise for resale, or supplies for use in the customer's business,
5) Agrees to do substantially all of this work personally,
6) Has no substantial investment in the facilities used to do the work, other than in facilities for transportation,
7) Maintains a continuing relationship with the person or company for which he or she works, and
8) Is not an employee under common-law rules.

3. Employees of Exempt Organizations

Many ***nonprofit organizations*** are exempt from income tax. Although they do not have to pay income tax themselves, they must still withhold income tax from the pay of their employees. However, there are special social security, Medicare, and Federal unemployment (FUTA) tax rules that apply to the wages they pay their employees.

Section 501(c)(3) organizations. Nonprofit organizations that are exempt from income tax under section 501(c)(3) of the Internal Revenue Code include any community chest, fund, or foundation organized and operated exclusively for religious, charitable, scientific, testing for public safety, literary or educational purposes, fostering national or international amateur sports competition, or for the prevention of cruelty to children or animals. These organizations are usually corporations and are exempt from income tax under section 501(a).

Social security and Medicare taxes. Wages paid to employees of section 501(c)(3) organizations are subject to social security and Medicare taxes unless one of the following situations applies:

1) The organization pays an employee less than $100 in a calendar year.
2) The organization is a church or church-controlled organization opposed for religious reasons to the payment of social security and Medicare taxes and has filed **Form 8274,** Certification by Churches and Qualified Church-Controlled Organizations Electing Exemption From Employer Social Security and Medicare Taxes, to elect exemption from social security and Medicare taxes. The organization must have filed for exemption before the first date on which a quarterly employment tax return (Form 941) would otherwise be due.

An employee of a church or church-controlled organization that is exempt from social security and Medicare taxes must pay self-employment tax if the employee is paid $108.28 or more in a year. However, an employee who is a member of a qualified religious sect can apply for an exemption from the self-employment tax by filing **Form 4029,** Application for Exemption From Social Security and Medicare Taxes and Waiver of Benefits. See **Members of recognized religious sects opposed to insurance** in section 4.

Federal unemployment tax. An organization that is exempt from income tax under section 501(c)(3) of the Internal Revenue Code is also exempt from the Federal unemployment (FUTA) tax. This exemption cannot be waived.

Note: *An organization wholly owned by a state or its political subdivision should contact the appropriate state official for information about reporting and getting social security and Medicare coverage for its employees.*

Other than section 501(c)(3) organizations. Nonprofit organizations that are not section 501(c)(3) organizations may also be exempt from income tax under section 501(a) or section 521. However, these organizations are not exempt from withholding income, social security, or Medicare tax from their employees' pay, or from paying FUTA tax. Two special rules for social security, Medicare, and FUTA taxes apply.

1) If an employee is paid less than $100 during a calendar year, his or her wages are not subject to social security and Medicare taxes.
2) If an employee is paid less than $50 in a calendar quarter, his or her wages are not subject to FUTA tax for the quarter.

The above rules do not apply to employees who work for pension plans and other similar organizations described in section 401(a).

4. Religious Exemptions

Special rules apply to the treatment of ministers for social security purposes. An exemption from social security is available for ministers and certain other religious workers and members of certain recognized religious sects. For more information on getting an exemption, see **Pub. 517,** Social Security and Other Information for Members of the Clergy and Religious Workers.

Ministers. Ministers are individuals who are duly ordained, commissioned, or licensed by a religious body constituting a church or church denomination. They are given the authority to conduct religious worship, perform sacerdotal functions, and administer ordinances and sacraments according to the prescribed tenets and practices of that religious organization.

A minister who performs services for you subject to your will and control is your employee. The common-law rules discussed in sections 1 and 2 should be applied to determine whether a minister is your employee or is self-employed. The earnings of a minister are not subject to income, social security, and Medicare tax withholding. They are subject to self-employment tax and income tax. You do not withhold these taxes from wages earned by a minister. However, you may agree with the minister to voluntarily withhold tax to cover the minister's liability for self-employment tax and income tax.

Form W-2. If your employee is an ordained minister, report all taxable compensation as wages in box 1 on Form W-2. Include in this amount expense allowances or reimbursements paid under a nonaccountable plan, discussed in section 5 of Circular E. Do not include a parsonage allowance (excludable housing allowance) in this amount. You may report a parsonage or rental allowance (housing allowance), utilities allowance, and the rental value of housing provided in a separate statement or in box 14 on Form W-2. Do not show on Form W-2 or 941 any amount as social security or Medicare wages, or any withholding for social security or Medicare taxes. If you withheld tax from the minister under a voluntary agreement, this amount should be shown in box 2 on Form W-2 as Federal income tax withheld. For more information on ministers, see Pub. 517.

Exemptions for ministers and others. Certain ordained ministers, Christian Science practitioners, and members of religious orders who have not taken a vow of poverty, who are subject to self-employment tax, may apply to exempt their earnings from the tax on religious grounds. The application must be based on conscientious opposition to public insurance because of personal religious considerations. The exemption applies only to qualified services performed for the religious organization. See Rev. Proc. 91-20, 1991-1 C.B. 524, for guidelines to determine whether an organization is a religious order or whether an individual is a member of a religious order.

To apply for the exemption, the employee should file **Form 4361,** Application for Exemption From Self-Employment Tax for Use by Ministers, Members of Religious Orders and Christian Science Practitioners. See Pub. 517 for more information about Form 4361.

Members of recognized religious sects opposed to insurance. If you belong to a recognized religious sect or a division of such sect that is opposed to insurance, you may qualify for an exemption from the self-employment tax. To qualify, you must be conscientiously opposed to accepting the benefits of any public or private insurance that makes payments because of death, disability, old age, or retirement, or makes payments toward the cost of, or provides services for, medical care (including social security and Medicare benefits). If you buy a retirement annuity from an insurance company, you will not be eligible for this exemption. Religious opposition based on the teachings of the sect is the only legal basis for the exemption. In addition, your religious sect (or division) must have existed since December 31, 1950.

Self-employed. If you are self-employed and a member of a recognized religious sect opposed to insurance, you can apply for exemption by filing **Form 4029,** Application for Exemption From Social Security and Medicare Taxes and Waiver of Benefits, and waive all social security benefits.

Employees. The social security and Medicare tax exemption available to the self-employed who are members of a recognized religious sect opposed to insurance is also available to their employees who are members of such a sect. This applies to partnerships only if each partner is a member of the sect. This exemption for employees applies only if both the employee and the employer are members of such a sect, and the employer has an exemption. To get the exemption, the employee must file Form 4029.

An employee of a church or church-controlled organization that is exempt from social security and Medicare taxes can also apply for an exemption on Form 4029.

5. Wages and Other Compensation

Circular E provides a general discussion of taxable wages. Publication 15-B discusses fringe benefits. The following topics supplement those discussions.

Employee Achievement Awards

Do not withhold income, social security, or Medicare taxes on the fair market value of an employee achievement award if it is excludable from your employee's gross income. To be excludable from your employee's gross income, the award must be tangible personal property (not cash or securities) given to an employee for length of service or safety achievement, awarded as part of a meaningful presentation, and awarded under circumstances that do not indicate that the payment is disguised compensation. Excludable employee achievement awards also are not subject to FUTA tax.

Limits. The most you can exclude for the cost of all employee achievement awards to the same employee for the year is $400. A higher limit of $1,600 applies to qualified plan awards. These awards are employee achievement awards under a written plan that does not discriminate in favor of highly compensated employees. An award cannot be treated as a qualified plan award if the average cost per recipient of all awards under all your qualified plans is more than $400.

If during the year an employee receives awards not made under a qualified plan and also receives awards under a qualified plan, the exclusion for the total cost of all awards to that employee cannot be more than $1,600. The $400 and $1,600 limits cannot be added together to exclude more than $1,600 for the cost of awards to any one employee during the year.

Scholarship and Fellowship Payments

Only amounts you pay as a qualified scholarship to a candidate for a degree may be excluded from the recipient's gross income. A qualified scholarship is any amount granted as a scholarship or fellowship that is used for:

- Tuition and fees required to enroll in, or to attend, an educational institution or
- Fees, books, supplies, and equipment that are required for courses at the educational institution.

Any amounts you pay for room and board, and any amounts you pay for teaching, research, or other services required as a condition of receiving the scholarship, are not excludable from the recipient's gross income. A qualified scholarship is not subject to social security, Medicare, and FUTA taxes, or income tax withholding. For more information, see **Pub. 520,** Scholarships and Fellowships.

Outplacement Services

If you provide outplacement services to your employees to help them find new employment (such as career counseling, resume assistance, or skills assessment), the value of these benefits may be income to them and subject to all withholding taxes. However, the value of these services will not be subject to any employment taxes if:

1) You derive a substantial business benefit from providing the services (such as improved employee morale or business image) separate from the benefit you would receive from the mere payment of additional compensation, and
2) The employee would be able to deduct the cost of the services as employee business expenses if he or she had paid for them.

However, if you receive no additional benefit from providing the services, or if the services are not provided on the basis of employee need, then the value of the services is treated as wages and is subject to income tax withholding and social security and Medicare taxes. Similarly, if an employee receives the outplacement services in exchange for reduced severance pay (or other taxable compensation), then the amount the severance pay is reduced is treated as wages for employment tax purposes.

Withholding for Idle Time

Payments made under a voluntary guarantee to employees for ***idle time*** (any time during which an employee performs no services) are wages for the purposes of social security, Medicare, and FUTA taxes, and income tax withholding.

Back Pay

Treat back pay as wages in the year paid and withhold and pay employment taxes as required. If back pay was awarded by a court or government agency to enforce a Federal or state statute protecting an employee's right to employment or wages, special rules apply for reporting those wages to the Social Security Administration. These rules also apply to litigation actions, and settlement agreements or agency directives that are resolved out of court and not under a court decree or order. Examples of pertinent statutes include, but are not limited to, the National Labor Relations Act, Fair Labor Standards Act, Equal Pay Act, and Age Discrimination in Employment Act. See **Pub. 957,** Reporting Back Pay and Special Wage Payments to the Social Security Administration, and **Form SSA-131,** Employer Report of Special Wage Payments, for details.

Supplemental Unemployment Benefits

If you pay, under a plan, supplemental unemployment benefits to a former employee, all or part of the payments may be taxable and subject to income tax withholding, depending on how the plan is funded. Amounts that represent a return to the employee of amounts previously subject to tax are not taxable and are not subject to withholding. You should withhold income tax on the taxable part of the payments made, under a plan, to an employee who is involuntarily separated because of a reduction in force, discontinuance of a plant or operation, or other similar condition. It does not matter whether the separation is temporary or permanent. There are special rules that apply in determining whether benefits qualify as supplemental unemployment benefits that are excluded from wages for social security, Medicare, and FUTA purposes. To qualify as supplemental unemployment benefits for these purposes, the benefits must meet the following requirements:

1) Benefits are paid only to unemployed former employees who are laid off by the employer.
2) Eligibility for benefits depends on meeting prescribed conditions after termination.
3) The amount of weekly benefits payable is based upon state unemployment benefits, other compensation allowable under state law, and the amount of regular weekly pay.
4) The duration of the benefits is affected by the fund level and employee seniority.
5) The right to benefits does not accrue until a prescribed period after termination.
6) Benefits are not attributable to the performance of particular services.
7) No employee has any right to the benefits until qualified and eligible to receive benefits.
8) Benefits may not be paid in a lump sum.

Withholding on taxable supplemental unemployment benefits must be based on the withholding certificate (Form W-4) the employee gave you.

Golden Parachutes (Excessive Termination Payments)

A golden parachute is a contract entered into by a corporation and key personnel under which the corporation agrees to pay certain amounts to the key personnel in the event of a change in ownership or control of the corporation. Payments under golden parachute contracts, like any termination pay, are subject to social security, Medicare, and FUTA taxes, and income tax withholding.

Beginning with payments under contracts entered into, significantly amended, or renewed after June 14, 1984, no deduction is allowed to the corporation for ***excess parachute payments.*** The employee is subject to a 20% nondeductible excise tax to be withheld by the corporation on all excess payments. The payment is generally considered an excess parachute payment if it equals or exceeds three times the average annual compensation of the recipient over the previous 5-year period. The amount over the average is the excess parachute payment.

Example. An officer of a corporation receives a golden parachute payment of $400,000. This is more than three times greater than his or her average compensation of $100,000 over the previous 5-year period. The excess parachute payment is $300,000 ($400,000 minus $100,000). The corporation cannot deduct the $300,000 and must withhold the excise tax of $60,000 (20% of $300,000).

Exempt payments. Most small business corporations are exempt from the golden parachute rules. See Code section 280G for more information.

Interest-Free and Below-Market-Interest-Rate Loans

If an employer lends an employee more than $10,000 at an interest rate less than the current applicable Federal rate (AFR), the difference between the interest paid and the interest that would be paid under the AFR is considered additional compensation to the employee. This rule applies to a loan of $10,000 or less if one of its principal purposes is the avoidance of Federal tax.

This additional compensation to the employee is subject to social security, Medicare, and FUTA taxes, but not to income tax withholding. Include it in compensation on Form W-2 (or Form 1099-MISC for an independent contractor). The AFR is established monthly and published by the IRS each month in the Internal Revenue Bulletin. You can get these rates by calling 1-800-829-1040 or by accessing the IRS's Web Site at **www.irs.gov.** For more information, see Pub. 15-B.

Workers' Compensation—Public Employees

State and local government employees, such as police officers and firefighters, sometimes receive payments due to injury in the line of duty under a statute that is ***not*** the general workers' compensation law of a state. If the statute limits benefits to work-related injuries or sickness and does not base payments on the employee's age, length of service, or prior contributions, the statute is "in the nature of" a workers' compensation law. Payments under the statute are not subject to FUTA tax or income tax withholding, but they are subject to social security and Medicare taxes to the same extent as the employee's regular wages. However, the payments are no longer subject to social security and Medicare taxes after the expiration of 6 months following the last calendar month in which the employee worked for the employer.

Leave Sharing Plans

If you establish a leave sharing plan for your employees that allows them to donate leave to other employees for medical emergencies, the amounts paid to the recipients of the leave are considered wages. These amounts are includible in the gross income of the recipients and are subject to social security, Medicare, and FUTA taxes, and income tax withholding. Do not include these amounts in the income of the donors.

Nonqualified Deferred Compensation Plans

Social security, Medicare, and FUTA taxes. Employer contributions to nonqualified deferred compensation or nonqualified pension plans are treated as social security, Medicare, and FUTA wages when the services are performed or the employee no longer has a substantial risk of forfeiting the right to the deferred compensation, whichever is later. This is true whether the plan is funded or unfunded.

Amounts deferred are subject to social security, Medicare, and FUTA taxes unless the value of the amount deferred cannot be determined; for example, if benefits are based on final pay. If the value of the future benefit is based on any factors that are not yet reasonably determinable, you may estimate the value of the future benefit and withhold and pay social security, Medicare, and FUTA taxes on that amount. If amounts that were not determinable in prior periods are now determinable, they are subject to social security, Medicare, and FUTA taxes on the amounts deferred plus the income attributable to those amounts deferred. For more information, see Regulations sections 31.3121(v)(2)-1 and 31.3306(r)(2)-1.

Income tax withholding. Amounts deferred under nonqualified deferred compensation plans are not subject to income taxes until benefit payments begin. Withhold income tax on nonqualified plans as follows:

- ***Funded plan.*** Withhold when the employees' rights to amounts are not subject to substantial risk of forfeiture or are transferable free of such risk. A funded plan is one in which an employer ***irrevocably*** contributes the deferred compensation to a separate fund, such as an irrevocable trust.
- ***Unfunded plan.*** Generally, withhold when you make payments to the employee, either constructively or actually.

Tax-Sheltered Annuities

Employer payments made by an educational institution or a tax-exempt organization to purchase a tax-sheltered annuity for an employee (annual deferrals) are included in the employee's social security and Medicare wages if the payments are made because of a salary reduction agreement. They are not included in box 1 on Form W-2 in the year the deferrals are made and are not subject to income tax withholding.

Contributions to a Simplified Employee Pension (SEP)

An employer's SEP contributions to an employee's individual retirement arrangement (IRA) are excluded from the employee's gross income. These excluded amounts are not subject to social security, Medicare, and FUTA taxes, or income tax withholding. However, any SEP contributions paid under a salary reduction agreement (SARSEP) are included in wages for purposes of social security and Medicare taxes and FUTA. See **Pub. 560,** for more information about SEPs.

Salary reduction simplified employee pensions (SARSEP) repealed. You may not establish a SARSEP after 1996. However, SARSEPs established before January 1, 1997, may continue to receive contributions.

SIMPLE Retirement Plans

Employer and employee contributions to a savings incentive match plan for employees (SIMPLE) retirement account (subject to limitations) are excludable from the employee's income and are exempt from Federal income tax withholding. An employer's nonelective (2%) or matching contributions are exempt from social security, Medicare, and FUTA taxes. However, an employee's salary reduction contributions to a SIMPLE are subject to social security, Medicare, and FUTA taxes. For more information about SIMPLE retirement plans, see Pub. 560.

6. Sick Pay Reporting

Special rules apply to the reporting of sick pay payments to employees. How these payments are reported depends on whether the payments are made by the employer or a third party, such as an insurance company.

Sick pay is usually subject to social security, Medicare, and FUTA taxes. For exceptions, see **Social Security, Medicare, and FUTA Taxes on Sick Pay** later. Sick pay also may be subject to either mandatory or voluntary Federal income tax withholding, depending on who pays it.

Sick Pay

Sick pay generally means any amount paid under a plan because of an employee's temporary absence from work due to injury, sickness, or disability. It may be paid by either the employer or a third party, such as an insurance company. Sick pay includes both short- and long-term benefits. It is often expressed as a percentage of the employee's regular wages.

Payments That Are Not Sick Pay

Sick pay does not include the following payments:

1) **Disability retirement payments.** Disability retirement payments are not sick pay and are not discussed in this section. Those payments are subject to the rules for income tax withholding from pensions and annuities. See section 9.
2) **Workers' compensation.** Payments because of a work-related injury or sickness that are made under a workers' compensation law are not sick pay and are not subject to employment taxes. But see **Workers' Compensation—Public Employees** in section 5.
3) **Medical expense payments.** Payments under a definite plan or system for medical and hospitalization expenses, or for insurance covering these expenses, are not sick pay and are not subject to employment taxes.
4) **Payments unrelated to absence from work.** Accident or health insurance payments unrelated to absence from work are not sick pay and are not subject to employment taxes. These include payments for:
   a) Permanent loss of a member or function of the body,
   b) Permanent loss of the use of a member or function of the body, or
   c) Permanent disfigurement of the body.

Example. Donald was injured in a car accident and lost an eye. Under a policy paid for by Donald's employer, Delta Insurance Co. paid Donald $5,000 as compensation for the loss of his eye. Because the payment was determined by the type of injury and was unrelated to Donald's absence from work, it is not sick pay and is not subject to employment taxes.

Sick Pay Plan

A sick pay plan is a plan or system established by an employer under which sick pay is available to employees generally or to a class or classes of employees. This does not include a situation in which benefits are provided on a discretionary or occasional basis with merely an intention to aid particular employees in time of need.

You have a sick pay plan or system if the plan is in writing or is otherwise made known to employees, such as by a bulletin board notice or your long and established practice. Some indications that you have a sick pay plan or system include references to the plan or system in the contract of employment, employer contributions to a plan, or segregated accounts for the payment of benefits.

Definition of employer. The ***employer for whom the employee normally works,*** a term used in the following discussion, is either the employer for whom the employee was working at the time the employee became sick or disabled or the last employer for whom the employee worked before becoming sick or disabled, if that employer made contributions to the sick pay plan on behalf of the sick or disabled employee.

Note: *Contributions to a sick pay plan through a cafeteria plan (by direct employer contributions or salary reduction) are employer contributions unless they are* ***aftertax*** *employee contributions (included in taxable wages).*

Third-Party Payers of Sick Pay

Employer's agent. An employer's agent is a third party that bears no insurance risk and is reimbursed on a cost-plus-fee basis for payment of sick pay and similar amounts. A third party may be your agent even if the third party is responsible for determining which employees are eligible to receive payments. For example, if a third party provides administrative services only, the third party is your agent. If the third party is paid an insurance premium and is not reimbursed on a cost-plus-fee basis, the third party is not your agent. Whether an insurance company or other third party is your agent depends on the terms of the agreement with you.

A third party that makes payments of sick pay as your agent is not considered the employer and generally has no responsibility for employment taxes. This responsibility remains with you. However, under an exception to this rule, the parties may enter into an agreement that makes the third-party agent responsible for employment taxes. In this situation, the third-party agent should use its own name and EIN (rather than your name and EIN) for the responsibilities it has assumed.

Third party not employer's agent. A third party that makes payments of sick pay ***other than as an agent of the employer*** is liable for income tax withholding (if requested by the employee) and the employee part of the social security and Medicare taxes. The third party is also liable for the employer part of the social security and Medicare taxes and the FUTA tax, unless the third party transfers this liability to the employer for whom the employee normally works. This liability is transferred if the third party takes the following steps:

1) Withholds the ***employee*** social security and Medicare taxes from the sick pay payments.
2) Makes timely deposits of the ***employee*** social security and Medicare taxes.
3) Notifies the employer for whom the employee normally works of the payments on which employee taxes were withheld and deposited. The third party must notify the employer within the time required for the third party's deposit of the employee part of the social security and Medicare taxes. For instance, if the third party is a monthly schedule depositor, it must notify the employer by the 15th day of the month following the month in which the sick pay payment is made because that is the day by which the deposit is required to be made. The third party should notify the employer as soon as information on payments is available so that an employer required to make electronic deposits can make them timely. For multi-employer plans, see the special rule discussed next.

Multi-employer plan timing rule. A special rule applies to sick pay payments made to employees by a third-party insurer under an insurance contract with a multi-employer plan established under a collectively bargained agreement. If the third-party insurer making the payments complies with steps 1 and 2 above and gives the plan (rather than the employer) the required timely notice described in step 3 above, then the plan (not the third-party insurer) must pay the employer part of the social security and Medicare taxes and the FUTA tax. Similarly, if, within 6 business days of the plan's receipt of notification, the plan gives notice to the employer for whom the employee normally works, the employer (not the plan) must pay the employer part of the social security and Medicare taxes and the FUTA tax.

Reliance on information supplied by the employer. A third party that pays sick pay should request information from the employer to determine amounts that are not subject to employment taxes. Unless the third party has reason not to believe the information, it may rely on that information as to the following items:

- The total wages paid the employee during the calendar year.
- The last month in which the employee worked for the employer.
- The employee contributions to the sick pay plan made with aftertax dollars.

The third party should not rely on statements regarding these items made by the employee.

Social Security, Medicare, and FUTA Taxes on Sick Pay

Employer. If you pay sick pay to your employee, you must generally withhold employee social security and Medicare taxes from the sick pay. You must timely deposit employee and employer social security and Medicare taxes and FUTA tax. There are no special deposit rules for sick pay. See section 11 of Circular E for more information on the deposit rules.

Amounts not subject to social security, Medicare, or FUTA taxes. The following payments, whether made by the employer or a third party, are not subject to social security, Medicare, or FUTA taxes (different rules apply to income tax withholding):

- **Payments after employee's death or disability retirement.** Social security, Medicare, and FUTA taxes do not apply to amounts paid under a definite plan or system, as defined under **Sick Pay Plan** earlier, on or after the termination of the employment relationship because of death or disability retirement.
  However, even if there is a definite plan or system, amounts paid to a former employee are subject to social security, Medicare, and FUTA taxes if they would have been paid even if the employment relationship had not terminated because of death or disability retirement. For example, a payment to a disabled former employee for unused vacation time would have been made whether or not the employee retired on disability. Therefore, the payment is wages and is subject to social security, Medicare, and FUTA taxes.

- **Payments after calendar year of employee's death.** Sick pay paid to the employee's estate or survivor after the calendar year of the employee's death is not subject to social security, Medicare, or FUTA taxes. (Also see **Amounts not subject to income tax withholding** under **Income Tax Withholding on Sick Pay** later.)
  Example. Sandra became entitled to sick pay on November 30, 2001, and died December 26, 2001. On January 14, 2002, Sandra's sick pay for the period from December 19 through December 26, 2001, was paid to her survivor. The payment is not subject to social security, Medicare, or FUTA taxes.

- **Payments to an employee entitled to disability insurance benefits.** Payments to an employee when the employee is entitled to disability insurance benefits under section 223(a) of the Social Security Act are not subject to social security and Medicare taxes. This rule applies only if the employee became entitled to the Social Security Act benefits before the calendar year in which the payments are made, and the employee performs no services for the employer during the period for which the payments are made. These payments **are** subject to FUTA tax.

- **Payments that exceed the applicable wage base.** Social security and FUTA taxes do not apply to payments of sick pay that, when combined with the regular wages and sick pay previously paid to the employee during the year, exceed the applicable wage base. Because there is no Medicare tax wage base, this exception does not apply to Medicare tax. The social security tax wage base for 2002 is $84,900. The FUTA tax wage base is $7,000.
  Example. If an employee receives $80,000 in wages from an employer in 2002 and then receives $10,000 of sick pay, only the first $4,900 of the sick pay is subject to social security tax. All of the sick pay is subject to Medicare tax. None of the sick pay is subject to FUTA tax. See **Example of Figuring and Reporting Sick Pay** later.

- **Payments after 6 months absence from work.** Social security, Medicare, and FUTA taxes do not apply to sick pay paid more than 6 calendar months after the last calendar month in which the employee worked.
  Example 1. Ralph's last day of work before he became entitled to receive sick pay was December 10, 2001. He was paid sick pay for 9 months before his return to work on September 9, 2002. Sick pay paid to Ralph after June 30, 2002, is not subject to social security, Medicare, or FUTA taxes.
  Example 2. The facts are the same as in Example 1, except that Ralph worked 1 day during the 9-month period, on February 15, 2002. Because the 6-month period begins again in March, only the sick pay paid to Ralph after August 31, 2002, is exempt from social security, Medicare, and FUTA taxes.

- **Payments attributable to employee contributions.** Social security, Medicare, and FUTA taxes do not apply to payments, or parts of payments, attributable to employee contributions to a sick pay plan made with ***aftertax*** dollars. (Contributions to a sick pay plan made on behalf of employees with employees' pretax dollars under a cafeteria plan are employer contributions.)
  Group policy. If both the employer and the employee contributed to the sick pay plan under a group insurance policy, figure the taxable sick pay by multiplying it by the percentage of the policy's cost that was contributed by the employer for the 3 policy years before the calendar year in which the sick pay is paid. If the policy has been in effect fewer than 3 years, use the cost for the policy years in effect or, if in effect less than 1 year, a reasonable estimate of the cost for the first policy year.
  Example. Alan is employed by Edgewood Corporation. Because of an illness, he was absent from work for 3 months during 2002. Key Insurance Company paid Alan $2,000 sick pay for each month of his absence under a policy paid for by contributions from both Edgewood and its employees. All the employees' contributions were paid with aftertax dollars. For the 3 policy years before 2002, Edgewood paid 70% of the policy's cost and its employees paid 30%. Because 70% of the sick pay paid under the policy is due to Edgewood's contributions, $1,400 ($2,000 × 70%) of each payment made to Alan is taxable sick

pay. The remaining $600 of each payment that is due to employee contributions is not taxable sick pay and is not subject to employment taxes. Also, see **Example of Figuring and Reporting Sick Pay** later.

Income Tax Withholding on Sick Pay

The requirements for income tax withholding on sick pay and the methods for figuring it differ depending on whether the sick pay is paid by:

- The employer,
- An agent of the employer (defined earlier), or
- A third party that is not the employer's agent.

Employer or employer's agent. Sick pay paid by you or your agent is subject to mandatory income tax withholding. An employer or agent paying sick pay generally determines the income tax to be withheld based on the employee's Form W-4. The employee cannot choose how much will be withheld by giving you or your agent a **Form W-4S,** Request for Federal Tax Withholding From Sick Pay. Sick pay paid by an agent is treated as supplemental wages. If the agent does not pay regular wages to the employee, the agent may choose to withhold income tax at a flat 27% rate, rather than at the wage withholding rate.

Third party not an agent. Sick pay paid by a third party that is not your agent is **not** subject to mandatory income tax withholding. However, an employee may elect to have income tax withheld by submitting Form W-4S to the third party.

If Form W-4S has been submitted, the third party should withhold income tax on all payments of sick pay made 8 or more days after receiving the form. The third party may, at its option, withhold income tax before 8 days have passed.

The employee may request on Form W-4S to have a specific whole dollar amount withheld. However, if the requested withholding would reduce any net payment below $10, the third party should not withhold any income tax from that payment. The minimum amount of withholding that the employee can specify is $20 a week.

Withhold from all payments at the same rate. For example, if $25 is withheld from a regular full payment of $100, then $20 (25%) should be withheld from a partial payment of $80.

Amounts not subject to income tax withholding. The following amounts, whether paid by you or a third party, are not wages subject to income tax withholding.

- **Payments after the employee's death.** Sick pay paid to the employee's estate or survivor at any time after the employee's death is not subject to income tax withholding, regardless of who pays it.
- **Payments attributable to employee contributions.** Payments, or parts of payments, attributable to employee contributions made to a sick pay plan with aftertax dollars are not subject to income tax withholding. For more information, see the corresponding discussion in **Amounts not subject to social security, Medicare, or FUTA taxes** earlier.

Depositing and Reporting

This section discusses who is liable for depositing social security, Medicare, FUTA, and withheld income taxes on sick pay. These taxes must be deposited under the same rules that apply to deposits of taxes on regular wage payments. See Circular E for information on the deposit rules.

This section also explains how sick pay should be reported on Forms W-2, W-3, 940 or 940-EZ, and 941.

Sick Pay Paid by Employer or Agent

If you or your agent (defined earlier) make sick pay payments, you deposit taxes and file Forms W-2, W-3, 940 or 940-EZ, and 941 under the same rules that apply to regular wage payments.

However, the agreement between the parties may require your agent to carry out responsibilities that would otherwise have been borne by you. In this situation, your agent should use its own name and EIN (rather than yours) for the responsibilities it has assumed.

Reporting sick pay on Form W-2. You may either combine the sick pay with other wages and prepare a single Form W-2 for each employee, or you may prepare separate Forms W-2 for each employee, one reporting sick pay and the other reporting regular wages. A Form W-2 must be prepared even if all the sick pay is nontaxable (see **Box 12** below in the list of information that must be included on Form W-2). All Forms W-2 must be given to the employees by January 31.

The Form W-2 filed for the sick pay must include the employer's name, address, and EIN; the employee's name, address, and SSN; and the following information:

Box 1 – Sick pay the employee must include in income.

Box 2 – Any Federal income tax withheld from the sick pay.

Box 3 – Sick pay subject to employee social security tax.

Box 4 – Employee social security tax withheld from the sick pay.

Box 5 – Sick pay subject to employee Medicare tax.

Box 6 – Employee Medicare tax withheld from the sick pay.

Box 12 – Any amount not subject to Federal income tax because the employee contributed to the sick pay plan (enter code J).

Sick Pay Paid by Third Party

The rules for a third party that is not your agent depend on whether liability has been transferred as discussed under **Third-Party Payers of Sick Pay** earlier.

To figure the due dates and amounts of its deposits of employment taxes, a third party should combine:

1) The liability for the wages paid to its own employees and
2) The liability for payments it made to all employees of all its clients. This does not include liability transferred to the employer.

Liability not transferred to the employer. If the third party does not satisfy the requirements for transferring liability for FUTA tax and the employer's part of the social security and Medicare taxes, the third party reports the sick pay on its own Forms 940 (or 940-EZ) and 941. In this situation, the employer has no tax responsibilities for sick pay.

The third party must deposit social security, Medicare, FUTA, and withheld income taxes using its own name and EIN. The third party must give each employee to whom it paid sick pay a Form W-2 by January 31 of the following year. The Form W-2 must include the third party's name, address, and EIN instead of the employer information. Otherwise, the third party must complete Form W-2 as shown in **Reporting sick pay on Form W-2** earlier.

Liability transferred to the employer. Generally, if a third party satisfies the requirements for transferring liability for the employer part of the social security and Medicare taxes and for the FUTA tax, the following rules apply.

Deposits. The third party must make deposits of withheld employee social security and Medicare taxes and withheld income tax using its own name and EIN. You must make deposits of the employer part of the social security and Medicare taxes and the FUTA tax using your name and EIN. In applying the deposit rules, your liability for these taxes begins when you receive the third party's notice of sick pay payments.

Form 941. The third party and you must each file Form 941. Line 9 of each Form 941 must contain a special adjusting entry for social security and Medicare taxes. These entries are required because the total tax liability for social security and Medicare taxes (employee and employer parts) is split between you and the third party.

- ***Employer.*** You must include third-party sick pay on lines 2, 6a, and 7a of Form 941. (There are no entries for sick pay on lines 3 through 5.) After completing line 8, subtract on line 9 the employee social security and Medicare taxes withheld and deposited by the third party. Enter that amount in the "Sick Pay" space provided. If line 9 includes adjustments unrelated to sick pay, show those amounts in the spaces provided and the total in the line 9 entry space on the right.
- ***Third party.*** The third party must include on Form 941 the employee part of the social security and Medicare taxes (and income tax, if any) it withheld. The third party does **not** include on line 2 any sick pay paid as a third party but does include on line 3 any income tax withheld. On line 6a, the third party enters the total amount it paid subject to social security taxes. This amount includes both wages paid to its own employees and sick pay paid as a third party. The third party completes line 7a in a similar manner. On line 9, the third party subtracts the employer part of the social security and Medicare taxes that you must pay. The third party enters the amount you must pay on line 9 in the "Sick Pay" space provided. If line 9 includes adjustments unrelated to sick pay, the third party shows those amounts in the spaces provided and the total of all adjustments in the line 9 entry space.

Form 940 or 940-EZ. You, not the third party, must prepare Form 940 or 940-EZ for sick pay.

Third-party sick pay recap Forms W-2 and W-3. The third party must prepare a "third-party sick pay recap" Form W-2 and a "third-party sick pay recap" Form W-3. These forms, previously called "dummy" forms, do not reflect sick pay paid to individual employees, but instead show the combined amount of sick pay paid to all employees of all clients of the third party. The recap forms provide a means of reconciling the wages shown on the third party's Form 941. However, see ***Optional rule for Form W-2*** later. Do not file the recap Form W-2 and W-3 electronically or on magnetic media.

The third party fills out the third-party sick pay recap Form W-2 as follows:

Box b – Third party's EIN.

Box c – Third party's name and address.

Box e – "Third-Party Sick Pay Recap" in place of the employee's name.

Box 1 – Total sick pay paid to all employees.

Box 2 – Any Federal income tax withheld from the sick pay.

Box 3 – Sick pay subject to employee social security tax.

Box 4 – Employee social security tax withheld from sick pay.

Box 5 – Sick pay subject to employee Medicare tax.

Box 6 – Employee Medicare tax withheld from the sick pay.

The third party attaches the third-party sick pay recap Form W-2 to a separate recap Form W-3, on which only boxes b, e, f, g, 1, 2, 3, 4, 5, 6, and 12 are completed. Enter "Third-Party Sick Pay Recap" in box 12. Only the employer makes an entry in box 14 of Form W-3.

Optional rule for Form W-2. You and the third party may choose to enter into a legally binding agreement designating the third party to be your agent for purposes of preparing Forms W-2 reporting sick pay. The agreement must specify what part, if any, of the payments under the sick pay plan is excludable from the employees' gross incomes because it is attributable to their contributions to the plan. If you enter into an agreement, the third party prepares the actual Forms W-2, not the third-party sick pay recap Form W-2 as discussed earlier, for each employee who receives sick pay from the third party. If the optional rule is used:

1) The third party does not provide you with the sick pay statement described below and
2) You (not the third party) prepare third-party sick pay recap Forms W-2 and W-3. These recap forms are needed to reconcile the sick pay shown on your Form 941.

Sick pay statement. The third party must furnish you with a sick pay statement by January 15 of the year following the year in which the sick pay was paid. The statement must show the following information about each employee who was paid sick pay:

- The employee's name.
- The employee's SSN (if social security, Medicare, or income tax was withheld).
- The sick pay paid to the employee.
- Any Federal income tax withheld.
- Any employee social security tax withheld.
- Any employee Medicare tax withheld.

Example of Figuring and Reporting Sick Pay

Dave, an employee, was seriously injured in a car accident on January 1, 2001. Dave's last day of work was December 31, 2000. The accident was not job related.

Key, an insurance company that was not an agent of the employer, paid Dave $2,000 each month for 10 months, beginning in January 2001. Dave submitted a Form W-4S to Key, requesting $210 be withheld from each payment for Federal income tax. Dave received no payments from Edgewood, his employer, from January 2001 through October 2001. Dave returned to work in November 2001.

For the policy year in which the car accident occurred, Dave paid a part of the premiums for his coverage, and Edgewood paid the remaining part. The plan was, therefore, a "contributory plan." During the 3 policy years before the calendar year of the accident, Edgewood paid 70% of the total of the net premiums for its employees' insurance coverage, and its employees paid 30%.

Social security and Medicare taxes. For social security and Medicare tax purposes, taxable sick pay was $8,400 ($2,000 per month × 70% = $1,400 taxable portion per payment; $1,400 × 6 months = $8,400 total taxable sick pay). Only the six $2,000 checks received by Dave from January through June are included in the calculation. The check received by Dave in July (the seventh check) was received more than 6 months after the month in which Dave last worked.

Of each $2,000 payment Dave received, 30% ($600) is not subject to social security and Medicare taxes because the plan is contributory and Dave's aftertax contribution is considered to be 30% of the premiums during the 3 policy years before the calendar year of the accident.

FUTA tax. Of the $8,400 taxable sick pay (figured the same as for social security and Medicare taxes), only $7,000 is subject to the FUTA tax because the FUTA contribution base is $7,000.

Income tax withholding. Of each $2,000 payment, $1,400 ($2,000 × 70%) is subject to voluntary income tax withholding. In accordance with Dave's Form W-4S, $210 was withheld from each payment ($2,100 for the 10 payments made during 2001).

Liability transferred. For the first 6 months following the last month in which Dave worked, Key was liable for social security, Medicare, and FUTA taxes on any payments that constituted taxable wages. However, Key could have shifted the liability for the employer part of the social security and Medicare taxes (and for the FUTA tax) during the first 6 months by withholding Dave's part of the social security and Medicare taxes, timely depositing the taxes, and notifying Edgewood of the payments.

If Key shifted liability for the employer part of the social security and Medicare taxes to Edgewood and provided Edgewood with a sick pay statement, Key would not prepare a Form W-2 for Dave. However, Key would prepare third-party sick pay recap Forms W-2 and W-3. Key and Edgewood must each prepare Form 941. Edgewood must also report the sick pay and withholding for Dave on Forms W-2, W-3, and 940.

As an alternative, the parties could have followed the optional rule described under ***Optional rule for Form W-2*** earlier. Under this rule, Key would prepare Form W-2 even though liability for the employer part of the social security and Medicare taxes had been shifted to Edgewood. Also, Key would not prepare a sick pay statement, and Edgewood, not Key, would prepare the recap Forms W-2 and W-3 reflecting the sick pay shown on Edgewood's Form 941.

Liability not transferred. If Key did not shift liability for the employer part of the social security and Medicare taxes to Edgewood, Key would prepare Forms W-2 and W-3 as well as Forms 941 and 940. In this situation, Edgewood would not report the sick pay.

Payments received after 6 months. The payments received by Dave in July through October are not subject to social security, Medicare, or FUTA taxes, because they were received more than 6 months after the last month in which Dave worked (December 2000). However, Key must continue to withhold income tax from each payment because Dave furnished Key a Form W-4S. Also, Key must prepare Forms W-2 and W-3, unless it has furnished Edgewood with a sick pay statement. If the sick pay statement was furnished, then Edgewood must prepare Forms W-2 and W-3.

7. Special Rules for Paying Taxes

Common Paymaster

If two or more related corporations employ the same individual ***at the same time*** and pay this individual through a common paymaster, which is one of the corporations, the corporations are considered a single employer. They have to pay, in total, no more in social security and Medicare taxes than a single employer would.

Each corporation must pay its own part of the employment taxes and may deduct only its own part of the wages. The deductions will not be allowed unless the corporation reimburses the common paymaster for the wage and tax payments. See Regulations section 31.3121(s)–1 for more information.

Reporting Agents

You must submit an application for authorization to act as an agent to the IRS service center where you will be filing returns. A **Form 2678,** Employer Appointment of Agent, properly completed by each employer, must be submitted with this application. See Rev. Proc. 70-6, 1970-1 C.B. 420, Rev. Proc. 84-33, 1984-1 C.B. 502, and the separate **Instructions for Forms W-2 and W-3** for procedures and reporting requirements. Form 2678 does not apply to FUTA taxes reportable on Form 940.

Magnetic tape filing of Forms 940 and 941. Reporting agents filing Forms 940 and 941 for a large number of employers may file them on magnetic tape. For authorization to file using this method, reporting agents must submit a **Form 8655,** Reporting Agent Authorization for Magnetic Tape/Electronic Filers, completed by each employer. See Rev. Proc. 96-18, 1996-1 C.B. 637, for the procedures for filing Forms 940 and 941 on magnetic tape. You can find Rev. Proc. 96-18 on page 73 of Internal Revenue Bulletin 1996-4 at **www. irs.gov.**Also, see **Pub. 1314** (Form 940) and **Pub. 1264** (Form 941) for the tape specifications.

Electronic filing of Form 941. The 941e-file program accepts and processes Form 941 electronically in the Electronic Data Interchange (EDI) format. The program allows a reporting agent taxpayer to electronically file Form 941 using a personal computer, modem, and commercial tax preparation software. See Rev. Proc. 96-17, 1996-1 C.B. 633 and Rev. Proc. 99-39, 1999-43 IRB 532 for procedural information. You can find Rev. Proc. 96-17 on page 69 of Internal Revenue Bulletin 1996-4, and Rev. Proc. 99-39 on page 532 of Internal Revenue Bulletin 1999-43, at **www.irs.gov.** Also see **Pub. 1855** for technical specifications.

Payment of Employment Taxes by Disregarded Entities

Employment taxes for employees of a qualified subchapter S subsidiary or other entity disregarded as an entity separate from its owner may be reported and paid under one of the following methods:

- By its owner (as if the employees of the disregarded entity are employed directly by the owner) using the owner's name and taxpayer identification number or
- By each entity recognized as a separate entity under state law using the entity's own name and taxpayer identification number.

If the second method is chosen, the owner retains responsibility for the employment tax obligations of the disregarded entity. For more information, see Notice 99-6, 1999-3 C.B. 321. You can find Notice 99-6 on page 12 of Internal Revenue Bulletin 1999-3 at **www.irs.gov.**

Lender, Surety, or Other Third-Party Payers

Any lender, surety, or other person who pays wages, or supplies funds specifically to pay wages directly to the employees of an employer, or to the employee's agent, is responsible for any required withholding on those wages. The third party is also liable for any interest and penalties accruing on these accounts. This includes the withholding of income, social security, Medicare, and railroad retirement taxes.

Note: *These rules do not apply to a person acting only as an agent, or to third-party payers of sick pay, discussed in section 6.*

If a third party supplies funds to an employer so that the employer can pay the employees' wages, and if the third party knows that the employer will not pay or deposit the taxes that are required to be withheld when due, then the third party must pay the taxes withheld from the employees' wages but not paid by the employer. However, the third party does not have to pay more than 25% of the amount that is specifically supplied for paying wages. The third-party supplier must also pay interest on the taxes if they are paid after the due date of the employer's return.

Third parties are liable only for payment of the employees' parts of payroll taxes. They are not liable for the employer's part. The employer must file an employment tax return for wages that he or she or a third party pays and must furnish Forms W-2 to employees for the wages paid and taxes withheld. The employer also remains liable for any withholding taxes not paid by the third party.

Liability of trustee in bankruptcy. A trustee in bankruptcy must withhold, report, and pay income, social security, and Medicare taxes from the payment of priority claims for employees' wages earned prior to, but unpaid at the time of, an employer's bankruptcy.

How to pay withheld tax. Third parties who pay employment taxes must file two copies of **Form 4219,** Statement of Liability of Lender, Surety, or Other Person for Withholding Taxes. A separate set of forms must be filed for each employer and calendar quarter.

Form 4219 must be filed with the IRS service center where the employer for whom wages were paid, or funds were supplied, files Federal employment tax returns.

Each Form 4219 must be accompanied by a check or money order made payable to the "United States Treasury." To avoid interest, full payment should be made on or before the due date of the employer's Federal employment tax return.

Employee's Portion of Taxes Paid by Employer

If you pay your employee's social security and Medicare taxes without deducting them from the employee's pay, you must include the amount of the payments in the employee's wages for income tax withholding and social security, Medicare, and FUTA taxes. This increase in the employee's wage payment for your payment of the employee's social security and Medicare taxes is also subject to employee social security and Medicare taxes. This again increases the amount of the additional taxes you must pay.

Note: *This discussion does not apply to household and agricultural employers. If you pay a household or agricultural employee's social security and Medicare taxes, these payments must be included in the employee's wages. However, this wage increase due to the tax payments made for the employee is not subject to social security or Medicare taxes as discussed in this section.*

To figure the employee's increased wages in this situation, divide the ***stated pay*** (the amount you pay without taking into account your payment of employee social security and Medicare taxes) by a factor for that year. This factor is determined by subtracting from 1 the combined employee social security and Medicare tax rate for the year the wages are paid. For 2002, the factor is .9235 (1 − .0765). If the stated pay is more than $78,405.15 (2002 wage base $84,900 × .9235), follow the procedure described under **Stated pay of more than $78,405.15 in 2002** below.

Stated pay of $78,405.15 or less in 2002. For an employee with stated pay of $78,405.15 or less in 2002, figure the correct wages (wages plus employer-paid employee taxes) and withholding to report by dividing the stated pay by .9235. This will give you the wages to report in box 1 and the social security and Medicare wages to report in boxes 3 and 5 of Form W-2.

To figure the correct social security tax to enter in box 4 and Medicare tax to enter in box 6, multiply the amounts in boxes 3 and 5 by the withholding rates (6.2% and 1.45%) for those taxes, and enter the results in boxes 4 and 6.

Example. Donald Devon hires Lydia Lone for only 1 week during 2002. He pays her $300 for that week. Donald agrees to pay Lydia's part of the social security and Medicare taxes. To figure her reportable wages, he divides $300 by .9235. The result, $324.85, is the amount he reports as wages in boxes 1, 3, and 5 of Form W-2. To figure the amount to report as social security tax, Donald multiplies $324.85 by the social security tax rate of 6.2% (.062). The result, $20.14, is entered in box 4 of Form W-2. To figure the amount to report as Medicare tax, Donald multiplies $324.85 by the Medicare tax rate of 1.45% (.0145). The result, $4.71, is entered in box 6 of Form W-2. Although he did not actually withhold these amounts from Lydia, he will report these amounts as taxes withheld on Form 941 and is responsible for matching these amounts with the employer share of these taxes.

For FUTA tax and income tax withholding, Lydia's weekly wages are $324.85.

Stated pay of more than $78,405.15 in 2002. For an employee with stated pay of more than $78,405.15 in 2002, the correct social security wage amount is $84,900 (the first $78,405.15 of wages ÷ .9235). The stated pay in excess of $78,405.15 is not subject to social security tax because the tax only applies to the first $84,900 of wages (stated pay plus employer-paid employee taxes). Enter $84,900 in box 3 of Form W-2. The social security tax to enter in box 4 is $5,263.80 (84,900 x .062).

To figure the correct Medicare wages to enter in box 5 of Form W-2, subtract $78,405.15 from the stated pay. Divide this amount by .9855 (1 − .0145) and add $84,900. For example, if stated pay is $100,000, the correct Medicare wages are figured as follows:

$100,000 − $78,405.15 = $21,594.85

$21,594.85 ÷ .9855 = $21,912.58

$21,912.58 + $84,900 = $106,812.58

The Medicare wages are $106,812.58. Enter this amount in box 5 of Form W-2. The Medicare tax to enter in box 6 is $1,548.78 ($106,812.58 × .0145).

Although these employment tax amounts are not actually withheld, report them as withheld on Form 941, and pay this amount as the employer's share of the social security and Medicare taxes. If the wages for income tax purposes in the preceding example are the same as for social security and Medicare purposes, the correct wage amount for income tax withholding is $106,812.58 ($100,000 + $5,263.80 + $1,548.78), which is included in box 1 of Form W-2.

International Social Security Agreements

The United States has social security agreements with many countries to eliminate dual taxation and coverage under two social security systems. Under these agreements, sometimes known as totalization agreements, you generally must pay social security taxes only to the country where you work. Employees and employers who are subject only to foreign social security taxes under these agreements are exempt from U.S. social security taxes, including the Medicare portion.

The United States has social security agreements with the following countries: Austria, Belgium, Canada, Finland, France, Germany, Greece, Ireland, Italy, Korea, Luxembourg, the Netherlands, Norway, Portugal, Spain, Sweden, Switzerland, and the United Kingdom. Additional agreements are expected in the future. For more information, see **Pub. 519,** U.S. Tax Guide for Aliens, or contact:

Social Security Administration
Office of International Programs
P.O. Box 17741
Baltimore, MD 21235–7741

If you have access to the Internet, you can get more information from the SSA at **www.ssa.gov/international.**

8. Pensions and Annuities

Generally, pension and annuity payments are subject to Federal income tax withholding. The withholding rules apply to the ***taxable*** part of payments from an employer pension, annuity, profit-sharing, stock bonus, or other ***deferred compensation plan.*** The rules also apply to payments from an individual retirement arrangement (IRA), an annuity, endowment, or life insurance contract issued by a life insurance company. There is no withholding on any part of a distribution that is not expected to be includible in the recipient's gross income.

Generally, recipients of payments described above can choose not to have withholding apply to their pensions or annuities (however, see **Mandatory Withholding** below). The election remains in effect until the recipient revokes it. The payer must notify the recipient that this election is available.

Withholding

Periodic Payments

Generally, periodic payments are pension or annuity payments made for more than 1 year that are not eligible rollover distributions (see discussion below). Periodic payments include substantially equal payments made at least once a year over the life of the employee and/or beneficiaries or for 10 years or more. For withholding purposes, these payments are treated as if they are wages. You can figure withholding by using the recipient's **Form W-4P,** Withholding Certificate for Pension or Annuity Payments, and the income tax withholding tables and methods in Circular E or the alternative tables and methods in this publication.

Recipients of periodic payments can give you a Form W-4P to specify the number of withholding allowances and any additional amount they want withheld. They may also claim exemption from withholding on Form W-4P or revoke a previously claimed exemption. If they do not submit a Form W-4P, you must figure withholding by treating a recipient as married with three withholding allowances. See Form W-4P for more information.

Nonperiodic Payments

Withhold 10% of the taxable part of a nonperiodic payment that is not an eligible rollover distribution. The recipient may request additional withholding on Form W-4P or claim exemption from withholding.

Mandatory Withholding

Payments delivered outside the United States. The election to be exempt from income tax withholding does not apply to any periodic or nonperiodic payment delivered outside the United States or its possessions to a U.S. citizen or resident alien. See Form W-4P for more information.

A nonresident alien can elect exemption from withholding only if he or she certifies to the payer that he or she is not (1) a U.S. citizen or resident alien or (2) an individual to whom Internal Revenue Code section 877 applies (concerning expatriation to avoid tax). The certification must be made in a statement to the payer under penalties of perjury. However, nonresident aliens who choose such exemption will be subject to withholding under Code section 1441. See **Pub. 515,** Withholding of Tax on Nonresident Aliens and Foreign Entities, and the **Instructions for Form 1042-S.**

Eligible rollover distributions. Withhold 20% of an eligible rollover distribution unless the recipient elected to have the distribution paid in a direct rollover to an eligible retirement plan, including an IRA. An eligible rollover distribution is the taxable part of any distribution from a qualified plan, governmental section 457 plan (for distributions after December 31, 2001), or tax-sheltered annuity ***(but not an IRA)*** except:

1) One of a series of substantially equal periodic payments (at least annually) made for the life or life expectancy of the employee and the employee's beneficiary or for a specified period of 10 years or more.
2) Any part of a distribution that is a minimum distribution required by Code section 401(a)(9).
3) A hardship distribution. A distribution will qualify for hardship if it is (a) made on account of immediate and heavy need and (b) necessary to satisfy the need. This includes medical and educational expenses and costs for purchasing a new residence, or to prevent eviction or foreclosure on a current residence.
4) Other exceptions apply. For details see the **Instructions for Forms 1099-R and 5498.**

You are not required to withhold 20% of an eligible rollover distribution that, when added to other rollover distributions made to one person during the year, is less than $200.

A recipient of an eligible rollover distribution cannot claim exemption from the 20% withholding. However, a recipient may elect to have more than 20% withheld using Form W-4P. Do not provide the recipient a Form W-4P for eligible rollover distributions unless he or she wishes to request additional withholding in excess of the mandatory 20%.

Notice to recipient (section 402(f) notice). Generally, you must provide a written explanation to the recipient at least 30 but no more than 90 days before making an eligible rollover distribution. You must explain the rollover rules, special tax treatment for lump-sum distributions, direct rollover option, and the mandatory 20% withholding rule. Notice 2000-11 (2006 IRB 572), contains a model notice you can use to satisfy this requirement. You can find Notice 2000-11 on page 572 of Internal Revenue Bulletin 2000-6 at **www.irs.gov.**

Similar rules apply to distributions from tax-sheltered annuities. The IRS has issued regulations on these requirements under sections 401(a)(31), 402, 403(b), and 3405.

Depositing and Reporting Withholding

Report income tax withholding from pensions and annuities on **Form 945,** Annual Return of Withheld Federal Income Tax. Do not report these liabilities on Form 941. You must furnish the recipients and the IRS with **Form 1099-R,** Distributions From Pensions, Annuities, Retirement or Profit-Sharing Plans, IRAs, Insurance Contracts, etc.

Deposit withholding from pensions and annuities combined with any other nonpayroll withholding reported on Form 945 (e.g., backup withholding). Do not combine the Form 945 deposits with deposits for payroll taxes. Circular E and the separate **Instructions for Form 945** include information on the deposit rules.

9. Alternative Methods for Figuring Withholding

You may use various methods of figuring income tax withholding. The methods described below may be used instead of the common payroll methods provided in Circular E. Use the method that best suits your payroll system and employees.

Annualized wages. Using your employee's annual wages, figure the withholding using the Percentage Method, Table 7–Annual Payroll Period, in Circular E. Divide the amount from the table by the number of payroll periods, and the result will be the amount of withholding for each payroll period.

Average estimated wages. You may withhold the tax for a payroll period based on estimated average wages, with necessary adjustments, for any quarter. For details, see Regulations section 31.3402(h)(1)-1.

Cumulative wages. An employee may ask you, in writing, to withhold tax on cumulative wages. If you agree to do so, and you have paid the employee for the same kind of payroll period (weekly, biweekly, etc.) since the beginning of the year, you may figure the tax as follows:

Add the wages you have paid the employee for the current calendar year to the current payroll period amount. Divide this amount by the number of payroll periods so far this year including the current period. Figure the withholding on this amount, and multiply the withholding by the number of payroll periods used above. Use the percentage method shown in Circular E. Subtract the total withholding calculated from the total tax withheld during the calendar year. The excess is the amount to withhold for the current payroll period. (See Rev. Proc. 78-8, 1978-1 C.B. 562, for an example of the cumulative method.)

Part-year employment. A part-year employee who figures income tax on a calendar-year basis may ask you to withhold tax by the part-year employment method. The request must be in writing and must contain the following information:

1) The last day of any employment during the calendar year with any prior employer.
2) A statement that the employee uses the calendar year accounting period.
3) A statement that the employee reasonably anticipates he or she will be employed by all employers for a total of no more than 245 days in all ***terms of continuous employment*** (defined below) during the current calendar year.

Complete the following steps to figure withholding tax by the part-year method:

1) Add the wages to be paid to the employee for the current payroll period to any wages you have already paid the employee in the current term of continuous employment.
2) Add the number of payroll periods used in step 1 to the number of payroll periods between the employee's last employment and current employment. To find the number of periods between the last employment and current employment, divide (a) the number of calendar days between the employee's last day of earlier employment (or the previous December 31, if later) and the first day of current employment by (b) the number of calendar days in the current payroll period.
3) Divide the step 1 amount by the total number of payroll periods from step 2.
4) Find the tax in the withholding tax tables on the step 3 amount. Be sure to use the correct payroll period table and to take into account the employee's withholding allowances.

5) Multiply the total number of payroll periods from step 2 by the step 4 amount.
6) Subtract from the step 5 amount the total tax already withheld during the current term of continuous employment. Any excess is the amount to withhold for the current payroll period.

(See Regulations section 31.3402(h)(4)-1(c)(4) for examples of the part-year method.)

Term of continuous employment. A term of continuous employment may be a single term or two or more following terms of employment with the same employer. A continuous term includes holidays, regular days off, and days off for illness or vacation. A continuous term begins on the first day an employee works for you and earns pay. It ends on the earlier of the employee's last day of work for you or, if the employee performs no services for you for more than 30 calendar days, the last workday before the 30-day period. If an employment relationship is ended, the term of continuous employment is ended, even if a new employment relationship is established with the same employer within 30 days.

Other methods. You may use other methods and tables for withholding taxes, as long as the amount of tax withheld is consistently about the same as it would be under the percentage method shown in Circular E. If you develop an alternative method or table, you should test the full range of wage and allowance situations to be sure that they meet the tolerances contained in Regulations section 31.3402(h)(4)-1 as shown in the chart below.

If the tax required to be withheld under the annual percentage rate is—	The annual tax withheld under your method may not differ by more than—
Less than $10	$9.99
$10 or more but under $100	$10 plus 10% of the excess over $10
$100 or more but under $1,000	$19 plus 3% of the excess over $100
$1,000 or more	$46 plus 1% of the excess over $1,000

Formula Tables for Percentage Method Withholding (for Automated Payroll Systems)

Two formula tables for percentage method withholding are on pages 23 and 24. The differences in the Alternative Percentage Method formulas and the steps for figuring withheld tax for different payroll systems are shown in this example.

MARRIED PERSON
(Weekly Payroll Period)

If wages exceeding the allowance amount are over $124 but not over $355:

Method:	Income Tax Withheld:
Percentage (Pub. 15) . .	10% of excess over $124
Alternative 1 (Page 23)	10% of such wages minus $12.40
Alternative 2 (Page 24)	Such wages minus $124, times 10% of remainder

When employers use the percentage method in Circular E or the formula tables for percentage method withholding in this publication, the tax for the pay period may be rounded to the nearest dollar. If rounding is used, it must be used consistently. Withheld tax amounts should be rounded to the nearest whole dollar by (1) dropping amounts under 50 cents and (2) increasing amounts from 50 to 99 cents to the next higher dollar. This rounding will be considered to meet the tolerances under section 3402(h)(4).

(For Wages Paid in 2002)

Alternative 1.—Tables for Percentage Method Withholding Computations

Table A(1)—WEEKLY PAYROLL PERIOD (Amount for each allowance claimed is $57.69)

Single Person

If the wage in excess of allowance amount is: Over—	But not over—	The income tax to be withheld is: Of such wage—	From product
$0	—$51	0%	$0
$51	—$164	10% less	$5.10
$164	—$570	15% less	$13.30
$570	—$1,247	27% less	$81.70
$1,247	—$2,749	30% less	$119.11
$2,749	—$5,938	35% less	$256.56
$5,938	—	38.6% less	$470.33

Married Person

If the wage in excess of allowance amount is: Over—	But not over—	The income tax to be withheld is: Of such wage—	From product
$124	—$0	0%	$0
$124	—$355	10% less	$12.40
$355	—$991	15% less	$30.15
$991	—$2,110	27% less	$149.07
$2,110	—$3,400	30% less	$212.37
$3,400	—$5,998	35% less	$382.37
$5,998	—	38.6% less	$598.30

Table B(1)—BIWEEKLY PAYROLL PERIOD (Amount for each allowance claimed is $115.38)

Single Person

If the wage in excess of allowance amount is: Over—	But not over—	The income tax to be withheld is: Of such wage—	From product
$0	—$102	0%	$0
$102	—$329	10% less	$10.20
$329	—$1,140	15% less	$26.65
$1,140	—$2,493	27% less	$163.45
$2,493	—$5,498	30% less	$238.24
$5,498	—$11,875	35% less	$513.14
$11,875	—	38.6% less	$940.64

Married Person

If the wage in excess of allowance amount is: Over—	But not over—	The income tax to be withheld is: Of such wage—	From product
$0	—$248	0%	$0
$248	—$710	10% less	$24.80
$710	—$1,983	15% less	$60.30
$1,983	—$4,219	27% less	$298.26
$4,219	—$6,800	30% less	$424.83
$6,800	—$11,996	35% less	$764.83
$11,996	—	38.6% less	$1,196.69

Table C(1)—SEMIMONTHLY PAYROLL PERIOD (Amount for each allowance claimed is $125.00)

Single Person

If the wage in excess of allowance amount is: Over—	But not over—	The income tax to be withheld is: Of such wage—	From product
$0	—$110	0%	$0
$110	—$356	10% less	$11.00
$356	—$1,235	15% less	$28.80
$1,235	—$2,701	27% less	$177.00
$2,701	—$5,956	30% less	$258.03
$5,956	—$12,865	35% less	$555.83
$12,865	—	38.6% less	1,018.97

Married Person

If the wage in excess of allowance amount is: Over—	But not over—	The income tax to be withheld is: Of such wage—	From product
$0	—$269	0%	$0
$269	—$769	10% less	$26.90
$769	—$2,148	15% less	$65.35
$2,148	—$4,571	27% less	$323.11
$4,571	—$7,367	30% less	$460.24
$7,367	—$12,996	35% less	$828.59
$12,996	—	38.6% less	$1,296.45

Table D(1)—MONTHLY PAYROLL PERIOD (Amount for each allowance claimed is $250.00)

Single Person

If the wage in excess of allowance amount is: Over—	But not over—	The income tax to be withheld is: Of such wage—	From product
$0	—$221	0%	$0
$221	—$713	10% less	$22.10
$713	—$2,471	15% less	$57.75
$2,471	—$5,402	27% less	$354.27
$5,402	—$11,913	30% less	$516.33
$11,913	—$25,729	35% less	1,111.98
$25,729	—	38.6% less	2,038.22

Married Person

If the wage in excess of allowance amount is: Over—	But not over—	The income tax to be withheld is: Of such wage—	From product
$0	—$538	0%	$0
$538	—$1,538	10% less	$53.80
$1,538	—$4,296	15% less	$130.70
$4,296	—$9,142	27% less	$646.22
$9,142	—$14,733	30% less	$920.48
$14,733	—$25,992	35% less	$1,657.13
$25,992	—	38.6% less	$2,592.84

Table E(1)—DAILY OR MISCELLANEOUS PAYROLL PERIOD (Amount for each allowance claimed per day for such period is $11.54)

Single Person

If the wage in excess of allowance amount divided by the number of days in the pay period is: Over—	But not over—	The income tax to be withheld multiplied by the number of days in such period is: Of such wage—	From product
$0.00	—$10.20	0%	$0
$10.20	—$32.90	10% less	$1.02
$32.90	—$114.00	15% less	$2.67
$114.00	—$249.30	27% less	$16.34
$249.30	—$549.80	30% less	$23.82
$549.80	—$1,187.50	35% less	$51.31
$1,187.50	—	38.6% less	$94.06

Married Person

If the wage in excess of allowance amount divided by the number of days in the pay period is: Over—	But not over—	The income tax to be withheld multiplied by the number of days in such period is: Of such wage—	From product
$0.00	—$24.80	0%	$0
$24.80	—$71.00	10% less	$2.48
$71.00	—$198.30	15% less	$6.03
$198.30	—$421.90	27% less	$29.82
$421.90	—$680.00	30% less	$42.48
$680.00	—$1,199.60	35% less	$76.48
$1,199.60	—	38.6% less	$119.67

Note.—The adjustment factors may be reduced by one-half cent (e.g., 7.50 to 7.495; 69.38 to 69.375) to eliminate separate half rounding operations.

The first two brackets of these tables may be combined, provided zero withholding is used to credit withholding amounts computed by the combined bracket rates, e.g., $0 to $51 and $51 to $536 combined to read, Over $0, But not over $536.

The employee's excess wage (gross wage less amount for allowances claimed) is used with the applicable percentage rates and subtraction factors to calculate the amount of income tax withheld.

Page 23

(For Wages Paid in 2002)

Alternative 2.—Tables for Percentage Method Withholding Computations

Table A(2)—WEEKLY PAYROLL PERIOD (Amount for each allowance claimed is $57.69)

Single Person

If the wage in excess of allowance amount is: Over—	But not over—	The income tax to be withheld is: Such wage—	Times
$0	—$51	$0.00	0%
$51	—$164	minus $51.00	10%
$164	—$570	minus $88.67	15%
$570	—$1,247	minus $302.59	27%
$1,247	—$2,749	minus $397.03	30%
$2,749	—$5,938	minus $733.03	35%
$5,938	—	minus $1,218.47	38.6%

Married Person

If the wage in excess of allowance amount is: Over—	But not over—	The income tax to be withheld is: Such wage—	Times
$0	—$124	$0.00	0%
$124	—$355	minus $124.00	10%
$355	—$991	minus $201.00	15%
$991	—$2,110	minus $552.11	27%
$2,110	—$3,400	minus $707.90	30%
$3,400	—$5,998	minus $1,092.49	35%
$5,998	—	minus $1,549.99	38.6%

Table B(2)—BIWEEKLY PAYROLL PERIOD (Amount for each allowance claimed is $115.38)

Single Person

If the wage in excess of allowance amount is: Over—	But not over—	The income tax to be withheld is: Such wage—	Times
$0	—$102	$0.00	0%
$102	—$329	minus $102.00	10%
$329	—$1,140	minus $177.67	15%
$1,140	—$2,493	minus $605.37	27%
$2,493	—$5,498	minus $794.13	30%
$5,498	—$11,875	minus $1,466.11	35%
$11,875	—	minus $2,436.89	38.6%

Married Person

If the wage in excess of allowance amount is: Over—	But not over—	The income tax to be withheld is: Such wage—	Times
$0	—$248	$0.00	0%
$248	—$710	minus $248.00	10%
$710	—$1,983	minus $402.00	15%
$1,983	—$4,219	minus $1,104.67	27%
$4,219	—$6,800	minus $1,416.10	30%
$6,800	—$11,996	minus $2,185.23	35%
$11,996	—	minus $3,100.22	38.6%

Table C(2)—SEMIMONTHLY PAYROLL PERIOD (Amount for each allowance claimed is $125.00)

Single Person

If the wage in excess of allowance amount is: Over—	But not over—	The income tax to be withheld is: Such wage—	Times
$0	—$110	$0.00	0%
$110	—$356	minus $110.00	10%
$356	—$1,235	minus $192.00	15%
$1,235	—$2,701	minus $655.56	27%
$2,701	—$5,956	minus $860.10	30%
$5,956	—$12,865	minus $1,588.09	35%
$12,865	—	minus $2,639.82	38.6%

Married Person

If the wage in excess of allowance amount is: Over—	But not over—	The income tax to be withheld is: Such wage—	Times
$0	—$269	$0.00	0%
$269	—$769	minus $269.00	10%
$769	—$2,148	minus $435.67	15%
$2,148	—$4,571	minus $1,196.70	27%
$4,571	—$7,367	minus $1,534.13	30%
$7,367	—$12,996	minus $2,367.40	35%
$12,996	—	minus $3,358.67	38.6%

Table D(2)—MONTHLY PAYROLL PERIOD (Amount for each allowance claimed is $250.00)

Single Person

If the wage in excess of allowance amount is: Over—	But not over—	The income tax to be withheld is: Such wage—	Times
$0	—$221	$0.00	0%
$221	—$713	minus $221.00	10%
$713	—$2,471	minus $385.00	15%
$2,471	—$5,402	minus $1,312.11	27%
$5,402	—$11,913	minus $1,721.10	30%
$11,913	—$25,729	minus $3,177.09	35%
$25,729	—	minus $5,280.37	38.6%

Married Person

If the wage in excess of allowance amount is: Over—	But not over—	The income tax to be withheld is: Such wage—	Times
$0	—$538	$0.00	0%
$538	—$1,538	minus $538.00	10%
$1,538	—$4,296	minus $871.33	15%
$4,296	—$9,142	minus $2,393.41	27%
$9,142	—$14,733	minus $3,068.27	30%
$14,733	—$25,992	minus $4,734.66	35%
$25,992	—	minus $6,717.21	38.6%

Table E(2)—DAILY OR MISCELLANEOUS PAYROLL PERIOD (Amount for each allowance claimed per day for such period is $11.54)

Single Person

If the wage in excess of allowance amount divided by the number of days in the pay period is: Over—	But not over—	The income tax to be withheld multiplied by the number of days in such period is: Such wage—	Times
$0.00	—$10.20	$0.00	0%
$10.20	—$32.90	minus $10.20	10%
$32.90	—$114.00	minus $17.77	15%
$114.00	—$249.30	minus $60.52	27%
$249.30	—$549.80	minus $79.40	30%
$549.80	—$1,187.50	minus $146.60	35%
$1,187.50	—	minus $243.67	38.6%

Married Person

If the wage in excess of allowance amount divided by the number of days in the pay period is: Over—	But not over—	The income tax to be withheld multiplied by the number of days in such period is: Such wage—	Times
$0.00	—$24.80	$0.00	0%
$24.80	—$71.00	minus $24.80	10%
$71.00	—$198.30	minus $40.20	15%
$198.30	—$421.90	minus $110.45	27%
$421.90	—$680.00	minus $141.60	30%
$680.00	—$1,199.60	minus $218.51	35%
$1,199.60	—	minus $310.01	38.6%

Note.—The first two brackets of these tables may be combined, provided zero withholding is used to credit withholding amounts computed by the combined bracket rates, e.g., $0 to $51 and $51 to $536 combined to read, Over $0, But not over $536.

The employee's excess wage (gross wage less amount for allowances claimed) is used with the applicable percentage rates and subtraction factors to calculate the amount of income tax withheld.

Page 24

Wage Bracket Percentage Method Tables (for Automated Payroll Systems)

The **Wage Bracket Percentage Method Tables** show the gross wage brackets that apply to each withholding percentage rate for employees with up to nine withholding allowances. These tables also show the computation factors for each number of withholding allowances and the applicable wage bracket. The computation factors are used to figure the amount of withholding tax by a percentage method.

Two kinds of **Wage Bracket Percentage Method Tables** are shown. Each has tables for married and single persons for weekly, biweekly, semimonthly, and monthly payroll periods.

The difference between the two kinds of tables is the reduction factor subtracted from wages before multiplying by the applicable percentage withholding rate. In the tables for **Computing Income Tax Withholding From Gross Wages** on pages 26–29, the reduction factor includes both the amount for withholding allowances claimed and a rate adjustment factor as shown in the **Alternative 2—Tables for Percentage Method Withholding Computations** on page 24. In the tables for **Computing Income Tax Withholding From Wages Exceeding Allowance Amount** on pages 30–33, the reduction factor does not include an amount for the number of allowances claimed.

Use the kind of wage bracket table that best suits your payroll system. For example, some pay systems automatically subtract from wages the allowance amount for each employee before finding the amount of tax to withhold. The tables for **Computing Income Tax Withholding From Wages Exceeding Allowance Amount** can be used in these systems. The reduction factors in these tables do not include the allowance amount that was automatically subtracted before applying the table factors in the calculation. For other systems that do not separately subtract the allowance amount, use the tables for **Computing Income Tax Withholding From Gross Wages.**

When employers use the **Wage Bracket Percentage Method Tables,** the tax for the period may be rounded to the nearest dollar. If rounding is used, it must be used consistently. Withheld tax amounts should be rounded to the nearest whole dollar by (1) dropping amounts under 50 cents and (2) increasing amounts from 50 to 99 cents to the next higher dollar. Such rounding will be deemed to meet the tolerances under section 3402(h)(4).

(For Wages Paid in 2002)

Wage Bracket Percentage Method Table for Computing
Income Tax Withholding From Gross Wages

Weekly Payroll Period

If the number of allowances is—	Single Persons				Married Persons			
	And gross wages are—		from gross wages[1]	Multiply result by—	And gross wages are—		from gross wages[1]	Multiply result by—
	Over	But not over			Over	But not over		
	A	**B**	**C**	**D**	**A**	**B**	**C**	**D**
0	$0.00	$164.00	subtract $51.00	10%	$0.00	$355.00	subtract $124.00	10%
	$164.00	$570.00	subtract $88.67	15%	$355.00	$991.00	subtract $201.00	15%
	$570.00	$1,247.00	subtract $302.59	27%	$991.00	$2,110.00	subtract $552.11	27%
	$1,247.00	$2,749.00	subtract $397.03	30%	$2,110.00	$3,400.00	subtract $707.90	30%
	$2,749.00	$5,938.00	subtract $733.03	35%	$3,400.00	$5,998.00	subtract $1,092.49	35%
	$5,938.00	......	subtract $1,218.47	38.6%	$5,998.00	......	subtract $1,549.99	38.6%
1	$0.00	$221.69	subtract $108.69	10%	$0.00	$412.69	subtract $181.69	10%
	$221.69	$627.69	subtract $146.36	15%	$412.69	$1,048.69	subtract $258.69	15%
	$627.69	$1,304.69	subtract $360.28	27%	$1,048.69	$2,167.69	subtract $609.80	27%
	$1,304.69	$2,806.69	subtract $454.72	30%	$2,167.69	$3,457.69	subtract $765.59	30%
	$2,806.69	$5,995.69	subtract $790.72	35%	$3,457.69	$6,055.69	subtract $1,150.18	35%
	$5,995.69	......	subtract $1,276.16	38.6%	$6,055.69	......	subtract $1,607.68	38.6%
2	$0.00	$279.38	subtract $166.38	10%	$0.00	$470.38	subtract $239.38	10%
	$279.38	$685.38	subtract $204.05	15%	$470.38	$1,106.38	subtract $316.38	15%
	$685.38	$1,362.38	subtract $417.97	27%	$1,106.38	$2,225.38	subtract $667.49	27%
	$1,362.38	$2,864.38	subtract $512.41	30%	$2,225.38	$3,515.38	subtract $823.28	30%
	$2,864.38	$6,053.38	subtract $848.41	35%	$3,515.38	$6,113.38	subtract $1,207.87	35%
	$6,053.38	......	subtract $1,333.85	38.6%	$6,113.38	......	subtract $1,665.37	38.6%
3	$0.00	$337.07	subtract $224.07	10%	$0.00	$528.07	subtract $297.07	10%
	$337.07	$743.07	subtract $261.74	15%	$528.07	$1,164.07	subtract $374.07	15%
	$743.07	$1,420.07	subtract $475.66	27%	$1,164.07	$2,283.07	subtract $725.18	27%
	$1,420.07	$2,922.07	subtract $570.10	30%	$2,283.07	$3,573.07	subtract $880.97	30%
	$2,922.07	$6,111.07	subtract $906.10	35%	$3,573.07	$6,171.07	subtract $1,265.56	35%
	$6,111.07	......	subtract $1,391.54	38.6%	$6,171.07	......	subtract $1,723.06	38.6%
4	$0.00	$394.76	subtract $281.76	10%	$0.00	$585.76	subtract $354.76	10%
	$394.76	$800.76	subtract $319.43	15%	$585.76	$1,221.76	subtract $431.76	15%
	$800.76	$1,477.76	subtract $533.35	27%	$1,221.76	$2,340.76	subtract $782.87	27%
	$1,477.76	$2,979.76	subtract $627.79	30%	$2,340.76	$3,630.76	subtract $938.66	30%
	$2,979.76	$6,168.76	subtract $963.79	35%	$3,630.76	$6,228.76	subtract $1,323.25	35%
	$6,168.76	......	subtract $1,449.23	38.6%	$6,228.76	......	subtract $1,780.75	38.6%
5	$0.00	$452.45	subtract $339.45	10%	$0.00	$643.45	subtract $412.45	10%
	$452.45	$858.45	subtract $377.12	15%	$643.45	$1,279.45	subtract $489.45	15%
	$858.45	$1,535.45	subtract $591.04	27%	$1,279.45	$2,398.45	subtract $840.56	27%
	$1,535.45	$3,037.45	subtract $685.48	30%	$2,398.45	$3,688.45	subtract $996.35	30%
	$3,037.45	$6,226.45	subtract $1,021.48	35%	$3,688.45	$6,286.45	subtract $1,380.94	35%
	$6,226.45	......	subtract $1,506.92	38.6%	$6,286.45	......	subtract $1,838.44	38.6%
6	$0.00	$510.14	subtract $397.14	10%	$0.00	$701.14	subtract $470.14	10%
	$510.14	$916.14	subtract $434.81	15%	$701.14	$1,337.14	subtract $547.14	15%
	$916.14	$1,593.14	subtract $648.73	27%	$1,337.14	$2,456.14	subtract $898.25	27%
	$1,593.14	$3,095.14	subtract $743.17	30%	$2,456.14	$3,746.14	subtract $1,054.04	30%
	$3,095.14	$6,284.14	subtract $1,079.17	35%	$3,746.14	$6,344.14	subtract $1,438.63	35%
	$6,284.14	......	subtract $1,564.61	38.6%	$6,344.14	......	subtract $1,896.13	38.6%
7	$0.00	$567.83	subtract $454.83	10%	$0.00	$758.83	subtract $527.83	10%
	$567.83	$973.83	subtract $492.50	15%	$758.83	$1,394.83	subtract $604.83	15%
	$973.83	$1,650.83	subtract $706.42	27%	$1,394.83	$2,513.83	subtract $955.94	27%
	$1,650.83	$3,152.83	subtract $800.86	30%	$2,513.83	$3,803.83	subtract $1,111.73	30%
	$3,152.83	$6,341.83	subtract $1,136.86	35%	$3,803.83	$6,401.83	subtract $1,496.32	35%
	$6,341.83	......	subtract $1,622.30	38.6%	$6,401.83	......	subtract $1,953.82	38.6%
8	$0.00	$625.52	subtract $512.52	10%	$0.00	$816.52	subtract $585.52	10%
	$625.52	$1,031.52	subtract $550.19	15%	$816.52	$1,452.52	subtract $662.52	15%
	$1,031.52	$1,708.52	subtract $764.11	27%	$1,452.52	$2,571.52	subtract $1,013.63	27%
	$1,708.52	$3,210.52	subtract $858.55	30%	$2,571.52	$3,861.52	subtract $1,169.42	30%
	$3,210.52	$6,399.52	subtract $1,194.55	35%	$3,861.52	$6,459.52	subtract $1,554.01	35%
	$6,399.52	......	subtract $1,679.99	38.6%	$6,459.52	......	subtract $2,011.51	38.6%
9[2]	$0.00	$683.21	subtract $570.21	10%	$0.00	$874.21	subtract $643.21	10%
	$683.21	$1,089.21	subtract $607.88	15%	$874.21	$1,510.21	subtract $720.21	15%
	$1,089.21	$1,766.21	subtract $821.80	27%	$1,510.21	$2,629.21	subtract $1,071.32	27%
	$1,766.21	$3,268.21	subtract $916.24	30%	$2,629.21	$3,919.21	subtract $1,227.11	30%
	$3,268.21	$6,457.21	subtract $1,252.24	35%	$3,919.21	$6,517.21	subtract $1,611.70	35%
	$6,457.21	......	subtract $1,737.68	38.6%	$6,517.21	......	subtract $2,069.20	38.6%

Instructions

A. For each employee, use the appropriate payroll period table and marital status section, and select the subsection showing the number of allowances claimed.

B. Read across the selected subsection and locate the bracket applicable to the employee's gross wages in columns A and B.

C. Subtract the amount shown in column C from the employee's gross wages.

D. Multiply the result by the withholding percentage rate shown in column D to obtain the amount of tax to be withheld.

[1] If the gross wages are less than the amount to be subtracted, the withholding is zero.

[2] You can expand these tables for additional allowances. To do this, increase the amounts in this subsection by $57.69 for each additional allowance claimed.

Page 26

(For Wages Paid in 2002)

Wage Bracket Percentage Method Table for Computing Income Tax Withholding From Gross Wages

Biweekly Payroll Period

If the number of allowances is—	Single Persons				Married Persons			
	And gross wages are—		from gross wages[1]	Multiply result by—	And gross wages are—		from gross wages[1]	Multiply result by—
	Over	But not over			Over	But not over		
	A	B	C	D	A	B	C	D
0	$0.00	$329.00	subtract $102.00	10%	$0.00	$710.00	subtract $248.00	10%
	$329.00	$1,140.00	subtract $177.67	15%	$710.00	$1,983.00	subtract $402.00	15%
	$1,140.00	$2,493.00	subtract $605.37	27%	$1,983.00	$4,219.00	subtract $1,104.67	27%
	$2,493.00	$5,498.00	subtract $794.13	30%	$4,219.00	$6,800.00	subtract $1,416.10	30%
	$5,498.00	$11,875.00	subtract $1,466.11	35%	$6,800.00	$11,996.00	subtract $2,185.23	35%
	$11,875.00	......	subtract $2,436.89	38.6%	$11,996.00	......	subtract $3,100.22	38.6%
1	$0.00	$444.38	subtract $217.38	10%	$0.00	$825.38	subtract $363.38	10%
	$444.38	$1,255.38	subtract $293.05	15%	$825.38	$2,098.38	subtract $517.38	15%
	$1,255.38	$2,608.38	subtract $720.75	27%	$2,098.38	$4,334.38	subtract $1,220.05	27%
	$2,608.38	$5,613.38	subtract $909.51	30%	$4,334.38	$6,915.38	subtract $1,531.48	30%
	$5,613.38	$11,990.38	subtract $1,581.49	35%	$6,915.38	$12,111.38	subtract $2,300.61	35%
	$11,990.38	......	subtract $2,552.27	38.6%	$12,111.38	......	subtract $3,215.60	38.6%
2	$0.00	$559.76	subtract $332.76	10%	$0.00	$940.76	subtract $478.76	10%
	$559.76	$1,370.76	subtract $408.43	15%	$940.76	$2,213.76	subtract $632.76	15%
	$1,370.76	$2,723.76	subtract $836.13	27%	$2,213.76	$4,449.76	subtract $1,335.43	27%
	$2,723.76	$5,728.76	subtract $1,024.89	30%	$4,449.76	$7,030.76	subtract $1,646.86	30%
	$5,728.76	$12,105.76	subtract $1,696.87	35%	$7,030.76	$12,226.76	subtract $2,415.99	35%
	$12,105.76	......	subtract $2,667.65	38.6%	$12,226.76	......	subtract $3,330.98	38.6%
3	$0.00	$675.14	subtract $448.14	10%	$0.00	$1,056.14	subtract $594.14	10%
	$675.14	$1,486.14	subtract $523.81	15%	$1,056.14	$2,329.14	subtract $748.14	15%
	$1,486.14	$2,839.14	subtract $951.51	27%	$2,329.14	$4,565.14	subtract $1,450.81	27%
	$2,839.14	$5,844.14	subtract $1,140.27	30%	$4,565.14	$7,146.14	subtract $1,762.24	30%
	$5,844.14	$12,221.14	subtract $1,812.25	35%	$7,146.14	$12,342.14	subtract $2,531.37	35%
	$12,221.14	......	subtract $2,783.03	38.6%	$12,342.14	......	subtract $3,446.36	38.6%
4	$0.00	$790.52	subtract $563.52	10%	$0.00	$1,171.52	subtract $709.52	10%
	$790.52	$1,601.52	subtract $639.19	15%	$1,171.52	$2,444.52	subtract $863.52	15%
	$1,601.52	$2,954.52	subtract $1,066.89	27%	$2,444.52	$4,680.52	subtract $1,566.19	27%
	$2,954.52	$5,959.52	subtract $1,255.65	30%	$4,680.52	$7,261.52	subtract $1,877.62	30%
	$5,959.52	$12,336.52	subtract $1,927.63	35%	$7,261.52	$12,457.52	subtract $2,646.75	35%
	$12,336.52	......	subtract $2,898.41	38.6%	$12,457.52	......	subtract $3,561.74	38.6%
5	$0.00	$905.90	subtract $678.90	10%	$0.00	$1,286.90	subtract $824.90	10%
	$905.90	$1,716.90	subtract $754.57	15%	$1,286.90	$2,559.90	subtract $978.90	15%
	$1,716.90	$3,069.90	subtract $1,182.27	27%	$2,559.90	$4,795.90	subtract $1,681.57	27%
	$3,069.90	$6,074.90	subtract $1,371.03	30%	$4,795.90	$7,376.90	subtract $1,993.00	30%
	$6,074.90	$12,451.90	subtract $2,043.01	35%	$7,376.90	$12,572.90	subtract $2,762.13	35%
	$12,451.90	......	subtract $3,013.79	38.6%	$12,572.90	......	subtract $3,677.12	38.6%
6	$0.00	$1,021.28	subtract $794.28	10%	$0.00	$1,402.28	subtract $940.28	10%
	$1,021.28	$1,832.28	subtract $869.95	15%	$1,402.28	$2,675.28	subtract $1,094.28	15%
	$1,832.28	$3,185.28	subtract $1,297.65	27%	$2,675.28	$4,911.28	subtract $1,796.95	27%
	$3,185.28	$6,190.28	subtract $1,486.41	30%	$4,911.28	$7,492.28	subtract $2,108.38	30%
	$6,190.28	$12,567.28	subtract $2,158.39	35%	$7,492.28	$12,688.28	subtract $2,877.51	35%
	$12,567.28	......	subtract $3,129.17	38.6%	$12,688.28	......	subtract $3,792.50	38.6%
7	$0.00	$1,136.66	subtract $909.66	10%	$0.00	$1,517.66	subtract $1,055.66	10%
	$1,136.66	$1,947.66	subtract $985.33	15%	$1,517.66	$2,790.66	subtract $1,209.66	15%
	$1,947.66	$3,300.66	subtract $1,413.03	27%	$2,790.66	$5,026.66	subtract $1,912.33	27%
	$3,300.66	$6,305.66	subtract $1,601.79	30%	$5,026.66	$7,607.66	subtract $2,223.76	30%
	$6,305.66	$12,682.66	subtract $2,273.77	35%	$7,607.66	$12,803.66	subtract $2,992.89	35%
	$12,682.66	......	subtract $3,244.55	38.6%	$12,803.66	......	subtract $3,907.88	38.6%
8	$0.00	$1,252.04	subtract $1,025.04	10%	$0.00	$1,633.04	subtract $1,171.04	10%
	$1,252.04	$2,063.04	subtract $1,100.71	15%	$1,633.04	$2,906.04	subtract $1,325.04	15%
	$2,063.04	$3,416.04	subtract $1,528.41	27%	$2,906.04	$5,142.04	subtract $2,027.71	27%
	$3,416.04	$6,421.04	subtract $1,717.17	30%	$5,142.04	$7,723.04	subtract $2,339.14	30%
	$6,421.04	$12,798.04	subtract $2,389.15	35%	$7,723.04	$12,919.04	subtract $3,108.27	35%
	$12,798.04	......	subtract $3,359.93	38.6%	$12,919.04	......	subtract $4,023.26	38.6%
9[2]	$0.00	$1,367.42	subtract $1,140.42	10%	$0.00	$1,748.42	subtract $1,286.42	10%
	$1,367.42	$2,178.42	subtract $1,216.09	15%	$1,748.42	$3,021.42	subtract $1,440.42	15%
	$2,178.42	$3,531.42	subtract $1,643.79	27%	$3,021.42	$5,257.42	subtract $2,143.09	27%
	$3,531.42	$6,536.42	subtract $1,832.55	30%	$5,257.42	$7,838.42	subtract $2,454.52	30%
	$6,536.42	$12,913.42	subtract $2,504.53	35%	$7,838.42	$13,034.42	subtract $3,223.65	35%
	$12,913.42	......	subtract $3,475.31	38.6%	$13,034.42	......	subtract $4,138.64	38.6%

Instructions

A. For each employee, use the appropriate payroll period table and marital status section, and select the subsection showing the number of allowances claimed.

B. Read across the selected subsection and locate the bracket applicable to the employee's gross wages in columns A and B.

C. Subtract the amount shown in column C from the employee's gross wages.

D. Multiply the result by the withholding percentage rate shown in column D to obtain the amount of tax to be withheld.

[1] If the gross wages are less than the amount to be subtracted, the withholding is zero.

[2] You can expand these tables for additional allowances. To do this, increase the amounts in this subsection by $115.38 for each additional allowance claimed.

Page 27

(For Wages Paid in 2002)

Wage Bracket Percentage Method Table for Computing Income Tax Withholding From Gross Wages

Semimonthly Payroll Period

If the number of allowances is—	Single Persons				Married Persons			
	And gross wages are—		from gross wages[1]	Multiply result by—	And gross wages are—		from gross wages[1]	Multiply result by—
	Over	But not over			Over	But not over		
	A	B	C	D	A	B	C	D
0	$0.00	$356.00	subtract $110.00	10%	$0.00	$769.00	subtract $269.00	10%
	$356.00	$1,235.00	subtract $192.00	15%	$769.00	$2,148.00	subtract $435.67	15%
	$1,235.00	$2,701.00	subtract $655.56	27%	$2,148.00	$4,571.00	subtract $1,196.70	27%
	$2,701.00	$5,956.00	subtract $860.10	30%	$4,571.00	$7,367.00	subtract $1,534.13	30%
	$5,956.00	$12,865.00	subtract $1,588.09	35%	$7,367.00	$12,996.00	subtract $2,367.40	35%
	$12,865.00	......	subtract $2,639.82	38.6%	$12,996.00	......	subtract $3,358.67	38.6%
1	$0.00	$481.00	subtract $235.00	10%	$0.00	$894.00	subtract $394.00	10%
	$481.00	$1,360.00	subtract $317.00	15%	$894.00	$2,273.00	subtract $560.67	15%
	$1,360.00	$2,826.00	subtract $780.56	27%	$2,273.00	$4,696.00	subtract $1,321.70	27%
	$2,826.00	$6,081.00	subtract $985.10	30%	$4,696.00	$7,492.00	subtract $1,659.13	30%
	$6,081.00	$12,990.00	subtract $1,713.09	35%	$7,492.00	$13,121.00	subtract $2,492.40	35%
	$12,990.00	......	subtract $2,764.82	38.6%	$13,121.00	......	subtract $3,483.67	38.6%
2	$0.00	$606.00	subtract $360.00	10%	$0.00	$1,019.00	subtract $519.00	10%
	$606.00	$1,485.00	subtract $442.00	15%	$1,019.00	$2,398.00	subtract $685.67	15%
	$1,485.00	$2,951.00	subtract $905.56	27%	$2,398.00	$4,821.00	subtract $1,446.70	27%
	$2,951.00	$6,206.00	subtract $1,110.10	30%	$4,821.00	$7,617.00	subtract $1,784.13	30%
	$6,206.00	$13,115.00	subtract $1,838.09	35%	$7,617.00	$13,246.00	subtract $2,617.40	35%
	$13,115.00	......	subtract $2,889.82	38.6%	$13,246.00	......	subtract $3,608.67	38.6%
3	$0.00	$731.00	subtract $485.00	10%	$0.00	$1,144.00	subtract $644.00	10%
	$731.00	$1,610.00	subtract $567.00	15%	$1,144.00	$2,523.00	subtract $810.67	15%
	$1,610.00	$3,076.00	subtract $1,030.56	27%	$2,523.00	$4,946.00	subtract $1,571.70	27%
	$3,076.00	$6,331.00	subtract $1,235.10	30%	$4,946.00	$7,742.00	subtract $1,909.13	30%
	$6,331.00	$13,240.00	subtract $1,963.09	35%	$7,742.00	$13,371.00	subtract $2,742.40	35%
	$13,240.00	......	subtract $3,014.82	38.6%	$13,371.00	......	subtract $3,733.67	38.6%
4	$0.00	$856.00	subtract $610.00	10%	$0.00	$1,269.00	subtract $769.00	10%
	$856.00	$1,735.00	subtract $692.00	15%	$1,269.00	$2,648.00	subtract $935.67	15%
	$1,735.00	$3,201.00	subtract $1,155.56	27%	$2,648.00	$5,071.00	subtract $1,696.70	27%
	$3,201.00	$6,456.00	subtract $1,360.10	30%	$5,071.00	$7,867.00	subtract $2,034.13	30%
	$6,456.00	$13,365.00	subtract $2,088.09	35%	$7,867.00	$13,496.00	subtract $2,867.40	35%
	$13,365.00	......	subtract $3,139.82	38.6%	$13,496.00	......	subtract $3,858.67	38.6%
5	$0.00	$981.00	subtract $735.00	10%	$0.00	$1,394.00	subtract $894.00	10%
	$981.00	$1,860.00	subtract $817.00	15%	$1,394.00	$2,773.00	subtract $1,060.67	15%
	$1,860.00	$3,326.00	subtract $1,280.56	27%	$2,773.00	$5,196.00	subtract $1,821.70	27%
	$3,326.00	$6,581.00	subtract $1,485.10	30%	$5,196.00	$7,992.00	subtract $2,159.13	30%
	$6,581.00	$13,490.00	subtract $2,213.09	35%	$7,992.00	$13,621.00	subtract $2,992.40	35%
	$13,490.00	......	subtract $3,264.82	38.6%	$13,621.00	......	subtract $3,983.67	38.6%
6	$0.00	$1,106.00	subtract $860.00	10%	$0.00	$1,519.00	subtract $1,019.00	10%
	$1,106.00	$1,985.00	subtract $942.00	15%	$1,519.00	$2,898.00	subtract $1,185.67	15%
	$1,985.00	$3,451.00	subtract $1,405.56	27%	$2,898.00	$5,321.00	subtract $1,946.70	27%
	$3,451.00	$6,706.00	subtract $1,610.10	30%	$5,321.00	$8,117.00	subtract $2,284.13	30%
	$6,706.00	$13,615.00	subtract $2,338.09	35%	$8,117.00	$13,746.00	subtract $3,117.40	35%
	$13,615.00	......	subtract $3,389.82	38.6%	$13,746.00	......	subtract $4,108.67	38.6%
7	$0.00	$1,231.00	subtract $985.00	10%	$0.00	$1,644.00	subtract $1,144.00	10%
	$1,231.00	$2,110.00	subtract $1,067.00	15%	$1,644.00	$3,023.00	subtract $1,310.67	15%
	$2,110.00	$3,576.00	subtract $1,530.56	27%	$3,023.00	$5,446.00	subtract $2,071.70	27%
	$3,576.00	$6,831.00	subtract $1,735.10	30%	$5,446.00	$8,242.00	subtract $2,409.13	30%
	$6,831.00	$13,740.00	subtract $2,463.09	35%	$8,242.00	$13,871.00	subtract $3,242.40	35%
	$13,740.00	......	subtract $3,514.82	38.6%	$13,871.00	......	subtract $4,233.67	38.6%
8	$0.00	$1,356.00	subtract $1,110.00	10%	$0.00	$1,769.00	subtract $1,269.00	10%
	$1,356.00	$2,235.00	subtract $1,192.00	15%	$1,769.00	$3,148.00	subtract $1,435.67	15%
	$2,235.00	$3,701.00	subtract $1,655.56	27%	$3,148.00	$5,571.00	subtract $2,196.70	27%
	$3,701.00	$6,956.00	subtract $1,860.10	30%	$5,571.00	$8,367.00	subtract $2,534.13	30%
	$6,956.00	$13,865.00	subtract $2,588.09	35%	$8,367.00	$13,996.00	subtract $3,367.40	35%
	$13,865.00	......	subtract $3,639.82	38.6%	$13,996.00	......	subtract $4,358.67	38.6%
9[2]	$0.00	$1,481.00	subtract $1,235.00	10%	$0.00	$1,894.00	subtract $1,394.00	10%
	$1,481.00	$2,360.00	subtract $1,317.00	15%	$1,894.00	$3,273.00	subtract $1,560.67	15%
	$2,360.00	$3,826.00	subtract $1,780.56	27%	$3,273.00	$5,696.00	subtract $2,321.70	27%
	$3,826.00	$7,081.00	subtract $1,985.10	30%	$5,696.00	$8,492.00	subtract $2,659.13	30%
	$7,081.00	$13,990.00	subtract $2,713.09	35%	$8,492.00	$14,121.00	subtract $3,492.40	35%
	$13,990.00	......	subtract $3,764.82	38.6%	$14,121.00	......	subtract $4,483.67	38.6%

Instructions

A. For each employee, use the appropriate payroll period table and marital status section, and select the subsection showing the number of allowances claimed.

B. Read across the selected subsection and locate the bracket applicable to the employee's gross wages in columns A and B.

C. Subtract the amount shown in column C from the employee's gross wages.

D. Multiply the result by the withholding percentage rate shown in column D to obtain the amount of tax to be withheld.

[1] If the gross wages are less than the amount to be subtracted, the withholding is zero.

[2] You can expand these tables for additional allowances. To do this, increase the amounts in this subsection by $125.00 for each additional allowance claimed.

Page 28

(For Wages Paid in 2002)

Wage Bracket Percentage Method Table for Computing
Income Tax Withholding From Gross Wages

Monthly Payroll Period

If the number of allowances is—	Single Persons: And gross wages are— Over	Single Persons: But not over	Single Persons: from gross wages[1]	Single Persons: Multiply result by—	Married Persons: And gross wages are— Over	Married Persons: But not over	Married Persons: from gross wages[1]	Married Persons: Multiply result by—
	A	B	C	D	A	B	C	D
0	$0.00	$713.00	subtract $221.00	10%	$0.00	$1,538.00	subtract $538.00	10%
	$713.00	$2,471.00	subtract $385.00	15%	$1,538.00	$4,296.00	subtract $871.33	15%
	$2,471.00	$5,402.00	subtract $1,312.11	27%	$4,296.00	$9,142.00	subtract $2,393.41	27%
	$5,402.00	$11,913.00	subtract $1,721.10	30%	$9,142.00	$14,733.00	subtract $3,068.27	30%
	$11,913.00	$25,729.00	subtract $3,177.09	35%	$14,733.00	$25,992.00	subtract $4,734.66	35%
	$25,729.00	......	subtract $5,280.37	38.6%	$25,992.00	......	subtract $6,717.21	38.6%
1	$0.00	$963.00	subtract $471.00	10%	$0.00	$1,788.00	subtract $788.00	10%
	$963.00	$2,721.00	subtract $635.00	15%	$1,788.00	$4,546.00	subtract $1,121.33	15%
	$2,721.00	$5,652.00	subtract $1,562.11	27%	$4,546.00	$9,392.00	subtract $2,643.41	27%
	$5,652.00	$12,163.00	subtract $1,971.10	30%	$9,392.00	$14,983.00	subtract $3,318.27	30%
	$12,163.00	$25,979.00	subtract $3,427.09	35%	$14,983.00	$26,242.00	subtract $4,984.66	35%
	$25,979.00	......	subtract $5,530.37	38.6%	$26,242.00	......	subtract $6,967.21	38.6%
2	$0.00	$1,213.00	subtract $721.00	10%	$0.00	$2,038.00	subtract $1,038.00	10%
	$1,213.00	$2,971.00	subtract $885.00	15%	$2,038.00	$4,796.00	subtract $1,371.33	15%
	$2,971.00	$5,902.00	subtract $1,812.11	27%	$4,796.00	$9,642.00	subtract $2,893.41	27%
	$5,902.00	$12,413.00	subtract $2,221.10	30%	$9,642.00	$15,233.00	subtract $3,568.27	30%
	$12,413.00	$26,229.00	subtract $3,677.09	35%	$15,233.00	$26,492.00	subtract $5,234.66	35%
	$26,229.00	......	subtract $5,780.37	38.6%	$26,492.00	......	subtract $7,217.21	38.6%
3	$0.00	$1,463.00	subtract $971.00	10%	$0.00	$2,288.00	subtract $1,288.00	10%
	$1,463.00	$3,221.00	subtract $1,135.00	15%	$2,288.00	$5,046.00	subtract $1,621.33	15%
	$3,221.00	$6,152.00	subtract $2,062.11	27%	$5,046.00	$9,892.00	subtract $3,143.41	27%
	$6,152.00	$12,663.00	subtract $2,471.10	30%	$9,892.00	$15,483.00	subtract $3,818.27	30%
	$12,663.00	$26,479.00	subtract $3,927.09	35%	$15,483.00	$26,742.00	subtract $5,484.66	35%
	$26,479.00	......	subtract $6,030.37	38.6%	$26,742.00	......	subtract $7,467.21	38.6%
4	$0.00	$1,713.00	subtract $1,221.00	10%	$0.00	$2,538.00	subtract $1,538.00	10%
	$1,713.00	$3,471.00	subtract $1,385.00	15%	$2,538.00	$5,296.00	subtract $1,871.33	15%
	$3,471.00	$6,402.00	subtract $2,312.11	27%	$5,296.00	$10,142.00	subtract $3,393.41	27%
	$6,402.00	$12,913.00	subtract $2,721.10	30%	$10,142.00	$15,733.00	subtract $4,068.27	30%
	$12,913.00	$26,729.00	subtract $4,177.09	35%	$15,733.00	$26,992.00	subtract $5,734.66	35%
	$26,729.00	......	subtract $6,280.37	38.6%	$26,992.00	......	subtract $7,717.21	38.6%
5	$0.00	$1,963.00	subtract $1,471.00	10%	$0.00	$2,788.00	subtract $1,788.00	10%
	$1,963.00	$3,721.00	subtract $1,635.00	15%	$2,788.00	$5,546.00	subtract $2,121.33	15%
	$3,721.00	$6,652.00	subtract $2,562.11	27%	$5,546.00	$10,392.00	subtract $3,643.41	27%
	$6,652.00	$13,163.00	subtract $2,971.10	30%	$10,392.00	$15,983.00	subtract $4,318.27	30%
	$13,163.00	$26,979.00	subtract $4,427.09	35%	$15,983.00	$27,242.00	subtract $5,984.66	35%
	$26,979.00	......	subtract $6,530.37	38.6%	$27,242.00	......	subtract $7,967.21	38.6%
6	$0.00	$2,213.00	subtract $1,721.00	10%	$0.00	$3,038.00	subtract $2,038.00	10%
	$2,213.00	$3,971.00	subtract $1,885.00	15%	$3,038.00	$5,796.00	subtract $2,371.33	15%
	$3,971.00	$6,902.00	subtract $2,812.11	27%	$5,796.00	$10,642.00	subtract $3,893.41	27%
	$6,902.00	$13,413.00	subtract $3,221.10	30%	$10,642.00	$16,233.00	subtract $4,568.27	30%
	$13,413.00	$27,229.00	subtract $4,677.09	35%	$16,233.00	$27,492.00	subtract $6,234.66	35%
	$27,229.00	......	subtract $6,780.37	38.6%	$27,492.00	......	subtract $8,217.21	38.6%
7	$0.00	$2,463.00	subtract $1,971.00	10%	$0.00	$3,288.00	subtract $2,288.00	10%
	$2,463.00	$4,221.00	subtract $2,135.00	15%	$3,288.00	$6,046.00	subtract $2,621.33	15%
	$4,221.00	$7,152.00	subtract $3,062.11	27%	$6,046.00	$10,892.00	subtract $4,143.41	27%
	$7,152.00	$13,663.00	subtract $3,471.10	30%	$10,892.00	$16,483.00	subtract $4,818.27	30%
	$13,663.00	$27,479.00	subtract $4,927.09	35%	$16,483.00	$27,742.00	subtract $6,484.66	35%
	$27,479.00	......	subtract $7,030.37	38.6%	$27,742.00	......	subtract $8,467.21	38.6%
8	$0.00	$2,713.00	subtract $2,221.00	10%	$0.00	$3,538.00	subtract $2,538.00	10%
	$2,713.00	$4,471.00	subtract $2,385.00	15%	$3,538.00	$6,296.00	subtract $2,871.33	15%
	$4,471.00	$7,402.00	subtract $3,312.11	27%	$6,296.00	$11,142.00	subtract $4,393.41	27%
	$7,402.00	$13,913.00	subtract $3,721.10	30%	$11,142.00	$16,733.00	subtract $5,068.27	30%
	$13,913.00	$27,729.00	subtract $5,177.09	35%	$16,733.00	$27,992.00	subtract $6,734.66	35%
	$27,729.00	......	subtract $7,280.37	38.6%	$27,992.00	......	subtract $8,717.21	38.6%
9[2]	$0.00	$2,963.00	subtract $2,471.00	10%	$0.00	$3,788.00	subtract $2,788.00	10%
	$2,963.00	$4,721.00	subtract $2,635.00	15%	$3,788.00	$6,546.00	subtract $3,121.33	15%
	$4,721.00	$7,652.00	subtract $3,562.11	27%	$6,546.00	$11,392.00	subtract $4,643.41	27%
	$7,652.00	$14,163.00	subtract $3,971.10	30%	$11,392.00	$16,983.00	subtract $5,318.27	30%
	$14,163.00	$27,979.00	subtract $5,427.09	35%	$16,983.00	$28,242.00	subtract $6,984.66	35%
	$27,979.00	......	subtract $7,530.37	38.6%	$28,242.00	......	subtract $8,967.21	38.6%

Instructions

A. For each employee, use the appropriate payroll period table and marital status section, and select the subsection showing the number of allowances claimed.

B. Read across the selected subsection and locate the bracket applicable to the employee's gross wages in columns A and B.

C. Subtract the amount shown in column C from the employee's gross wages.

D. Multiply the result by the withholding percentage rate shown in column D to obtain the amount of tax to be withheld.

[1] If the gross wages are less than the amount to be subtracted, the withholding is zero.

[2] You can expand these tables for additional allowances. To do this, increase the amounts in this subsection by $250.00 for each additional allowance claimed.

Page 29

(For Wages Paid in 2002)

Wage Bracket Percentage Method Table for Computing Income Tax Withholding From Wages Exceeding Allowance Amount

Weekly Payroll Period

If the number of allowances is—	Single Persons				Married Persons			
	And gross wages are— Over	But not over	from excess wages[1]	Multiply result by—	And gross wages are— Over	But not over	from excess wages[1]	Multiply result by—
	A	B	C	D	A	B	C	D
0	$0	$164.00	subtract $51.00	10%	$0	$355.00	subtract $124.00	10%
	$164.00	$570.00	subtract $88.67	15%	$355.00	$991.00	subtract $201.00	15%
	$570.00	$1,247.00	subtract $302.59	27%	$991.00	$2,110.00	subtract $552.11	27%
	$1,247.00	$2,749.00	subtract $397.03	30%	$2,110.00	$3,400.00	subtract $707.90	30%
	$2,749.00	$5,938.00	subtract $733.03	35%	$3,400.00	$5,998.00	subtract $1,092.49	35%
	$5,938.00	......	subtract $1,218.47	38.6%	$5,998.00	......	subtract $1,549.99	38.6%
1	$0	$221.69	subtract $51.00	10%	$0	$412.69	subtract $124.00	10%
	$221.69	$627.69	subtract $88.67	15%	$412.69	$1,048.69	subtract $201.00	15%
	$627.69	$1,304.69	subtract $302.59	27%	$1,048.69	$2,167.69	subtract $552.11	27%
	$1,304.69	$2,806.69	subtract $397.03	30%	$2,167.69	$3,457.69	subtract $707.90	30%
	$2,806.69	$5,995.69	subtract $733.03	35%	$3,457.69	$6,055.69	subtract $1,092.49	35%
	$5,995.69	......	subtract $1,218.47	38.6%	$6,055.69	......	subtract $1,549.99	38.6%
2	$0	$279.38	subtract $51.00	10%	$0	$470.38	subtract $124.00	10%
	$279.38	$685.38	subtract $88.67	15%	$470.38	$1,106.38	subtract $201.00	15%
	$685.38	$1,362.38	subtract $302.59	27%	$1,106.38	$2,225.38	subtract $552.11	27%
	$1,362.38	$2,864.38	subtract $397.03	30%	$2,225.38	$3,515.38	subtract $707.90	30%
	$2,864.38	$6,053.38	subtract $733.03	35%	$3,515.38	$6,113.38	subtract $1,092.49	35%
	$6,053.38	......	subtract $1,218.47	38.6%	$6,113.38	......	subtract $1,549.99	38.6%
3	$0	$337.07	subtract $51.00	10%	$0	$528.07	subtract $124.00	10%
	$337.07	$743.07	subtract $88.67	15%	$528.07	$1,164.07	subtract $201.00	15%
	$743.07	$1,420.07	subtract $302.59	27%	$1,164.07	$2,283.07	subtract $552.11	27%
	$1,420.07	$2,922.07	subtract $397.03	30%	$2,283.07	$3,573.07	subtract $707.90	30%
	$2,922.07	$6,111.07	subtract $733.03	35%	$3,573.07	$6,171.07	subtract $1,092.49	35%
	$6,111.07	......	subtract $1,218.47	38.6%	$6,171.07	......	subtract $1,549.99	38.6%
4	$0	$394.76	subtract $51.00	10%	$0	$585.76	subtract $124.00	10%
	$394.76	$800.76	subtract $88.67	15%	$585.76	$1,221.76	subtract $201.00	15%
	$800.76	$1,477.76	subtract $302.59	27%	$1,221.76	$2,340.76	subtract $552.11	27%
	$1,477.76	$2,979.76	subtract $397.03	30%	$2,340.76	$3,630.76	subtract $707.90	30%
	$2,979.76	$6,168.76	subtract $733.03	35%	$3,630.76	$6,228.76	subtract $1,092.49	35%
	$6,168.76	......	subtract $1,218.47	38.6%	$6,228.76	......	subtract $1,549.99	38.6%
5	$0	$452.45	subtract $51.00	10%	$0	$643.45	subtract $124.00	10%
	$452.45	$858.45	subtract $88.67	15%	$643.45	$1,279.45	subtract $201.00	15%
	$858.45	$1,535.45	subtract $302.59	27%	$1,279.45	$2,398.45	subtract $552.11	27%
	$1,535.45	$3,037.45	subtract $397.03	30%	$2,398.45	$3,688.45	subtract $707.90	30%
	$3,037.45	$6,226.45	subtract $733.03	35%	$3,688.45	$6,286.45	subtract $1,092.49	35%
	$6,226.45	......	subtract $1,218.47	38.6%	$6,286.45	......	subtract $1,549.99	38.6%
6	$0	$510.14	subtract $51.00	10%	$0	$701.14	subtract $124.00	10%
	$510.14	$916.14	subtract $88.67	15%	$701.14	$1,337.14	subtract $201.00	15%
	$916.14	$1,593.14	subtract $302.59	27%	$1,337.14	$2,456.14	subtract $552.11	27%
	$1,593.14	$3,095.14	subtract $397.03	30%	$2,456.14	$3,746.14	subtract $707.90	30%
	$3,095.14	$6,284.14	subtract $733.03	35%	$3,746.14	$6,344.14	subtract $1,092.49	35%
	$6,284.14	......	subtract $1,218.47	38.6%	$6,344.14	......	subtract $1,549.99	38.6%
7	$0	$567.83	subtract $51.00	10%	$0	$758.83	subtract $124.00	10%
	$567.83	$973.83	subtract $88.67	15%	$758.83	$1,394.83	subtract $201.00	15%
	$973.83	$1,650.83	subtract $302.59	27%	$1,394.83	$2,513.83	subtract $552.11	27%
	$1,650.83	$3,152.83	subtract $397.03	30%	$2,513.83	$3,803.83	subtract $707.90	30%
	$3,152.83	$6,341.83	subtract $733.03	35%	$3,803.83	$6,401.83	subtract $1,092.49	35%
	$6,341.83	......	subtract $1,218.47	38.6%	$6,401.83	......	subtract $1,549.99	38.6%
8	$0	$625.52	subtract $51.00	10%	$0	$816.52	subtract $124.00	10%
	$625.52	$1,031.52	subtract $88.67	15%	$816.52	$1,452.52	subtract $201.00	15%
	$1,031.52	$1,708.52	subtract $302.59	27%	$1,452.52	$2,571.52	subtract $552.11	27%
	$1,708.52	$3,210.52	subtract $397.03	30%	$2,571.52	$3,861.52	subtract $707.90	30%
	$3,210.52	$6,399.52	subtract $733.03	35%	$3,861.52	$6,459.52	subtract $1,092.49	35%
	$6,399.52	......	subtract $1,218.47	38.6%	$6,459.52	......	subtract $1,549.99	38.6%
9[2]	$0	$683.21	subtract $51.00	10%	$0	$874.21	subtract $124.00	10%
	$683.21	$1,089.21	subtract $88.67	15%	$874.21	$1,510.21	subtract $201.00	15%
	$1,089.21	$1,766.21	subtract $302.59	27%	$1,510.21	$2,629.21	subtract $552.11	27%
	$1,766.21	$3,268.21	subtract $397.03	30%	$2,629.21	$3,919.21	subtract $707.90	30%
	$3,268.21	$6,457.21	subtract $733.03	35%	$3,919.21	$6,517.21	subtract $1,092.49	35%
	$6,457.21	......	subtract $1,218.47	38.6%	$6,517.21	......	subtract $1,549.99	38.6%

Instructions

A. For each employee, use the appropriate payroll period table and marital status section, and select the subsection showing the number of allowances claimed.

B. Read across the selected subsection and locate the bracket applicable to the employee's gross wages in columns A and B.

C. Subtract the amount shown in column C from the employee's excess wages (gross wages less amount for allowances claimed).

Caution.—The adjustment (subtraction) factors shown in this table (instruction C) do not include an amount for the number of allowances claimed by the employee on Form W-4. The amount for allowances claimed must be deducted from gross wages before withholding tax is computed.

D. Multiply the result by the withholding percentage rate shown in column D to obtain the amount of tax to be withheld.

[1] If the excess wages are less than the amount to be subtracted, the withholding is zero.

[2] You can expand these tables for additional allowances. To do this, increase the wage bracket amounts in this subsection by $57.69 for each additional allowance claimed.

Page 30

(For Wages Paid in 2002)

Wage Bracket Percentage Method Table for Computing
Income Tax Withholding From Wages Exceeding Allowance Amount

Biweekly Payroll Period

If the number of allowances is—	Single Persons				Married Persons			
	And gross wages are—		from excess wages[1]	Multiply result by—	And gross wages are—		from excess wages[1]	Multiply result by—
	Over	But not over			Over	But not over		
	A	B	C	D	A	B	C	D
0	$0	$329.00	subtract $102.00	10%	$0	$710.00	subtract $248.00	10%
	$329.00	$1,140.00	subtract $177.67	15%	$710.00	$1,983.00	subtract $402.00	15%
	$1,140.00	$2,493.00	subtract $605.37	27%	$1,983.00	$4,219.00	subtract $1,104.67	27%
	$2,493.00	$5,498.00	subtract $794.13	30%	$4,219.00	$6,800.00	subtract $1,416.10	30%
	$5,498.00	$11,875.00	subtract $1,466.11	35%	$6,800.00	$11,996.00	subtract $2,185.23	35%
	$11,875.00	......	subtract $2,436.89	38.6%	$11,996.00	......	subtract $3,100.22	38.6%
1	$0	$444.38	subtract $102.00	10%	$0	$825.38	subtract $248.00	10%
	$444.38	$1,255.38	subtract $177.67	15%	$825.38	$2,098.38	subtract $402.00	15%
	$1,255.38	$2,608.38	subtract $605.37	27%	$2,098.38	$4,334.38	subtract $1,104.67	27%
	$2,608.38	$5,613.38	subtract $794.13	30%	$4,334.38	$6,915.38	subtract $1,416.10	30%
	$5,613.38	$11,990.38	subtract $1,466.11	35%	$6,915.38	$12,111.38	subtract $2,185.23	35%
	$11,990.38	......	subtract $2,436.89	38.6%	$12,111.38	......	subtract $3,100.22	38.6%
2	$0	$559.76	subtract $102.00	10%	$0	$940.76	subtract $248.00	10%
	$559.76	$1,370.76	subtract $177.67	15%	$940.76	$2,213.76	subtract $402.00	15%
	$1,370.76	$2,723.76	subtract $605.37	27%	$2,213.76	$4,449.76	subtract $1,104.67	27%
	$2,723.76	$5,728.76	subtract $794.13	30%	$4,449.76	$7,030.76	subtract $1,416.10	30%
	$5,728.76	$12,105.76	subtract $1,466.11	35%	$7,030.76	$12,226.76	subtract $2,185.23	35%
	$12,105.76	......	subtract $2,436.89	38.6%	$12,226.76	......	subtract $3,100.22	38.6%
3	$0	$675.14	subtract $102.00	10%	$0	$1,056.14	subtract $248.00	10%
	$675.14	$1,486.14	subtract $177.67	15%	$1,056.14	$2,329.14	subtract $402.00	15%
	$1,486.14	$2,839.14	subtract $605.37	27%	$2,329.14	$4,565.14	subtract $1,104.67	27%
	$2,839.14	$5,844.14	subtract $794.13	30%	$4,565.14	$7,146.14	subtract $1,416.10	30%
	$5,844.14	$12,221.14	subtract $1,466.11	35%	$7,146.14	$12,342.14	subtract $2,185.23	35%
	$12,221.14	......	subtract $2,436.89	38.6%	$12,342.14	......	subtract $3,100.22	38.6%
4	$0	$790.52	subtract $102.00	10%	$0	$1,171.52	subtract $248.00	10%
	$790.52	$1,601.52	subtract $177.67	15%	$1,171.52	$2,444.52	subtract $402.00	15%
	$1,601.52	$2,954.52	subtract $605.37	27%	$2,444.52	$4,680.52	subtract $1,104.67	27%
	$2,954.52	$5,959.52	subtract $794.13	30%	$4,680.52	$7,261.52	subtract $1,416.10	30%
	$5,959.52	$12,336.52	subtract $1,466.11	35%	$7,261.52	$12,457.52	subtract $2,185.23	35%
	$12,336.52	......	subtract $2,436.89	38.6%	$12,457.52	......	subtract $3,100.22	38.6%
5	$0	$905.90	subtract $102.00	10%	$0	$1,286.90	subtract $248.00	10%
	$905.90	$1,716.90	subtract $177.67	15%	$1,286.90	$2,559.90	subtract $402.00	15%
	$1,716.90	$3,069.90	subtract $605.37	27%	$2,559.90	$4,795.90	subtract $1,104.67	27%
	$3,069.90	$6,074.90	subtract $794.13	30%	$4,795.90	$7,376.90	subtract $1,416.10	30%
	$6,074.90	$12,451.90	subtract $1,466.11	35%	$7,376.90	$12,572.90	subtract $2,185.23	35%
	$12,451.90	......	subtract $2,436.89	38.6%	$12,572.90	......	subtract $3,100.22	38.6%
6	$0	$1,021.28	subtract $102.00	10%	$0	$1,402.28	subtract $248.00	10%
	$1,021.28	$1,832.28	subtract $177.67	15%	$1,402.28	$2,675.28	subtract $402.00	15%
	$1,832.28	$3,185.28	subtract $605.37	27%	$2,675.28	$4,911.28	subtract $1,104.67	27%
	$3,185.28	$6,190.28	subtract $794.13	30%	$4,911.28	$7,492.28	subtract $1,416.10	30%
	$6,190.28	$12,567.28	subtract $1,466.11	35%	$7,492.28	$12,688.28	subtract $2,185.23	35%
	$12,567.28	......	subtract $2,436.89	38.6%	$12,688.28	......	subtract $3,100.22	38.6%
7	$0	$1,136.66	subtract $102.00	10%	$0	$1,517.66	subtract $248.00	10%
	$1,136.66	$1,947.66	subtract $177.67	15%	$1,517.66	$2,790.66	subtract $402.00	15%
	$1,947.66	$3,300.66	subtract $605.37	27%	$2,790.66	$5,026.66	subtract $1,104.67	27%
	$3,300.66	$6,305.66	subtract $794.13	30%	$5,026.66	$7,607.66	subtract $1,416.10	30%
	$6,305.66	$12,682.66	subtract $1,466.11	35%	$7,607.66	$12,803.66	subtract $2,185.23	35%
	$12,682.66	......	subtract $2,436.89	38.6%	$12,803.66	......	subtract $3,100.22	38.6%
8	$0	$1,252.04	subtract $102.00	10%	$0	$1,633.04	subtract $248.00	10%
	$1,252.04	$2,063.04	subtract $177.67	15%	$1,633.04	$2,906.04	subtract $402.00	15%
	$2,063.04	$3,416.04	subtract $605.37	27%	$2,906.04	$5,142.04	subtract $1,104.67	27%
	$3,416.04	$6,421.04	subtract $794.13	30%	$5,142.04	$7,723.04	subtract $1,416.10	30%
	$6,421.04	$12,798.04	subtract $1,466.11	35%	$7,723.04	$12,919.04	subtract $2,185.23	35%
	$12,798.04	......	subtract $2,436.89	38.6%	$12,919.04	......	subtract $3,100.22	38.6%
9[2]	$0	$1,367.42	subtract $102.00	10%	$0	$1,748.42	subtract $248.00	10%
	$1,367.42	$2,178.42	subtract $177.67	15%	$1,748.42	$3,021.42	subtract $402.00	15%
	$2,178.42	$3,531.42	subtract $605.37	27%	$3,021.42	$5,257.42	subtract $1,104.67	27%
	$3,531.42	$6,536.42	subtract $794.13	30%	$5,257.42	$7,838.42	subtract $1,416.10	30%
	$6,536.42	$12,913.42	subtract $1,466.11	35%	$7,838.42	$13,034.42	subtract $2,185.23	35%
	$12,913.42	......	subtract $2,436.89	38.6%	$13,034.42	......	subtract $3,100.22	38.6%

Instructions

A. For each employee, use the appropriate payroll period table and marital status section, and select the subsection showing the number of allowances claimed.

B. Read across the selected subsection and locate the bracket applicable to the employee's gross wages in columns A and B.

C. Subtract the amount shown in column C from the employee's excess wages (gross wages less amount for allowances claimed).

Caution.—The adjustment (subtraction) factors shown in this table (instruction C) do not include an amount for the number of allowances claimed by the employee on Form W-4. The amount for allowances claimed must be deducted from gross wages before withholding tax is computed.

D. Multiply the result by the withholding percentage rate shown in column D to obtain the amount of tax to be withheld.

[1] If the excess wages are less than the amount to be subtracted, the withholding is zero.

[2] You can expand these tables for additional allowances. To do this, increase the wage bracket amounts in this subsection by $115.38 for each additional allowance claimed.

Page 31

(For Wages Paid in 2002)

Wage Bracket Percentage Method Table for Computing
Income Tax Withholding From Wages Exceeding Allowance Amount

Semimonthly Payroll Period

If the number of allowances is—	Single Persons				Married Persons			
	And gross wages are—		from excess wages[1]	Multiply result by—	And gross wages are—		from excess wages[1]	Multiply result by—
	Over	But not over			Over	But not over		
	A	B	C	D	A	B	C	D
0	$0	$356.00	subtract $110.00	10%	$0	$769.00	subtract $269.00	10%
	$356.00	$1,235.00	subtract $192.00	15%	$769.00	$2,148.00	subtract $435.67	15%
	$1,235.00	$2,701.00	subtract $655.56	27%	$2,148.00	$4,571.00	subtract $1,196.70	27%
	$2,701.00	$5,956.00	subtract $860.10	30%	$4,571.00	$7,367.00	subtract $1,534.13	30%
	$5,956.00	$12,865.00	subtract $1,588.09	35%	$7,367.00	$12,996.00	subtract $2,367.40	35%
	$12,865.00	......	subtract $2,639.82	38.6%	$12,996.00	......	subtract $3,358.67	38.6%
1	$0	$481.00	subtract $110.00	10%	$0	$894.00	subtract $269.00	10%
	$481.00	$1,360.00	subtract $192.00	15%	$894.00	$2,273.00	subtract $435.67	15%
	$1,360.00	$2,826.00	subtract $655.56	27%	$2,273.00	$4,696.00	subtract $1,196.70	27%
	$2,826.00	$6,081.00	subtract $860.10	30%	$4,696.00	$7,492.00	subtract $1,534.13	30%
	$6,081.00	$12,990.00	subtract $1,588.09	35%	$7,492.00	$13,121.00	subtract $2,367.40	35%
	$12,990.00	......	subtract $2,639.82	38.6%	$13,121.00	......	subtract $3,358.67	38.6%
2	$0	$606.00	subtract $110.00	10%	$0	$1,019.00	subtract $269.00	10%
	$606.00	$1,485.00	subtract $192.00	15%	$1,019.00	$2,398.00	subtract $435.67	15%
	$1,485.00	$2,951.00	subtract $655.56	27%	$2,398.00	$4,821.00	subtract $1,196.70	27%
	$2,951.00	$6,206.00	subtract $860.10	30%	$4,821.00	$7,617.00	subtract $1,534.13	30%
	$6,206.00	$13,115.00	subtract $1,588.09	35%	$7,617.00	$13,246.00	subtract $2,367.40	35%
	$13,115.00	......	subtract $2,639.82	38.6%	$13,246.00	......	subtract $3,358.67	38.6%
3	$0	$731.00	subtract $110.00	10%	$0	$1,144.00	subtract $269.00	10%
	$731.00	$1,610.00	subtract $192.00	15%	$1,144.00	$2,523.00	subtract $435.67	15%
	$1,610.00	$3,076.00	subtract $655.56	27%	$2,523.00	$4,946.00	subtract $1,196.70	27%
	$3,076.00	$6,331.00	subtract $860.10	30%	$4,946.00	$7,742.00	subtract $1,534.13	30%
	$6,331.00	$13,240.00	subtract $1,588.09	35%	$7,742.00	$13,371.00	subtract $2,367.40	35%
	$13,240.00	......	subtract $2,639.82	38.6%	$13,371.00	......	subtract $3,358.67	38.6%
4	$0	$856.00	subtract $110.00	10%	$0	$1,269.00	subtract $269.00	10%
	$856.00	$1,735.00	subtract $192.00	15%	$1,269.00	$2,648.00	subtract $435.67	15%
	$1,735.00	$3,201.00	subtract $655.56	27%	$2,648.00	$5,071.00	subtract $1,196.70	27%
	$3,201.00	$6,456.00	subtract $860.10	30%	$5,071.00	$7,867.00	subtract $1,534.13	30%
	$6,456.00	$13,365.00	subtract $1,588.09	35%	$7,867.00	$13,496.00	subtract $2,367.40	35%
	$13,365.00	......	subtract $2,639.82	38.6%	$13,496.00	......	subtract $3,358.67	38.6%
5	$0	$981.00	subtract $110.00	10%	$0	$1,394.00	subtract $269.00	10%
	$981.00	$1,860.00	subtract $192.00	15%	$1,394.00	$2,773.00	subtract $435.67	15%
	$1,860.00	$3,326.00	subtract $655.56	27%	$2,773.00	$5,196.00	subtract $1,196.70	27%
	$3,326.00	$6,581.00	subtract $860.10	30%	$5,196.00	$7,992.00	subtract $1,534.13	30%
	$6,581.00	$13,490.00	subtract $1,588.09	35%	$7,992.00	$13,621.00	subtract $2,367.40	35%
	$13,490.00	......	subtract $2,639.82	38.6%	$13,621.00	......	subtract $3,358.67	38.6%
6	$0	$1,106.00	subtract $110.00	10%	$0	$1,519.00	subtract $269.00	10%
	$1,106.00	$1,985.00	subtract $192.00	15%	$1,519.00	$2,898.00	subtract $435.67	15%
	$1,985.00	$3,451.00	subtract $655.56	27%	$2,898.00	$5,321.00	subtract $1,196.70	27%
	$3,451.00	$6,706.00	subtract $860.10	30%	$5,321.00	$8,117.00	subtract $1,534.13	30%
	$6,706.00	$13,615.00	subtract $1,588.09	35%	$8,117.00	$13,746.00	subtract $2,367.40	35%
	$13,615.00	......	subtract $2,639.82	38.6%	$13,746.00	......	subtract $3,358.67	38.6%
7	$0	$1,231.00	subtract $110.00	10%	$0	$1,644.00	subtract $269.00	10%
	$1,231.00	$2,110.00	subtract $192.00	15%	$1,644.00	$3,023.00	subtract $435.67	15%
	$2,110.00	$3,576.00	subtract $655.56	27%	$3,023.00	$5,446.00	subtract $1,196.70	27%
	$3,576.00	$6,831.00	subtract $860.10	30%	$5,446.00	$8,242.00	subtract $1,534.13	30%
	$6,831.00	$13,740.00	subtract $1,588.09	35%	$8,242.00	$13,871.00	subtract $2,367.40	35%
	$13,740.00	......	subtract $2,639.82	38.6%	$13,871.00	......	subtract $3,358.67	38.6%
8	$0	$1,356.00	subtract $110.00	10%	$0	$1,769.00	subtract $269.00	10%
	$1,356.00	$2,235.00	subtract $192.00	15%	$1,769.00	$3,148.00	subtract $435.67	15%
	$2,235.00	$3,701.00	subtract $655.56	27%	$3,148.00	$5,571.00	subtract $1,196.70	27%
	$3,701.00	$6,956.00	subtract $860.10	30%	$5,571.00	$8,367.00	subtract $1,534.13	30%
	$6,956.00	$13,865.00	subtract $1,588.09	35%	$8,367.00	$13,996.00	subtract $2,367.40	35%
	$13,865.00	......	subtract $2,639.82	38.6%	$13,996.00	......	subtract $3,358.67	38.6%
9[2]	$0	$1,481.00	subtract $110.00	10%	$0	$1,894.00	subtract $269.00	10%
	$1,481.00	$2,360.00	subtract $192.00	15%	$1,894.00	$3,273.00	subtract $435.67	15%
	$2,360.00	$3,826.00	subtract $655.56	27%	$3,273.00	$5,696.00	subtract $1,196.70	27%
	$3,826.00	$7,081.00	subtract $860.10	30%	$5,696.00	$8,492.00	subtract $1,534.13	30%
	$7,081.00	$13,990.00	subtract $1,588.09	35%	$8,492.00	$14,121.00	subtract $2,367.40	35%
	$13,990.00	......	subtract $2,639.82	38.6%	$14,121.00	......	subtract $3,358.67	38.6%

Instructions

A. For each employee, use the appropriate payroll period table and marital status section, and select the subsection showing the number of allowances claimed.

B. Read across the selected subsection and locate the bracket applicable to the employee's gross wages in columns A and B.

C. Subtract the amount shown in column C from the employee's excess wages (gross wages less amount for allowances claimed).

Caution.—The adjustment (subtraction) factors shown in this table (instruction C) do not include an amount for the number of allowances claimed by the employee on Form W-4. The amount for allowances claimed must be deducted from gross wages before withholding tax is computed.

D. Multiply the result by the withholding percentage rate shown in column D to obtain the amount of tax to be withheld.

[1] If the excess wages are less than the amount to be subtracted, the withholding is zero.

[2] You can expand these tables for additional allowances. To do this, increase the wage bracket amounts in this subsection by $125.00 for each additional allowance claimed.

Page 32

(For Wages Paid in 2002)

Wage Bracket Percentage Method Table for Computing Income Tax Withholding From Wages Exceeding Allowance Amount

Monthly Payroll Period

If the number of allowances is—	Single Persons				Married Persons			
	And gross wages are— Over	But not over	from excess wages[1]	Multiply result by—	And gross wages are— Over	But not over	from excess wages[1]	Multiply result by—
	A	B	C	D	A	B	C	D
0	$0	$713.00	subtract $221.00	10%	$0	$1,538.00	subtract $538.00	10%
	$713.00	$2,471.00	subtract $385.00	15%	$1,538.00	$4,296.00	subtract $871.33	15%
	$2,471.00	$5,402.00	subtract $1,312.11	27%	$4,296.00	$9,142.00	subtract $2,393.41	27%
	$5,402.00	$11,913.00	subtract $1,721.10	30%	$9,142.00	$14,733.00	subtract $3,068.27	30%
	$11,913.00	$25,729.00	subtract $3,177.09	35%	$14,733.00	$25,992.00	subtract $4,734.66	35%
	$25,729.00	......	subtract $5,280.37	38.6%	$25,992.00	......	subtract $6,717.21	38.6%
1	$0	$963.00	subtract $221.00	10%	$0	$1,788.00	subtract $538.00	10%
	$963.00	$2,721.00	subtract $385.00	15%	$1,788.00	$4,546.00	subtract $871.33	15%
	$2,721.00	$5,652.00	subtract $1,312.11	27%	$4,546.00	$9,392.00	subtract $2,393.41	27%
	$5,652.00	$12,163.00	subtract $1,721.10	30%	$9,392.00	$14,983.00	subtract $3,068.27	30%
	$12,163.00	$25,979.00	subtract $3,177.09	35%	$14,983.00	$26,242.00	subtract $4,734.66	35%
	$25,979.00	......	subtract $5,280.37	38.6%	$26,242.00	......	subtract $6,717.21	38.6%
2	$0	$1,213.00	subtract $221.00	10%	$0	$2,038.00	subtract $538.00	10%
	$1,213.00	$2,971.00	subtract $385.00	15%	$2,038.00	$4,796.00	subtract $871.33	15%
	$2,971.00	$5,902.00	subtract $1,312.11	27%	$4,796.00	$9,642.00	subtract $2,393.41	27%
	$5,902.00	$12,413.00	subtract $1,721.10	30%	$9,642.00	$15,233.00	subtract $3,068.27	30%
	$12,413.00	$26,229.00	subtract $3,177.09	35%	$15,233.00	$26,492.00	subtract $4,734.66	35%
	$26,229.00	......	subtract $5,280.37	38.6%	$26,492.00	......	subtract $6,717.21	38.6%
3	$0	$1,463.00	subtract $221.00	10%	$0	$2,288.00	subtract $538.00	10%
	$1,463.00	$3,221.00	subtract $385.00	15%	$2,288.00	$5,046.00	subtract $871.33	15%
	$3,221.00	$6,152.00	subtract $1,312.11	27%	$5,046.00	$9,892.00	subtract $2,393.41	27%
	$6,152.00	$12,663.00	subtract $1,721.10	30%	$9,892.00	$15,483.00	subtract $3,068.27	30%
	$12,663.00	$26,479.00	subtract $3,177.09	35%	$15,483.00	$26,742.00	subtract $4,734.66	35%
	$26,479.00	......	subtract $5,280.37	38.6%	$26,742.00	......	subtract $6,717.21	38.6%
4	$0	$1,713.00	subtract $221.00	10%	$0	$2,538.00	subtract $538.00	10%
	$1,713.00	$3,471.00	subtract $385.00	15%	$2,538.00	$5,296.00	subtract $871.33	15%
	$3,471.00	$6,402.00	subtract $1,312.11	27%	$5,296.00	$10,142.00	subtract $2,393.41	27%
	$6,402.00	$12,913.00	subtract $1,721.10	30%	$10,142.00	$15,733.00	subtract $3,068.27	30%
	$12,913.00	$26,729.00	subtract $3,177.09	35%	$15,733.00	$26,992.00	subtract $4,734.66	35%
	$26,729.00	......	subtract $5,280.37	38.6%	$26,992.00	......	subtract $6,717.21	38.6%
5	$0	$1,963.00	subtract $221.00	10%	$0	$2,788.00	subtract $538.00	10%
	$1,963.00	$3,721.00	subtract $385.00	15%	$2,788.00	$5,546.00	subtract $871.33	15%
	$3,721.00	$6,652.00	subtract $1,312.11	27%	$5,546.00	$10,392.00	subtract $2,393.41	27%
	$6,652.00	$13,163.00	subtract $1,721.10	30%	$10,392.00	$15,983.00	subtract $3,068.27	30%
	$13,163.00	$26,979.00	subtract $3,177.09	35%	$15,983.00	$27,242.00	subtract $4,734.66	35%
	$26,979.00	......	subtract $5,280.37	38.6%	$27,242.00	......	subtract $6,717.21	38.6%
6	$0	$2,213.00	subtract $221.00	10%	$0	$3,038.00	subtract $538.00	10%
	$2,213.00	$3,971.00	subtract $385.00	15%	$3,038.00	$5,796.00	subtract $871.33	15%
	$3,971.00	$6,902.00	subtract $1,312.11	27%	$5,796.00	$10,642.00	subtract $2,393.41	27%
	$6,902.00	$13,413.00	subtract $1,721.10	30%	$10,642.00	$16,233.00	subtract $3,068.27	30%
	$13,413.00	$27,229.00	subtract $3,177.09	35%	$16,233.00	$27,492.00	subtract $4,734.66	35%
	$27,229.00	......	subtract $5,280.37	38.6%	$27,492.00	......	subtract $6,717.21	38.6%
7	$0	$2,463.00	subtract $221.00	10%	$0	$3,288.00	subtract $538.00	10%
	$2,463.00	$4,221.00	subtract $385.00	15%	$3,288.00	$6,046.00	subtract $871.33	15%
	$4,221.00	$7,152.00	subtract $1,312.11	27%	$6,046.00	$10,892.00	subtract $2,393.41	27%
	$7,152.00	$13,663.00	subtract $1,721.10	30%	$10,892.00	$16,483.00	subtract $3,068.27	30%
	$13,663.00	$27,479.00	subtract $3,177.09	35%	$16,483.00	$27,742.00	subtract $4,734.66	35%
	$27,479.00	......	subtract $5,280.37	38.6%	$27,742.00	......	subtract $6,717.21	38.6%
8	$0	$2,713.00	subtract $221.00	10%	$0	$3,538.00	subtract $538.00	10%
	$2,713.00	$4,471.00	subtract $385.00	15%	$3,538.00	$6,296.00	subtract $871.33	15%
	$4,471.00	$7,402.00	subtract $1,312.11	27%	$6,296.00	$11,142.00	subtract $2,393.41	27%
	$7,402.00	$13,913.00	subtract $1,721.10	30%	$11,142.00	$16,733.00	subtract $3,068.27	30%
	$13,913.00	$27,729.00	subtract $3,177.09	35%	$16,733.00	$27,992.00	subtract $4,734.66	35%
	$27,729.00	......	subtract $5,280.37	38.6%	$27,992.00	......	subtract $6,717.21	38.6%
9[2]	$0	$2,963.00	subtract $221.00	10%	$0	$3,788.00	subtract $538.00	10%
	$2,963.00	$4,721.00	subtract $385.00	15%	$3,788.00	$6,546.00	subtract $871.33	15%
	$4,721.00	$7,652.00	subtract $1,312.11	27%	$6,546.00	$11,392.00	subtract $2,393.41	27%
	$7,652.00	$14,163.00	subtract $1,721.10	30%	$11,392.00	$16,983.00	subtract $3,068.27	30%
	$14,163.00	$27,979.00	subtract $3,177.09	35%	$16,983.00	$28,242.00	subtract $4,734.66	35%
	$27,979.00	......	subtract $5,280.37	38.6%	$28,242.00	......	subtract $6,717.21	38.6%

Instructions

A. For each employee, use the appropriate payroll period table and marital status section, and select the subsection showing the number of allowances claimed.

B. Read across the selected subsection and locate the bracket applicable to the employee's gross wages in columns A and B.

C. Subtract the amount shown in column C from the employee's excess wages (gross wages less amount for allowances claimed).

Caution.—The adjustment (subtraction) factors shown in this table (instruction C) do not include an amount for the number of allowances claimed by the employee on Form W-4. The amount for allowances claimed must be deducted from gross wages before withholding tax is computed.

D. Multiply the result by the withholding percentage rate shown in column D to obtain the amount of tax to be withheld.

[1] If the excess wages are less than the amount to be subtracted, the withholding is zero.

[2] You can expand these tables for additional allowances. To do this, increase the wage bracket amounts in this subsection by $250.00 for each additional allowance claimed.

Page 33

Combined Income Tax, Employee Social Security Tax, and Employee Medicare Tax Withholding Tables

If you want to combine amounts to be withheld as income tax, employee social security tax, and employee Medicare tax, you may use the combined tables on pages 35–54.

Combined withholding tables for single and married taxpayers are shown for weekly, biweekly, semimonthly, monthly, and daily or miscellaneous payroll periods. The payroll period and marital status of the employee determine the table to be used.

If the wages are greater than the highest wage bracket in the applicable table, you will have to use one of the other methods for figuring income tax withholding described in this publication or in Circular E. For wages that do not exceed $84,900, the combined social security tax rate and Medicare tax rate is 7.65% each for both the employee and the employer for wages paid in 2002. You can figure the employee social security tax by multiplying the wages by 6.2%, and you can figure the employee Medicare tax by multiplying the wages by 1.45%.

The combined tables give the correct total withholding only if wages for social security and Medicare taxes and income tax withholding are the same. When you have paid more than the maximum amount of wages subject to social security tax ($84,900 in 2002) in a calendar year, you may no longer use the combined tables.

If you use the combined withholding tables, use the following steps to find the amounts to report on your **Form 941,** Employer's Quarterly Federal Tax Return.

1) Employee social security tax withheld. Multiply the wages by 6.2%.
2) Employee Medicare tax withheld. Multiply the wages by 1.45%.
3) Income tax withheld. Subtract the amounts from steps 1 and 2 from the total tax withheld.

You can figure the amounts to be shown on **Form W-2,** Wage and Tax Statement, in the same way.

SINGLE Persons—**WEEKLY** Payroll Period

(For Wages Paid in 2002)

And the wages are–		And the number of withholding allowances claimed is—										
At least	But less than	0	1	2	3	4	5	6	7	8	9	10
		The amount of income, social security, and Medicare taxes to be withheld is—										
$0	**$55**	7.65%	7.65%	7.65%	7.65%	7.65%	7.65%	7.65%	7.65%	7.65%	7.65%	7.65%
55	**60**	$5.40	$4.40	$4.40	$4.40	$4.40	$4.40	$4.40	$4.40	$4.40	$4.40	$4.40
60	**65**	5.78	4.78	4.78	4.78	4.78	4.78	4.78	4.78	4.78	4.78	4.78
65	**70**	7.16	5.16	5.16	5.16	5.16	5.16	5.16	5.16	5.16	5.16	5.16
70	**75**	7.55	5.55	5.55	5.55	5.55	5.55	5.55	5.55	5.55	5.55	5.55
75	**80**	8.93	5.93	5.93	5.93	5.93	5.93	5.93	5.93	5.93	5.93	5.93
80	**85**	9.31	6.31	6.31	6.31	6.31	6.31	6.31	6.31	6.31	6.31	6.31
85	**90**	10.69	6.69	6.69	6.69	6.69	6.69	6.69	6.69	6.69	6.69	6.69
90	**95**	11.08	7.08	7.08	7.08	7.08	7.08	7.08	7.08	7.08	7.08	7.08
95	**100**	12.46	7.46	7.46	7.46	7.46	7.46	7.46	7.46	7.46	7.46	7.46
100	**105**	12.84	7.84	7.84	7.84	7.84	7.84	7.84	7.84	7.84	7.84	7.84
105	**110**	14.22	8.22	8.22	8.22	8.22	8.22	8.22	8.22	8.22	8.22	8.22
110	**115**	14.61	8.61	8.61	8.61	8.61	8.61	8.61	8.61	8.61	8.61	8.61
115	**120**	15.99	9.99	8.99	8.99	8.99	8.99	8.99	8.99	8.99	8.99	8.99
120	**125**	16.37	10.37	9.37	9.37	9.37	9.37	9.37	9.37	9.37	9.37	9.37
125	**130**	17.75	11.75	9.75	9.75	9.75	9.75	9.75	9.75	9.75	9.75	9.75
130	**135**	18.14	12.14	10.14	10.14	10.14	10.14	10.14	10.14	10.14	10.14	10.14
135	**140**	19.52	13.52	10.52	10.52	10.52	10.52	10.52	10.52	10.52	10.52	10.52
140	**145**	19.90	13.90	10.90	10.90	10.90	10.90	10.90	10.90	10.90	10.90	10.90
145	**150**	21.28	15.28	11.28	11.28	11.28	11.28	11.28	11.28	11.28	11.28	11.28
150	**155**	21.67	15.67	11.67	11.67	11.67	11.67	11.67	11.67	11.67	11.67	11.67
155	**160**	23.05	17.05	12.05	12.05	12.05	12.05	12.05	12.05	12.05	12.05	12.05
160	**165**	23.43	17.43	12.43	12.43	12.43	12.43	12.43	12.43	12.43	12.43	12.43
165	**170**	24.81	18.81	12.81	12.81	12.81	12.81	12.81	12.81	12.81	12.81	12.81
170	**175**	26.20	19.20	14.20	13.20	13.20	13.20	13.20	13.20	13.20	13.20	13.20
175	**180**	26.58	20.58	14.58	13.58	13.58	13.58	13.58	13.58	13.58	13.58	13.58
180	**185**	27.96	20.96	15.96	13.96	13.96	13.96	13.96	13.96	13.96	13.96	13.96
185	**190**	29.34	22.34	16.34	14.34	14.34	14.34	14.34	14.34	14.34	14.34	14.34
190	**195**	30.73	22.73	17.73	14.73	14.73	14.73	14.73	14.73	14.73	14.73	14.73
195	**200**	31.11	24.11	18.11	15.11	15.11	15.11	15.11	15.11	15.11	15.11	15.11
200	**210**	32.68	25.68	19.68	15.68	15.68	15.68	15.68	15.68	15.68	15.68	15.68
210	**220**	35.45	27.45	21.45	16.45	16.45	16.45	16.45	16.45	16.45	16.45	16.45
220	**230**	37.21	29.21	23.21	17.21	17.21	17.21	17.21	17.21	17.21	17.21	17.21
230	**240**	39.98	30.98	24.98	18.98	17.98	17.98	17.98	17.98	17.98	17.98	17.98
240	**250**	41.74	33.74	26.74	20.74	18.74	18.74	18.74	18.74	18.74	18.74	18.74
250	**260**	44.51	35.51	28.51	22.51	19.51	19.51	19.51	19.51	19.51	19.51	19.51
260	**270**	46.27	38.27	30.27	24.27	20.27	20.27	20.27	20.27	20.27	20.27	20.27
270	**280**	49.04	40.04	32.04	26.04	21.04	21.04	21.04	21.04	21.04	21.04	21.04
280	**290**	50.80	42.80	33.80	27.80	21.80	21.80	21.80	21.80	21.80	21.80	21.80
290	**300**	53.57	44.57	36.57	29.57	23.57	22.57	22.57	22.57	22.57	22.57	22.57
300	**310**	55.33	47.33	38.33	31.33	25.33	23.33	23.33	23.33	23.33	23.33	23.33
310	**320**	58.10	49.10	41.10	33.10	27.10	24.10	24.10	24.10	24.10	24.10	24.10
320	**330**	59.86	51.86	42.86	34.86	28.86	24.86	24.86	24.86	24.86	24.86	24.86
330	**340**	62.63	53.63	45.63	36.63	30.63	25.63	25.63	25.63	25.63	25.63	25.63
340	**350**	64.39	56.39	47.39	38.39	32.39	27.39	26.39	26.39	26.39	26.39	26.39
350	**360**	67.16	58.16	50.16	41.16	34.16	29.16	27.16	27.16	27.16	27.16	27.16
360	**370**	68.92	60.92	51.92	42.92	35.92	30.92	27.92	27.92	27.92	27.92	27.92
370	**380**	71.69	62.69	54.69	45.69	37.69	32.69	28.69	28.69	28.69	28.69	28.69
380	**390**	73.45	65.45	56.45	47.45	39.45	34.45	29.45	29.45	29.45	29.45	29.45
390	**400**	76.22	67.22	59.22	50.22	41.22	36.22	30.22	30.22	30.22	30.22	30.22
400	**410**	77.98	69.98	60.98	51.98	43.98	37.98	31.98	30.98	30.98	30.98	30.98
410	**420**	80.75	71.75	63.75	54.75	45.75	39.75	33.75	31.75	31.75	31.75	31.75
420	**430**	82.51	74.51	65.51	56.51	48.51	41.51	35.51	32.51	32.51	32.51	32.51
430	**440**	85.28	76.28	68.28	59.28	50.28	43.28	37.28	33.28	33.28	33.28	33.28
440	**450**	87.04	79.04	70.04	61.04	53.04	45.04	39.04	34.04	34.04	34.04	34.04
450	**460**	89.81	80.81	72.81	63.81	54.81	46.81	40.81	34.81	34.81	34.81	34.81
460	**470**	91.57	83.57	74.57	65.57	57.57	48.57	42.57	36.57	35.57	35.57	35.57
470	**480**	94.34	85.34	77.34	68.34	59.34	51.34	44.34	38.34	36.34	36.34	36.34
480	**490**	96.10	88.10	79.10	70.10	62.10	53.10	46.10	40.10	37.10	37.10	37.10
490	**500**	98.87	89.87	81.87	72.87	63.87	55.87	47.87	41.87	37.87	37.87	37.87
500	**510**	100.63	92.63	83.63	74.63	66.63	57.63	49.63	43.63	38.63	38.63	38.63
510	**520**	103.40	94.40	86.40	77.40	68.40	60.40	51.40	45.40	39.40	39.40	39.40
520	**530**	105.16	97.16	88.16	79.16	71.16	62.16	54.16	47.16	41.16	40.16	40.16
530	**540**	107.93	98.93	90.93	81.93	72.93	64.93	55.93	48.93	42.93	40.93	40.93
540	**550**	109.69	101.69	92.69	83.69	75.69	66.69	58.69	50.69	44.69	41.69	41.69
550	**560**	112.46	103.46	95.46	86.46	77.46	69.46	60.46	52.46	46.46	42.46	42.46
560	**570**	114.22	106.22	97.22	88.22	80.22	71.22	63.22	54.22	48.22	43.22	43.22
570	**580**	117.99	107.99	99.99	90.99	81.99	73.99	64.99	55.99	49.99	43.99	43.99
580	**590**	120.75	110.75	101.75	92.75	84.75	75.75	67.75	58.75	51.75	45.75	44.75
590	**600**	124.52	112.52	104.52	95.52	86.52	78.52	69.52	60.52	53.52	47.52	45.52

(Continued on next page)

SINGLE Persons—**WEEKLY** Payroll Period

(For Wages Paid in 2002)

And the wages are–		And the number of withholding allowances claimed is—										
At least	But less than	0	1	2	3	4	5	6	7	8	9	10
		The amount of income, social security, and Medicare taxes to be withheld is—										
$600	**$610**	$128.28	$115.28	$106.28	$97.28	$89.28	$80.28	$72.28	$63.28	$55.28	$49.28	$46.28
610	**620**	131.05	117.05	109.05	100.05	91.05	83.05	74.05	65.05	57.05	51.05	47.05
620	**630**	134.81	119.81	110.81	101.81	93.81	84.81	76.81	67.81	58.81	52.81	47.81
630	**640**	138.58	122.58	113.58	104.58	95.58	87.58	78.58	69.58	61.58	54.58	49.58
640	**650**	141.34	126.34	115.34	106.34	98.34	89.34	81.34	72.34	63.34	56.34	51.34
650	**660**	145.11	130.11	118.11	109.11	100.11	92.11	83.11	74.11	66.11	58.11	53.11
660	**670**	148.87	132.87	119.87	110.87	102.87	93.87	85.87	76.87	67.87	59.87	54.87
670	**680**	152.64	136.64	122.64	113.64	104.64	96.64	87.64	78.64	70.64	61.64	56.64
680	**690**	155.40	140.40	124.40	115.40	107.40	98.40	90.40	81.40	72.40	64.40	58.40
690	**700**	159.17	143.17	128.17	118.17	109.17	101.17	92.17	83.17	75.17	66.17	60.17
700	**710**	162.93	146.93	130.93	119.93	111.93	102.93	94.93	85.93	76.93	68.93	61.93
710	**720**	165.70	150.70	134.70	122.70	113.70	105.70	96.70	87.70	79.70	70.70	63.70
720	**730**	169.46	153.46	138.46	124.46	116.46	107.46	99.46	90.46	81.46	73.46	65.46
730	**740**	173.23	157.23	142.23	127.23	118.23	110.23	101.23	92.23	84.23	75.23	67.23
740	**750**	175.99	160.99	144.99	129.99	120.99	111.99	103.99	94.99	85.99	77.99	68.99
750	**760**	179.76	164.76	148.76	132.76	122.76	114.76	105.76	96.76	88.76	79.76	70.76
760	**770**	183.52	167.52	152.52	136.52	125.52	116.52	108.52	99.52	90.52	82.52	73.52
770	**780**	187.29	171.29	155.29	140.29	127.29	119.29	110.29	101.29	93.29	84.29	75.29
780	**790**	190.05	175.05	159.05	143.05	130.05	121.05	113.05	104.05	95.05	87.05	78.05
790	**800**	193.82	177.82	162.82	146.82	131.82	123.82	114.82	105.82	97.82	88.82	79.82
800	**810**	197.58	181.58	165.58	150.58	134.58	125.58	117.58	108.58	99.58	91.58	82.58
810	**820**	200.35	185.35	169.35	154.35	138.35	128.35	119.35	110.35	102.35	93.35	84.35
820	**830**	204.11	188.11	173.11	157.11	142.11	130.11	122.11	113.11	104.11	96.11	87.11
830	**840**	207.88	191.88	176.88	160.88	144.88	132.88	123.88	114.88	106.88	97.88	88.88
840	**850**	210.64	195.64	179.64	164.64	148.64	134.64	126.64	117.64	108.64	100.64	91.64
850	**860**	214.41	199.41	183.41	167.41	152.41	137.41	128.41	119.41	111.41	102.41	93.41
860	**870**	218.17	202.17	187.17	171.17	156.17	140.17	131.17	122.17	113.17	105.17	96.17
870	**880**	221.94	205.94	189.94	174.94	158.94	143.94	132.94	123.94	115.94	106.94	97.94
880	**890**	224.70	209.70	193.70	177.70	162.70	146.70	135.70	126.70	117.70	109.70	100.70
890	**900**	228.47	212.47	197.47	181.47	166.47	150.47	137.47	128.47	120.47	111.47	102.47
900	**910**	232.23	216.23	200.23	185.23	169.23	154.23	140.23	131.23	122.23	114.23	105.23
910	**920**	235.00	220.00	204.00	189.00	173.00	157.00	142.00	133.00	125.00	116.00	107.00
920	**930**	238.76	222.76	207.76	191.76	176.76	160.76	145.76	135.76	126.76	118.76	109.76
930	**940**	242.53	226.53	211.53	195.53	179.53	164.53	148.53	137.53	129.53	120.53	111.53
940	**950**	245.29	230.29	214.29	199.29	183.29	168.29	152.29	140.29	131.29	123.29	114.29
950	**960**	249.06	234.06	218.06	202.06	187.06	171.06	156.06	142.06	134.06	125.06	116.06
960	**970**	252.82	236.82	221.82	205.82	190.82	174.82	158.82	144.82	135.82	127.82	118.82
970	**980**	256.59	240.59	224.59	209.59	193.59	178.59	162.59	146.59	138.59	129.59	120.59
980	**990**	259.35	244.35	228.35	212.35	197.35	181.35	166.35	150.35	140.35	132.35	123.35
990	**1,000**	263.12	247.12	232.12	216.12	201.12	185.12	169.12	154.12	143.12	134.12	125.12
1,000	**1,010**	266.88	250.88	234.88	219.88	203.88	188.88	172.88	157.88	144.88	136.88	127.88
1,010	**1,020**	269.65	254.65	238.65	223.65	207.65	191.65	176.65	160.65	147.65	138.65	129.65
1,020	**1,030**	273.41	257.41	242.41	226.41	211.41	195.41	180.41	164.41	149.41	141.41	132.41
1,030	**1,040**	277.18	261.18	246.18	230.18	214.18	199.18	183.18	168.18	152.18	143.18	134.18
1,040	**1,050**	279.94	264.94	248.94	233.94	217.94	202.94	186.94	170.94	155.94	145.94	136.94
1,050	**1,060**	283.71	268.71	252.71	236.71	221.71	205.71	190.71	174.71	158.71	147.71	138.71
1,060	**1,070**	287.47	271.47	256.47	240.47	225.47	209.47	193.47	178.47	162.47	150.47	141.47
1,070	**1,080**	291.24	275.24	259.24	244.24	228.24	213.24	197.24	181.24	166.24	152.24	143.24
1,080	**1,090**	294.00	279.00	263.00	247.00	232.00	216.00	201.00	185.00	170.00	155.00	146.00
1,090	**1,100**	297.77	281.77	266.77	250.77	235.77	219.77	203.77	188.77	172.77	157.77	147.77
1,100	**1,110**	301.53	285.53	269.53	254.53	238.53	223.53	207.53	192.53	176.53	160.53	150.53
1,110	**1,120**	304.30	289.30	273.30	258.30	242.30	226.30	211.30	195.30	180.30	164.30	152.30
1,120	**1,130**	308.06	292.06	277.06	261.06	246.06	230.06	215.06	199.06	183.06	168.06	155.06
1,130	**1,140**	311.83	295.83	280.83	264.83	248.83	233.83	217.83	202.83	186.83	171.83	156.83
1,140	**1,150**	314.59	299.59	283.59	268.59	252.59	237.59	221.59	205.59	190.59	174.59	159.59
1,150	**1,160**	318.36	303.36	287.36	271.36	256.36	240.36	225.36	209.36	193.36	178.36	162.36
1,160	**1,170**	322.12	306.12	291.12	275.12	260.12	244.12	228.12	213.12	197.12	182.12	166.12
1,170	**1,180**	325.89	309.89	293.89	278.89	262.89	247.89	231.89	215.89	200.89	184.89	169.89
1,180	**1,190**	328.65	313.65	297.65	281.65	266.65	250.65	235.65	219.65	204.65	188.65	172.65
1,190	**1,200**	332.42	316.42	301.42	285.42	270.42	254.42	238.42	223.42	207.42	192.42	176.42
1,200	**1,210**	336.18	320.18	304.18	289.18	273.18	258.18	242.18	227.18	211.18	195.18	180.18
1,210	**1,220**	338.95	323.95	307.95	292.95	276.95	260.95	245.95	229.95	214.95	198.95	183.95
1,220	**1,230**	342.71	326.71	311.71	295.71	280.71	264.71	249.71	233.71	217.71	202.71	186.71
1,230	**1,240**	346.48	330.48	315.48	299.48	283.48	268.48	252.48	237.48	221.48	206.48	190.48
1,240	**1,250**	349.24	334.24	318.24	303.24	287.24	272.24	256.24	240.24	225.24	209.24	194.24

$1,250 and over — Do not use this table. See page 34 for instructions.

Page 36

APPENDIX A-1

MARRIED Persons—**WEEKLY** Payroll Period

(For Wages Paid in 2002)

And the wages are–		And the number of withholding allowances claimed is—										
At least	But less than	0	1	2	3	4	5	6	7	8	9	10
		The amount of income, social security, and Medicare taxes to be withheld is—										
$0	**$125**	7.65%	7.65%	7.65%	7.65%	7.65%	7.65%	7.65%	7.65%	7.65%	7.65%	7.65%
125	**130**	$9.75	$9.75	$9.75	$9.75	$9.75	$9.75	$9.75	$9.75	$9.75	$9.75	$9.75
130	**135**	11.14	10.14	10.14	10.14	10.14	10.14	10.14	10.14	10.14	10.14	10.14
135	**140**	11.52	10.52	10.52	10.52	10.52	10.52	10.52	10.52	10.52	10.52	10.52
140	**145**	12.90	10.90	10.90	10.90	10.90	10.90	10.90	10.90	10.90	10.90	10.90
145	**150**	13.28	11.28	11.28	11.28	11.28	11.28	11.28	11.28	11.28	11.28	11.28
150	**155**	14.67	11.67	11.67	11.67	11.67	11.67	11.67	11.67	11.67	11.67	11.67
155	**160**	15.05	12.05	12.05	12.05	12.05	12.05	12.05	12.05	12.05	12.05	12.05
160	**165**	16.43	12.43	12.43	12.43	12.43	12.43	12.43	12.43	12.43	12.43	12.43
165	**170**	16.81	12.81	12.81	12.81	12.81	12.81	12.81	12.81	12.81	12.81	12.81
170	**175**	18.20	13.20	13.20	13.20	13.20	13.20	13.20	13.20	13.20	13.20	13.20
175	**180**	18.58	13.58	13.58	13.58	13.58	13.58	13.58	13.58	13.58	13.58	13.58
180	**185**	19.96	13.96	13.96	13.96	13.96	13.96	13.96	13.96	13.96	13.96	13.96
185	**190**	20.34	15.34	14.34	14.34	14.34	14.34	14.34	14.34	14.34	14.34	14.34
190	**195**	21.73	15.73	14.73	14.73	14.73	14.73	14.73	14.73	14.73	14.73	14.73
195	**200**	22.11	17.11	15.11	15.11	15.11	15.11	15.11	15.11	15.11	15.11	15.11
200	**210**	23.68	17.68	15.68	15.68	15.68	15.68	15.68	15.68	15.68	15.68	15.68
210	**220**	25.45	19.45	16.45	16.45	16.45	16.45	16.45	16.45	16.45	16.45	16.45
220	**230**	27.21	21.21	17.21	17.21	17.21	17.21	17.21	17.21	17.21	17.21	17.21
230	**240**	28.98	22.98	17.98	17.98	17.98	17.98	17.98	17.98	17.98	17.98	17.98
240	**250**	30.74	24.74	19.74	18.74	18.74	18.74	18.74	18.74	18.74	18.74	18.74
250	**260**	32.51	26.51	21.51	19.51	19.51	19.51	19.51	19.51	19.51	19.51	19.51
260	**270**	34.27	28.27	23.27	20.27	20.27	20.27	20.27	20.27	20.27	20.27	20.27
270	**280**	36.04	30.04	25.04	21.04	21.04	21.04	21.04	21.04	21.04	21.04	21.04
280	**290**	37.80	31.80	26.80	21.80	21.80	21.80	21.80	21.80	21.80	21.80	21.80
290	**300**	39.57	33.57	28.57	22.57	22.57	22.57	22.57	22.57	22.57	22.57	22.57
300	**310**	41.33	35.33	30.33	24.33	23.33	23.33	23.33	23.33	23.33	23.33	23.33
310	**320**	43.10	37.10	32.10	26.10	24.10	24.10	24.10	24.10	24.10	24.10	24.10
320	**330**	44.86	38.86	33.86	27.86	24.86	24.86	24.86	24.86	24.86	24.86	24.86
330	**340**	46.63	40.63	35.63	29.63	25.63	25.63	25.63	25.63	25.63	25.63	25.63
340	**350**	48.39	42.39	37.39	31.39	26.39	26.39	26.39	26.39	26.39	26.39	26.39
350	**360**	50.16	44.16	39.16	33.16	27.16	27.16	27.16	27.16	27.16	27.16	27.16
360	**370**	52.92	45.92	40.92	34.92	28.92	27.92	27.92	27.92	27.92	27.92	27.92
370	**380**	54.69	47.69	42.69	36.69	30.69	28.69	28.69	28.69	28.69	28.69	28.69
380	**390**	57.45	49.45	44.45	38.45	32.45	29.45	29.45	29.45	29.45	29.45	29.45
390	**400**	59.22	51.22	46.22	40.22	34.22	30.22	30.22	30.22	30.22	30.22	30.22
400	**410**	61.98	52.98	47.98	41.98	35.98	30.98	30.98	30.98	30.98	30.98	30.98
410	**420**	63.75	54.75	49.75	43.75	37.75	31.75	31.75	31.75	31.75	31.75	31.75
420	**430**	66.51	57.51	51.51	45.51	39.51	33.51	32.51	32.51	32.51	32.51	32.51
430	**440**	68.28	59.28	53.28	47.28	41.28	35.28	33.28	33.28	33.28	33.28	33.28
440	**450**	71.04	62.04	55.04	49.04	43.04	37.04	34.04	34.04	34.04	34.04	34.04
450	**460**	72.81	63.81	56.81	50.81	44.81	38.81	34.81	34.81	34.81	34.81	34.81
460	**470**	75.57	66.57	58.57	52.57	46.57	40.57	35.57	35.57	35.57	35.57	35.57
470	**480**	77.34	68.34	60.34	54.34	48.34	42.34	36.34	36.34	36.34	36.34	36.34
480	**490**	80.10	71.10	62.10	56.10	50.10	44.10	38.10	37.10	37.10	37.10	37.10
490	**500**	81.87	72.87	64.87	57.87	51.87	45.87	39.87	37.87	37.87	37.87	37.87
500	**510**	84.63	75.63	66.63	59.63	53.63	47.63	41.63	38.63	38.63	38.63	38.63
510	**520**	86.40	77.40	69.40	61.40	55.40	49.40	43.40	39.40	39.40	39.40	39.40
520	**530**	89.16	80.16	71.16	63.16	57.16	51.16	45.16	40.16	40.16	40.16	40.16
530	**540**	90.93	81.93	73.93	64.93	58.93	52.93	46.93	41.93	40.93	40.93	40.93
540	**550**	93.69	84.69	75.69	67.69	60.69	54.69	48.69	43.69	41.69	41.69	41.69
550	**560**	95.46	86.46	78.46	69.46	62.46	56.46	50.46	45.46	42.46	42.46	42.46
560	**570**	98.22	89.22	80.22	72.22	64.22	58.22	52.22	47.22	43.22	43.22	43.22
570	**580**	99.99	90.99	82.99	73.99	65.99	59.99	53.99	48.99	43.99	43.99	43.99
580	**590**	102.75	93.75	84.75	76.75	67.75	61.75	55.75	50.75	44.75	44.75	44.75
590	**600**	104.52	95.52	87.52	78.52	69.52	63.52	57.52	52.52	46.52	45.52	45.52
600	**610**	107.28	98.28	89.28	81.28	72.28	65.28	59.28	54.28	48.28	46.28	46.28
610	**620**	109.05	100.05	92.05	83.05	74.05	67.05	61.05	56.05	50.05	47.05	47.05
620	**630**	111.81	102.81	93.81	85.81	76.81	68.81	62.81	57.81	51.81	47.81	47.81
630	**640**	113.58	104.58	96.58	87.58	78.58	70.58	64.58	59.58	53.58	48.58	48.58
640	**650**	116.34	107.34	98.34	90.34	81.34	72.34	66.34	61.34	55.34	49.34	49.34
650	**660**	118.11	109.11	101.11	92.11	83.11	75.11	68.11	63.11	57.11	51.11	50.11
660	**670**	120.87	111.87	102.87	94.87	85.87	76.87	69.87	64.87	58.87	52.87	50.87
670	**680**	122.64	113.64	105.64	96.64	87.64	79.64	71.64	66.64	60.64	54.64	51.64
680	**690**	125.40	116.40	107.40	99.40	90.40	81.40	73.40	68.40	62.40	56.40	52.40
690	**700**	127.17	118.17	110.17	101.17	92.17	84.17	75.17	70.17	64.17	58.17	53.17
700	**710**	129.93	120.93	111.93	103.93	94.93	85.93	77.93	71.93	65.93	59.93	53.93
710	**720**	131.70	122.70	114.70	105.70	96.70	88.70	79.70	73.70	67.70	61.70	55.70
720	**730**	134.46	125.46	116.46	108.46	99.46	90.46	82.46	75.46	69.46	63.46	57.46
730	**740**	136.23	127.23	119.23	110.23	101.23	93.23	84.23	77.23	71.23	65.23	59.23

(Continued on next page)

MARRIED Persons—**WEEKLY** Payroll Period

(For Wages Paid in 2002)

And the wages are–		And the number of withholding allowances claimed is—										
At least	But less than	0	1	2	3	4	5	6	7	8	9	10
		The amount of income, social security, and Medicare taxes to be withheld is—										
$740	**$750**	$138.99	$129.99	$120.99	$112.99	$103.99	$94.99	$86.99	$78.99	$72.99	$66.99	$60.99
750	**760**	140.76	131.76	123.76	114.76	105.76	97.76	88.76	80.76	74.76	68.76	62.76
760	**770**	143.52	134.52	125.52	117.52	108.52	99.52	91.52	82.52	76.52	70.52	64.52
770	**780**	145.29	136.29	128.29	119.29	110.29	102.29	93.29	85.29	78.29	72.29	66.29
780	**790**	148.05	139.05	130.05	122.05	113.05	104.05	96.05	87.05	80.05	74.05	68.05
790	**800**	149.82	140.82	132.82	123.82	114.82	106.82	97.82	89.82	81.82	75.82	69.82
800	**810**	152.58	143.58	134.58	126.58	117.58	108.58	100.58	91.58	83.58	77.58	71.58
810	**820**	154.35	145.35	137.35	128.35	119.35	111.35	102.35	94.35	85.35	79.35	73.35
820	**830**	157.11	148.11	139.11	131.11	122.11	113.11	105.11	96.11	87.11	81.11	75.11
830	**840**	158.88	149.88	141.88	132.88	123.88	115.88	106.88	98.88	89.88	82.88	76.88
840	**850**	161.64	152.64	143.64	135.64	126.64	117.64	109.64	100.64	91.64	84.64	78.64
850	**860**	163.41	154.41	146.41	137.41	128.41	120.41	111.41	103.41	94.41	86.41	80.41
860	**870**	166.17	157.17	148.17	140.17	131.17	122.17	114.17	105.17	96.17	88.17	82.17
870	**880**	167.94	158.94	150.94	141.94	132.94	124.94	115.94	107.94	98.94	89.94	83.94
880	**890**	170.70	161.70	152.70	144.70	135.70	126.70	118.70	109.70	100.70	92.70	85.70
890	**900**	172.47	163.47	155.47	146.47	137.47	129.47	120.47	112.47	103.47	94.47	87.47
900	**910**	175.23	166.23	157.23	149.23	140.23	131.23	123.23	114.23	105.23	97.23	89.23
910	**920**	177.00	168.00	160.00	151.00	142.00	134.00	125.00	117.00	108.00	99.00	91.00
920	**930**	179.76	170.76	161.76	153.76	144.76	135.76	127.76	118.76	109.76	101.76	92.76
930	**940**	181.53	172.53	164.53	155.53	146.53	138.53	129.53	121.53	112.53	103.53	95.53
940	**950**	184.29	175.29	166.29	158.29	149.29	140.29	132.29	123.29	114.29	106.29	97.29
950	**960**	186.06	177.06	169.06	160.06	151.06	143.06	134.06	126.06	117.06	108.06	100.06
960	**970**	188.82	179.82	170.82	162.82	153.82	144.82	136.82	127.82	118.82	110.82	101.82
970	**980**	190.59	181.59	173.59	164.59	155.59	147.59	138.59	130.59	121.59	112.59	104.59
980	**990**	193.35	184.35	175.35	167.35	158.35	149.35	141.35	132.35	123.35	115.35	106.35
990	**1,000**	196.12	186.12	178.12	169.12	160.12	152.12	143.12	135.12	126.12	117.12	109.12
1,000	**1,010**	198.88	188.88	179.88	171.88	162.88	153.88	145.88	136.88	127.88	119.88	110.88
1,010	**1,020**	202.65	190.65	182.65	173.65	164.65	156.65	147.65	139.65	130.65	121.65	113.65
1,020	**1,030**	206.41	193.41	184.41	176.41	167.41	158.41	150.41	141.41	132.41	124.41	115.41
1,030	**1,040**	209.18	195.18	187.18	178.18	169.18	161.18	152.18	144.18	135.18	126.18	118.18
1,040	**1,050**	212.94	197.94	188.94	180.94	171.94	162.94	154.94	145.94	136.94	128.94	119.94
1,050	**1,060**	216.71	200.71	191.71	182.71	173.71	165.71	156.71	148.71	139.71	130.71	122.71
1,060	**1,070**	219.47	204.47	193.47	185.47	176.47	167.47	159.47	150.47	141.47	133.47	124.47
1,070	**1,080**	223.24	208.24	196.24	187.24	178.24	170.24	161.24	153.24	144.24	135.24	127.24
1,080	**1,090**	227.00	211.00	198.00	190.00	181.00	172.00	164.00	155.00	146.00	138.00	129.00
1,090	**1,100**	230.77	214.77	200.77	191.77	182.77	174.77	165.77	157.77	148.77	139.77	131.77
1,100	**1,110**	233.53	218.53	202.53	194.53	185.53	176.53	168.53	159.53	150.53	142.53	133.53
1,110	**1,120**	237.30	221.30	206.30	196.30	187.30	179.30	170.30	162.30	153.30	144.30	136.30
1,120	**1,130**	241.06	225.06	209.06	199.06	190.06	181.06	173.06	164.06	155.06	147.06	138.06
1,130	**1,140**	243.83	228.83	212.83	200.83	191.83	183.83	174.83	166.83	157.83	148.83	140.83
1,140	**1,150**	247.59	231.59	216.59	203.59	194.59	185.59	177.59	168.59	159.59	151.59	142.59
1,150	**1,160**	251.36	235.36	220.36	205.36	196.36	188.36	179.36	171.36	162.36	153.36	145.36
1,160	**1,170**	254.12	239.12	223.12	208.12	199.12	190.12	182.12	173.12	164.12	156.12	147.12
1,170	**1,180**	257.89	242.89	226.89	210.89	200.89	192.89	183.89	175.89	166.89	157.89	149.89
1,180	**1,190**	261.65	245.65	230.65	214.65	203.65	194.65	186.65	177.65	168.65	160.65	151.65
1,190	**1,200**	265.42	249.42	233.42	218.42	205.42	197.42	188.42	180.42	171.42	162.42	154.42
1,200	**1,210**	268.18	253.18	237.18	222.18	208.18	199.18	191.18	182.18	173.18	165.18	156.18
1,210	**1,220**	271.95	255.95	240.95	224.95	209.95	201.95	192.95	184.95	175.95	166.95	158.95
1,220	**1,230**	275.71	259.71	243.71	228.71	212.71	203.71	195.71	186.71	177.71	169.71	160.71
1,230	**1,240**	278.48	263.48	247.48	232.48	216.48	206.48	197.48	189.48	180.48	171.48	163.48
1,240	**1,250**	282.24	266.24	251.24	235.24	220.24	208.24	200.24	191.24	182.24	174.24	165.24
1,250	**1,260**	286.01	270.01	255.01	239.01	223.01	211.01	202.01	194.01	185.01	176.01	168.01
1,260	**1,270**	288.77	273.77	257.77	242.77	226.77	212.77	204.77	195.77	186.77	178.77	169.77
1,270	**1,280**	292.54	277.54	261.54	245.54	230.54	215.54	206.54	198.54	189.54	180.54	172.54
1,280	**1,290**	296.30	280.30	265.30	249.30	234.30	218.30	209.30	200.30	191.30	183.30	174.30
1,290	**1,300**	300.07	284.07	268.07	253.07	237.07	222.07	211.07	203.07	194.07	185.07	177.07
1,300	**1,310**	302.83	287.83	271.83	256.83	240.83	224.83	213.83	204.83	195.83	187.83	178.83
1,310	**1,320**	306.60	290.60	275.60	259.60	244.60	228.60	215.60	207.60	198.60	189.60	181.60
1,320	**1,330**	310.36	294.36	278.36	263.36	247.36	232.36	218.36	209.36	200.36	192.36	183.36
1,330	**1,340**	313.13	298.13	282.13	267.13	251.13	235.13	220.13	212.13	203.13	194.13	186.13
1,340	**1,350**	316.89	300.89	285.89	269.89	254.89	238.89	223.89	213.89	204.89	196.89	187.89
1,350	**1,360**	320.66	304.66	289.66	273.66	257.66	242.66	226.66	216.66	207.66	198.66	190.66
1,360	**1,370**	323.42	308.42	292.42	277.42	261.42	246.42	230.42	218.42	209.42	201.42	192.42
1,370	**1,380**	327.19	312.19	296.19	280.19	265.19	249.19	234.19	221.19	212.19	203.19	195.19
1,380	**1,390**	330.95	314.95	299.95	283.95	268.95	252.95	236.95	222.95	213.95	205.95	196.95

$1,390 and over — Do not use this table. See page 34 for instructions.

SINGLE Persons—BIWEEKLY Payroll Period

(For Wages Paid in 2002)

And the wages are–		And the number of withholding allowances claimed is—										
At least	But less than	0	1	2	3	4	5	6	7	8	9	10
		The amount of income, social security, and Medicare taxes to be withheld is—										
$0	**$105**	7.65%	7.65%	7.65%	7.65%	7.65%	7.65%	7.65%	7.65%	7.65%	7.65%	7.65%
105	**110**	$9.22	$8.22	$8.22	$8.22	$8.22	$8.22	$8.22	$8.22	$8.22	$8.22	$8.22
110	**115**	9.61	8.61	8.61	8.61	8.61	8.61	8.61	8.61	8.61	8.61	8.61
115	**120**	10.99	8.99	8.99	8.99	8.99	8.99	8.99	8.99	8.99	8.99	8.99
120	**125**	11.37	9.37	9.37	9.37	9.37	9.37	9.37	9.37	9.37	9.37	9.37
125	**130**	12.75	9.75	9.75	9.75	9.75	9.75	9.75	9.75	9.75	9.75	9.75
130	**135**	13.14	10.14	10.14	10.14	10.14	10.14	10.14	10.14	10.14	10.14	10.14
135	**140**	14.52	10.52	10.52	10.52	10.52	10.52	10.52	10.52	10.52	10.52	10.52
140	**145**	14.90	10.90	10.90	10.90	10.90	10.90	10.90	10.90	10.90	10.90	10.90
145	**150**	16.28	11.28	11.28	11.28	11.28	11.28	11.28	11.28	11.28	11.28	11.28
150	**155**	16.67	11.67	11.67	11.67	11.67	11.67	11.67	11.67	11.67	11.67	11.67
155	**160**	18.05	12.05	12.05	12.05	12.05	12.05	12.05	12.05	12.05	12.05	12.05
160	**165**	18.43	12.43	12.43	12.43	12.43	12.43	12.43	12.43	12.43	12.43	12.43
165	**170**	19.81	12.81	12.81	12.81	12.81	12.81	12.81	12.81	12.81	12.81	12.81
170	**175**	20.20	13.20	13.20	13.20	13.20	13.20	13.20	13.20	13.20	13.20	13.20
175	**180**	21.58	13.58	13.58	13.58	13.58	13.58	13.58	13.58	13.58	13.58	13.58
180	**185**	21.96	13.96	13.96	13.96	13.96	13.96	13.96	13.96	13.96	13.96	13.96
185	**190**	23.34	14.34	14.34	14.34	14.34	14.34	14.34	14.34	14.34	14.34	14.34
190	**195**	23.73	14.73	14.73	14.73	14.73	14.73	14.73	14.73	14.73	14.73	14.73
195	**200**	25.11	15.11	15.11	15.11	15.11	15.11	15.11	15.11	15.11	15.11	15.11
200	**205**	25.49	15.49	15.49	15.49	15.49	15.49	15.49	15.49	15.49	15.49	15.49
205	**210**	26.87	15.87	15.87	15.87	15.87	15.87	15.87	15.87	15.87	15.87	15.87
210	**215**	27.26	16.26	16.26	16.26	16.26	16.26	16.26	16.26	16.26	16.26	16.26
215	**220**	28.64	16.64	16.64	16.64	16.64	16.64	16.64	16.64	16.64	16.64	16.64
220	**225**	29.02	18.02	17.02	17.02	17.02	17.02	17.02	17.02	17.02	17.02	17.02
225	**230**	30.40	18.40	17.40	17.40	17.40	17.40	17.40	17.40	17.40	17.40	17.40
230	**235**	30.79	19.79	17.79	17.79	17.79	17.79	17.79	17.79	17.79	17.79	17.79
235	**240**	32.17	20.17	18.17	18.17	18.17	18.17	18.17	18.17	18.17	18.17	18.17
240	**245**	32.55	21.55	18.55	18.55	18.55	18.55	18.55	18.55	18.55	18.55	18.55
245	**250**	33.93	21.93	18.93	18.93	18.93	18.93	18.93	18.93	18.93	18.93	18.93
250	**260**	34.51	23.51	19.51	19.51	19.51	19.51	19.51	19.51	19.51	19.51	19.51
260	**270**	36.27	25.27	20.27	20.27	20.27	20.27	20.27	20.27	20.27	20.27	20.27
270	**280**	38.04	27.04	21.04	21.04	21.04	21.04	21.04	21.04	21.04	21.04	21.04
280	**290**	39.80	28.80	21.80	21.80	21.80	21.80	21.80	21.80	21.80	21.80	21.80
290	**300**	41.57	30.57	22.57	22.57	22.57	22.57	22.57	22.57	22.57	22.57	22.57
300	**310**	43.33	32.33	23.33	23.33	23.33	23.33	23.33	23.33	23.33	23.33	23.33
310	**320**	45.10	34.10	24.10	24.10	24.10	24.10	24.10	24.10	24.10	24.10	24.10
320	**330**	46.86	35.86	24.86	24.86	24.86	24.86	24.86	24.86	24.86	24.86	24.86
330	**340**	49.63	37.63	25.63	25.63	25.63	25.63	25.63	25.63	25.63	25.63	25.63
340	**350**	51.39	39.39	27.39	26.39	26.39	26.39	26.39	26.39	26.39	26.39	26.39
350	**360**	54.16	41.16	29.16	27.16	27.16	27.16	27.16	27.16	27.16	27.16	27.16
360	**370**	55.92	42.92	30.92	27.92	27.92	27.92	27.92	27.92	27.92	27.92	27.92
370	**380**	58.69	44.69	32.69	28.69	28.69	28.69	28.69	28.69	28.69	28.69	28.69
380	**390**	60.45	46.45	34.45	29.45	29.45	29.45	29.45	29.45	29.45	29.45	29.45
390	**400**	63.22	48.22	36.22	30.22	30.22	30.22	30.22	30.22	30.22	30.22	30.22
400	**410**	64.98	49.98	37.98	30.98	30.98	30.98	30.98	30.98	30.98	30.98	30.98
410	**420**	67.75	51.75	39.75	31.75	31.75	31.75	31.75	31.75	31.75	31.75	31.75
420	**430**	69.51	53.51	41.51	32.51	32.51	32.51	32.51	32.51	32.51	32.51	32.51
430	**440**	72.28	55.28	43.28	33.28	33.28	33.28	33.28	33.28	33.28	33.28	33.28
440	**450**	74.04	57.04	45.04	34.04	34.04	34.04	34.04	34.04	34.04	34.04	34.04
450	**460**	76.81	58.81	46.81	35.81	34.81	34.81	34.81	34.81	34.81	34.81	34.81
460	**470**	78.57	61.57	48.57	37.57	35.57	35.57	35.57	35.57	35.57	35.57	35.57
470	**480**	81.34	63.34	50.34	39.34	36.34	36.34	36.34	36.34	36.34	36.34	36.34
480	**490**	83.10	66.10	52.10	41.10	37.10	37.10	37.10	37.10	37.10	37.10	37.10
490	**500**	85.87	67.87	53.87	42.87	37.87	37.87	37.87	37.87	37.87	37.87	37.87
500	**520**	89.02	72.02	57.02	45.02	39.02	39.02	39.02	39.02	39.02	39.02	39.02
520	**540**	93.55	76.55	60.55	48.55	40.55	40.55	40.55	40.55	40.55	40.55	40.55
540	**560**	98.08	81.08	64.08	52.08	42.08	42.08	42.08	42.08	42.08	42.08	42.08
560	**580**	102.61	85.61	67.61	55.61	44.61	43.61	43.61	43.61	43.61	43.61	43.61
580	**600**	107.14	90.14	72.14	59.14	48.14	45.14	45.14	45.14	45.14	45.14	45.14
600	**620**	111.67	94.67	76.67	62.67	51.67	46.67	46.67	46.67	46.67	46.67	46.67
620	**640**	116.20	99.20	81.20	66.20	55.20	48.20	48.20	48.20	48.20	48.20	48.20
640	**660**	120.73	103.73	85.73	69.73	58.73	49.73	49.73	49.73	49.73	49.73	49.73
660	**680**	125.26	108.26	90.26	73.26	62.26	51.26	51.26	51.26	51.26	51.26	51.26
680	**700**	129.79	112.79	94.79	77.79	65.79	53.79	52.79	52.79	52.79	52.79	52.79
700	**720**	134.32	117.32	99.32	82.32	69.32	57.32	54.32	54.32	54.32	54.32	54.32
720	**740**	138.85	121.85	103.85	86.85	72.85	60.85	55.85	55.85	55.85	55.85	55.85
740	**760**	143.38	126.38	108.38	91.38	76.38	64.38	57.38	57.38	57.38	57.38	57.38
760	**780**	147.91	130.91	112.91	95.91	79.91	67.91	58.91	58.91	58.91	58.91	58.91
780	**800**	152.44	135.44	117.44	100.44	83.44	71.44	60.44	60.44	60.44	60.44	60.44

(Continued on next page)

SINGLE Persons—**BIWEEKLY** Payroll Period

(For Wages Paid in 2002)

And the wages are—		And the number of withholding allowances claimed is—										
At least	But less than	0	1	2	3	4	5	6	7	8	9	10
		The amount of income, social security, and Medicare taxes to be withheld is—										
$800	**$820**	$156.97	$139.97	$121.97	$104.97	$87.97	$74.97	$63.97	$61.97	$61.97	$61.97	$61.97
820	**840**	161.50	144.50	126.50	109.50	92.50	78.50	67.50	63.50	63.50	63.50	63.50
840	**860**	166.03	149.03	131.03	114.03	97.03	82.03	71.03	65.03	65.03	65.03	65.03
860	**880**	170.56	153.56	135.56	118.56	101.56	85.56	74.56	66.56	66.56	66.56	66.56
880	**900**	175.09	158.09	140.09	123.09	106.09	89.09	78.09	68.09	68.09	68.09	68.09
900	**920**	179.62	162.62	144.62	127.62	110.62	92.62	81.62	69.62	69.62	69.62	69.62
920	**940**	184.15	167.15	149.15	132.15	115.15	97.15	85.15	73.15	71.15	71.15	71.15
940	**960**	188.68	171.68	153.68	136.68	119.68	101.68	88.68	76.68	72.68	72.68	72.68
960	**980**	193.21	176.21	158.21	141.21	124.21	106.21	92.21	80.21	74.21	74.21	74.21
980	**1,000**	197.74	180.74	162.74	145.74	128.74	110.74	95.74	83.74	75.74	75.74	75.74
1,000	**1,020**	202.27	185.27	167.27	150.27	133.27	115.27	99.27	87.27	77.27	77.27	77.27
1,020	**1,040**	206.80	189.80	171.80	154.80	137.80	119.80	102.80	90.80	79.80	78.80	78.80
1,040	**1,060**	211.33	194.33	176.33	159.33	142.33	124.33	107.33	94.33	83.33	80.33	80.33
1,060	**1,080**	215.86	198.86	180.86	163.86	146.86	128.86	111.86	97.86	86.86	81.86	81.86
1,080	**1,100**	220.39	203.39	185.39	168.39	151.39	133.39	116.39	101.39	90.39	83.39	83.39
1,100	**1,120**	224.92	207.92	189.92	172.92	155.92	137.92	120.92	104.92	93.92	84.92	84.92
1,120	**1,140**	229.45	212.45	194.45	177.45	160.45	142.45	125.45	108.45	97.45	86.45	86.45
1,140	**1,160**	234.98	216.98	198.98	181.98	164.98	146.98	129.98	112.98	100.98	88.98	87.98
1,160	**1,180**	241.51	221.51	203.51	186.51	169.51	151.51	134.51	117.51	104.51	92.51	89.51
1,180	**1,200**	249.04	226.04	208.04	191.04	174.04	156.04	139.04	122.04	108.04	96.04	91.04
1,200	**1,220**	255.57	230.57	212.57	195.57	178.57	160.57	143.57	126.57	111.57	99.57	92.57
1,220	**1,240**	263.10	235.10	217.10	200.10	183.10	165.10	148.10	131.10	115.10	103.10	94.10
1,240	**1,260**	269.63	239.63	221.63	204.63	187.63	169.63	152.63	135.63	118.63	106.63	95.63
1,260	**1,280**	276.16	245.16	226.16	209.16	192.16	174.16	157.16	140.16	122.16	110.16	98.16
1,280	**1,300**	283.69	252.69	230.69	213.69	196.69	178.69	161.69	144.69	126.69	113.69	101.69
1,300	**1,320**	290.22	259.22	235.22	218.22	201.22	183.22	166.22	149.22	131.22	117.22	105.22
1,320	**1,340**	297.75	265.75	239.75	222.75	205.75	187.75	170.75	153.75	135.75	120.75	108.75
1,340	**1,360**	304.28	273.28	244.28	227.28	210.28	192.28	175.28	158.28	140.28	124.28	112.28
1,360	**1,380**	310.81	279.81	248.81	231.81	214.81	196.81	179.81	162.81	144.81	127.81	115.81
1,380	**1,400**	318.34	287.34	256.34	236.34	219.34	201.34	184.34	167.34	149.34	132.34	119.34
1,400	**1,420**	324.87	293.87	262.87	240.87	223.87	205.87	188.87	171.87	153.87	136.87	122.87
1,420	**1,440**	332.40	300.40	269.40	245.40	228.40	210.40	193.40	176.40	158.40	141.40	126.40
1,440	**1,460**	338.93	307.93	276.93	249.93	232.93	214.93	197.93	180.93	162.93	145.93	129.93
1,460	**1,480**	345.46	314.46	283.46	254.46	237.46	219.46	202.46	185.46	167.46	150.46	133.46
1,480	**1,500**	352.99	321.99	290.99	258.99	241.99	223.99	206.99	189.99	171.99	154.99	137.99
1,500	**1,520**	359.52	328.52	297.52	266.52	246.52	228.52	211.52	194.52	176.52	159.52	142.52
1,520	**1,540**	367.05	335.05	304.05	273.05	251.05	233.05	216.05	199.05	181.05	164.05	147.05
1,540	**1,560**	373.58	342.58	311.58	280.58	255.58	237.58	220.58	203.58	185.58	168.58	151.58
1,560	**1,580**	380.11	349.11	318.11	287.11	260.11	242.11	225.11	208.11	190.11	173.11	156.11
1,580	**1,600**	387.64	356.64	325.64	293.64	264.64	246.64	229.64	212.64	194.64	177.64	160.64
1,600	**1,620**	394.17	363.17	332.17	301.17	270.17	251.17	234.17	217.17	199.17	182.17	165.17
1,620	**1,640**	401.70	369.70	338.70	307.70	276.70	255.70	238.70	221.70	203.70	186.70	169.70
1,640	**1,660**	408.23	377.23	346.23	315.23	283.23	260.23	243.23	226.23	208.23	191.23	174.23
1,660	**1,680**	414.76	383.76	352.76	321.76	290.76	264.76	247.76	230.76	212.76	195.76	178.76
1,680	**1,700**	422.29	391.29	360.29	328.29	297.29	269.29	252.29	235.29	217.29	200.29	183.29
1,700	**1,720**	428.82	397.82	366.82	335.82	304.82	273.82	256.82	239.82	221.82	204.82	187.82
1,720	**1,740**	436.35	404.35	373.35	342.35	311.35	280.35	261.35	244.35	226.35	209.35	192.35
1,740	**1,760**	442.88	411.88	380.88	349.88	317.88	286.88	265.88	248.88	230.88	213.88	196.88
1,760	**1,780**	449.41	418.41	387.41	356.41	325.41	294.41	270.41	253.41	235.41	218.41	201.41
1,780	**1,800**	456.94	425.94	394.94	362.94	331.94	300.94	274.94	257.94	239.94	222.94	205.94
1,800	**1,820**	463.47	432.47	401.47	370.47	339.47	307.47	279.47	262.47	244.47	227.47	210.47
1,820	**1,840**	471.00	439.00	408.00	377.00	346.00	315.00	284.00	267.00	249.00	232.00	215.00
1,840	**1,860**	477.53	446.53	415.53	384.53	352.53	321.53	290.53	271.53	253.53	236.53	219.53
1,860	**1,880**	484.06	453.06	422.06	391.06	360.06	329.06	297.06	276.06	258.06	241.06	224.06
1,880	**1,900**	491.59	460.59	429.59	397.59	366.59	335.59	304.59	280.59	262.59	245.59	228.59
1,900	**1,920**	498.12	467.12	436.12	405.12	374.12	342.12	311.12	285.12	267.12	250.12	233.12
1,920	**1,940**	505.65	473.65	442.65	411.65	380.65	349.65	318.65	289.65	271.65	254.65	237.65
1,940	**1,960**	512.18	481.18	450.18	419.18	387.18	356.18	325.18	294.18	276.18	259.18	242.18
1,960	**1,980**	518.71	487.71	456.71	425.71	394.71	363.71	331.71	300.71	280.71	263.71	246.71
1,980	**2,000**	526.24	495.24	464.24	432.24	401.24	370.24	339.24	308.24	285.24	268.24	251.24
2,000	**2,020**	532.77	501.77	470.77	439.77	408.77	376.77	345.77	314.77	289.77	272.77	255.77
2,020	**2,040**	540.30	508.30	477.30	446.30	415.30	384.30	353.30	322.30	294.30	277.30	260.30
2,040	**2,060**	546.83	515.83	484.83	453.83	421.83	390.83	359.83	328.83	298.83	281.83	264.83
2,060	**2,080**	553.36	522.36	491.36	460.36	429.36	398.36	366.36	335.36	304.36	286.36	269.36
2,080	**2,100**	560.89	529.89	498.89	466.89	435.89	404.89	373.89	342.89	311.89	290.89	273.89

$2,100 and over — Do not use this table. See page 34 for instructions.

Page 40

MARRIED Persons—BIWEEKLY Payroll Period

(For Wages Paid in 2002)

And the wages are–		And the number of withholding allowances claimed is—										
At least	But less than	0	1	2	3	4	5	6	7	8	9	10
		The amount of income, social security, and Medicare taxes to be withheld is—										
$0	$250	7.65%	7.65%	7.65%	7.65%	7.65%	7.65%	7.65%	7.65%	7.65%	7.65%	7.65%
250	260	$20.51	$19.51	$19.51	$19.51	$19.51	$19.51	$19.51	$19.51	$19.51	$19.51	$19.51
260	270	22.27	20.27	20.27	20.27	20.27	20.27	20.27	20.27	20.27	20.27	20.27
270	280	24.04	21.04	21.04	21.04	21.04	21.04	21.04	21.04	21.04	21.04	21.04
280	290	25.80	21.80	21.80	21.80	21.80	21.80	21.80	21.80	21.80	21.80	21.80
290	300	27.57	22.57	22.57	22.57	22.57	22.57	22.57	22.57	22.57	22.57	22.57
300	310	29.33	23.33	23.33	23.33	23.33	23.33	23.33	23.33	23.33	23.33	23.33
310	320	31.10	24.10	24.10	24.10	24.10	24.10	24.10	24.10	24.10	24.10	24.10
320	330	32.86	24.86	24.86	24.86	24.86	24.86	24.86	24.86	24.86	24.86	24.86
330	340	34.63	25.63	25.63	25.63	25.63	25.63	25.63	25.63	25.63	25.63	25.63
340	350	36.39	26.39	26.39	26.39	26.39	26.39	26.39	26.39	26.39	26.39	26.39
350	360	38.16	27.16	27.16	27.16	27.16	27.16	27.16	27.16	27.16	27.16	27.16
360	370	39.92	27.92	27.92	27.92	27.92	27.92	27.92	27.92	27.92	27.92	27.92
370	380	41.69	29.69	28.69	28.69	28.69	28.69	28.69	28.69	28.69	28.69	28.69
380	390	43.45	31.45	29.45	29.45	29.45	29.45	29.45	29.45	29.45	29.45	29.45
390	400	45.22	33.22	30.22	30.22	30.22	30.22	30.22	30.22	30.22	30.22	30.22
400	410	46.98	34.98	30.98	30.98	30.98	30.98	30.98	30.98	30.98	30.98	30.98
410	420	48.75	36.75	31.75	31.75	31.75	31.75	31.75	31.75	31.75	31.75	31.75
420	430	50.51	38.51	32.51	32.51	32.51	32.51	32.51	32.51	32.51	32.51	32.51
430	440	52.28	40.28	33.28	33.28	33.28	33.28	33.28	33.28	33.28	33.28	33.28
440	450	54.04	42.04	34.04	34.04	34.04	34.04	34.04	34.04	34.04	34.04	34.04
450	460	55.81	43.81	34.81	34.81	34.81	34.81	34.81	34.81	34.81	34.81	34.81
460	470	57.57	45.57	35.57	35.57	35.57	35.57	35.57	35.57	35.57	35.57	35.57
470	480	59.34	47.34	36.34	36.34	36.34	36.34	36.34	36.34	36.34	36.34	36.34
480	490	61.10	49.10	38.10	37.10	37.10	37.10	37.10	37.10	37.10	37.10	37.10
490	500	62.87	50.87	39.87	37.87	37.87	37.87	37.87	37.87	37.87	37.87	37.87
500	520	65.02	54.02	42.02	39.02	39.02	39.02	39.02	39.02	39.02	39.02	39.02
520	540	68.55	57.55	45.55	40.55	40.55	40.55	40.55	40.55	40.55	40.55	40.55
540	560	72.08	61.08	49.08	42.08	42.08	42.08	42.08	42.08	42.08	42.08	42.08
560	580	75.61	64.61	52.61	43.61	43.61	43.61	43.61	43.61	43.61	43.61	43.61
580	600	79.14	68.14	56.14	45.14	45.14	45.14	45.14	45.14	45.14	45.14	45.14
600	620	82.67	71.67	59.67	48.67	46.67	46.67	46.67	46.67	46.67	46.67	46.67
620	640	86.20	75.20	63.20	52.20	48.20	48.20	48.20	48.20	48.20	48.20	48.20
640	660	89.73	78.73	66.73	55.73	49.73	49.73	49.73	49.73	49.73	49.73	49.73
660	680	93.26	82.26	70.26	59.26	51.26	51.26	51.26	51.26	51.26	51.26	51.26
680	700	96.79	85.79	73.79	62.79	52.79	52.79	52.79	52.79	52.79	52.79	52.79
700	720	100.32	89.32	77.32	66.32	54.32	54.32	54.32	54.32	54.32	54.32	54.32
720	740	104.85	92.85	80.85	69.85	57.85	55.85	55.85	55.85	55.85	55.85	55.85
740	760	109.38	96.38	84.38	73.38	61.38	57.38	57.38	57.38	57.38	57.38	57.38
760	780	113.91	99.91	87.91	76.91	64.91	58.91	58.91	58.91	58.91	58.91	58.91
780	800	118.44	103.44	91.44	80.44	68.44	60.44	60.44	60.44	60.44	60.44	60.44
800	820	122.97	106.97	94.97	83.97	71.97	61.97	61.97	61.97	61.97	61.97	61.97
820	840	127.50	110.50	98.50	87.50	75.50	64.50	63.50	63.50	63.50	63.50	63.50
840	860	132.03	115.03	102.03	91.03	79.03	68.03	65.03	65.03	65.03	65.03	65.03
860	880	136.56	119.56	105.56	94.56	82.56	71.56	66.56	66.56	66.56	66.56	66.56
880	900	141.09	124.09	109.09	98.09	86.09	75.09	68.09	68.09	68.09	68.09	68.09
900	920	145.62	128.62	112.62	101.62	89.62	78.62	69.62	69.62	69.62	69.62	69.62
920	940	150.15	133.15	116.15	105.15	93.15	82.15	71.15	71.15	71.15	71.15	71.15
940	960	154.68	137.68	120.68	108.68	96.68	85.68	73.68	72.68	72.68	72.68	72.68
960	980	159.21	142.21	125.21	112.21	100.21	89.21	77.21	74.21	74.21	74.21	74.21
980	1,000	163.74	146.74	129.74	115.74	103.74	92.74	80.74	75.74	75.74	75.74	75.74
1,000	1,020	168.27	151.27	134.27	119.27	107.27	96.27	84.27	77.27	77.27	77.27	77.27
1,020	1,040	172.80	155.80	138.80	122.80	110.80	99.80	87.80	78.80	78.80	78.80	78.80
1,040	1,060	177.33	160.33	143.33	126.33	114.33	103.33	91.33	80.33	80.33	80.33	80.33
1,060	1,080	181.86	164.86	147.86	129.86	117.86	106.86	94.86	82.86	81.86	81.86	81.86
1,080	1,100	186.39	169.39	152.39	134.39	121.39	110.39	98.39	86.39	83.39	83.39	83.39
1,100	1,120	190.92	173.92	156.92	138.92	124.92	113.92	101.92	89.92	84.92	84.92	84.92
1,120	1,140	195.45	178.45	161.45	143.45	128.45	117.45	105.45	93.45	86.45	86.45	86.45
1,140	1,160	199.98	182.98	165.98	147.98	131.98	120.98	108.98	96.98	87.98	87.98	87.98
1,160	1,180	204.51	187.51	170.51	152.51	135.51	124.51	112.51	100.51	89.51	89.51	89.51
1,180	1,200	209.04	192.04	175.04	157.04	140.04	128.04	116.04	104.04	93.04	91.04	91.04
1,200	1,220	213.57	196.57	179.57	161.57	144.57	131.57	119.57	107.57	96.57	92.57	92.57
1,220	1,240	218.10	201.10	184.10	166.10	149.10	135.10	123.10	111.10	100.10	94.10	94.10
1,240	1,260	222.63	205.63	188.63	170.63	153.63	138.63	126.63	114.63	103.63	95.63	95.63
1,260	1,280	227.16	210.16	193.16	175.16	158.16	142.16	130.16	118.16	107.16	97.16	97.16
1,280	1,300	231.69	214.69	197.69	179.69	162.69	145.69	133.69	121.69	110.69	98.69	98.69
1,300	1,320	236.22	219.22	202.22	184.22	167.22	150.22	137.22	125.22	114.22	102.22	100.22
1,320	1,340	240.75	223.75	206.75	188.75	171.75	154.75	140.75	128.75	117.75	105.75	101.75
1,340	1,360	245.28	228.28	211.28	193.28	176.28	159.28	144.28	132.28	121.28	109.28	103.28
1,360	1,380	249.81	232.81	215.81	197.81	180.81	163.81	147.81	135.81	124.81	112.81	104.81

(Continued on next page)

MARRIED Persons—**BIWEEKLY** Payroll Period

(For Wages Paid in 2002)

And the wages are–		And the number of withholding allowances claimed is—										
At least	But less than	0	1	2	3	4	5	6	7	8	9	10
		The amount of income, social security, and Medicare taxes to be withheld is—										
$1,380	**$1,400**	$254.34	$237.34	$220.34	$202.34	$185.34	$168.34	$151.34	$139.34	$128.34	$116.34	$106.34
1,400	**1,420**	258.87	241.87	224.87	206.87	189.87	172.87	154.87	142.87	131.87	119.87	108.87
1,420	**1,440**	263.40	246.40	229.40	211.40	194.40	177.40	159.40	146.40	135.40	123.40	112.40
1,440	**1,460**	267.93	250.93	233.93	215.93	198.93	181.93	163.93	149.93	138.93	126.93	115.93
1,460	**1,480**	272.46	255.46	238.46	220.46	203.46	186.46	168.46	153.46	142.46	130.46	119.46
1,480	**1,500**	276.99	259.99	242.99	224.99	207.99	190.99	172.99	156.99	145.99	133.99	122.99
1,500	**1,520**	281.52	264.52	247.52	229.52	212.52	195.52	177.52	160.52	149.52	137.52	126.52
1,520	**1,540**	286.05	269.05	252.05	234.05	217.05	200.05	182.05	165.05	153.05	141.05	130.05
1,540	**1,560**	290.58	273.58	256.58	238.58	221.58	204.58	186.58	169.58	156.58	144.58	133.58
1,560	**1,580**	295.11	278.11	261.11	243.11	226.11	209.11	191.11	174.11	160.11	148.11	137.11
1,580	**1,600**	299.64	282.64	265.64	247.64	230.64	213.64	195.64	178.64	163.64	151.64	140.64
1,600	**1,620**	304.17	287.17	270.17	252.17	235.17	218.17	200.17	183.17	167.17	155.17	144.17
1,620	**1,640**	308.70	291.70	274.70	256.70	239.70	222.70	204.70	187.70	170.70	158.70	147.70
1,640	**1,660**	313.23	296.23	279.23	261.23	244.23	227.23	209.23	192.23	175.23	162.23	151.23
1,660	**1,680**	317.76	300.76	283.76	265.76	248.76	231.76	213.76	196.76	179.76	165.76	154.76
1,680	**1,700**	322.29	305.29	288.29	270.29	253.29	236.29	218.29	201.29	184.29	169.29	158.29
1,700	**1,720**	326.82	309.82	292.82	274.82	257.82	240.82	222.82	205.82	188.82	172.82	161.82
1,720	**1,740**	331.35	314.35	297.35	279.35	262.35	245.35	227.35	210.35	193.35	176.35	165.35
1,740	**1,760**	335.88	318.88	301.88	283.88	266.88	249.88	231.88	214.88	197.88	179.88	168.88
1,760	**1,780**	340.41	323.41	306.41	288.41	271.41	254.41	236.41	219.41	202.41	184.41	172.41
1,780	**1,800**	344.94	327.94	310.94	292.94	275.94	258.94	240.94	223.94	206.94	188.94	175.94
1,800	**1,820**	349.47	332.47	315.47	297.47	280.47	263.47	245.47	228.47	211.47	193.47	179.47
1,820	**1,840**	354.00	337.00	320.00	302.00	285.00	268.00	250.00	233.00	216.00	198.00	183.00
1,840	**1,860**	358.53	341.53	324.53	306.53	289.53	272.53	254.53	237.53	220.53	202.53	186.53
1,860	**1,880**	363.06	346.06	329.06	311.06	294.06	277.06	259.06	242.06	225.06	207.06	190.06
1,880	**1,900**	367.59	350.59	333.59	315.59	298.59	281.59	263.59	246.59	229.59	211.59	194.59
1,900	**1,920**	372.12	355.12	338.12	320.12	303.12	286.12	268.12	251.12	234.12	216.12	199.12
1,920	**1,940**	376.65	359.65	342.65	324.65	307.65	290.65	272.65	255.65	238.65	220.65	203.65
1,940	**1,960**	381.18	364.18	347.18	329.18	312.18	295.18	277.18	260.18	243.18	225.18	208.18
1,960	**1,980**	385.71	368.71	351.71	333.71	316.71	299.71	281.71	264.71	247.71	229.71	212.71
1,980	**2,000**	391.24	373.24	356.24	338.24	321.24	304.24	286.24	269.24	252.24	234.24	217.24
2,000	**2,020**	397.77	377.77	360.77	342.77	325.77	308.77	290.77	273.77	256.77	238.77	221.77
2,020	**2,040**	405.30	382.30	365.30	347.30	330.30	313.30	295.30	278.30	261.30	243.30	226.30
2,040	**2,060**	411.83	386.83	369.83	351.83	334.83	317.83	299.83	282.83	265.83	247.83	230.83
2,060	**2,080**	419.36	391.36	374.36	356.36	339.36	322.36	304.36	287.36	270.36	252.36	235.36
2,080	**2,100**	425.89	395.89	378.89	360.89	343.89	326.89	308.89	291.89	274.89	256.89	239.89
2,100	**2,120**	432.42	401.42	383.42	365.42	348.42	331.42	313.42	296.42	279.42	261.42	244.42
2,120	**2,140**	439.95	408.95	387.95	369.95	352.95	335.95	317.95	300.95	283.95	265.95	248.95
2,140	**2,160**	446.48	415.48	392.48	374.48	357.48	340.48	322.48	305.48	288.48	270.48	253.48
2,160	**2,180**	454.01	423.01	397.01	379.01	362.01	345.01	327.01	310.01	293.01	275.01	258.01
2,180	**2,200**	460.54	429.54	401.54	383.54	366.54	349.54	331.54	314.54	297.54	279.54	262.54
2,200	**2,220**	467.07	436.07	406.07	388.07	371.07	354.07	336.07	319.07	302.07	284.07	267.07
2,220	**2,240**	474.60	443.60	412.60	392.60	375.60	358.60	340.60	323.60	306.60	288.60	271.60
2,240	**2,260**	481.13	450.13	419.13	397.13	380.13	363.13	345.13	328.13	311.13	293.13	276.13
2,260	**2,280**	488.66	457.66	425.66	401.66	384.66	367.66	349.66	332.66	315.66	297.66	280.66
2,280	**2,300**	495.19	464.19	433.19	406.19	389.19	372.19	354.19	337.19	320.19	302.19	285.19
2,300	**2,320**	501.72	470.72	439.72	410.72	393.72	376.72	358.72	341.72	324.72	306.72	289.72
2,320	**2,340**	509.25	478.25	447.25	415.25	398.25	381.25	363.25	346.25	329.25	311.25	294.25
2,340	**2,360**	515.78	484.78	453.78	422.78	402.78	385.78	367.78	350.78	333.78	315.78	298.78
2,360	**2,380**	523.31	492.31	460.31	429.31	407.31	390.31	372.31	355.31	338.31	320.31	303.31
2,380	**2,400**	529.84	498.84	467.84	436.84	411.84	394.84	376.84	359.84	342.84	324.84	307.84
2,400	**2,420**	536.37	505.37	474.37	443.37	416.37	399.37	381.37	364.37	347.37	329.37	312.37
2,420	**2,440**	543.90	512.90	481.90	449.90	420.90	403.90	385.90	368.90	351.90	333.90	316.90
2,440	**2,460**	550.43	519.43	488.43	457.43	426.43	408.43	390.43	373.43	356.43	338.43	321.43
2,460	**2,480**	557.96	526.96	494.96	463.96	432.96	412.96	394.96	377.96	360.96	342.96	325.96
2,480	**2,500**	564.49	533.49	502.49	471.49	439.49	417.49	399.49	382.49	365.49	347.49	330.49
2,500	**2,520**	571.02	540.02	509.02	478.02	447.02	422.02	404.02	387.02	370.02	352.02	335.02
2,520	**2,540**	578.55	547.55	516.55	484.55	453.55	426.55	408.55	391.55	374.55	356.55	339.55
2,540	**2,560**	585.08	554.08	523.08	492.08	461.08	431.08	413.08	396.08	379.08	361.08	344.08
2,560	**2,580**	592.61	561.61	529.61	498.61	467.61	436.61	417.61	400.61	383.61	365.61	348.61
2,580	**2,600**	599.14	568.14	537.14	506.14	474.14	443.14	422.14	405.14	388.14	370.14	353.14
2,600	**2,620**	605.67	574.67	543.67	512.67	481.67	450.67	426.67	409.67	392.67	374.67	357.67
2,620	**2,640**	613.20	582.20	551.20	519.20	488.20	457.20	431.20	414.20	397.20	379.20	362.20
2,640	**2,660**	619.73	588.73	557.73	526.73	495.73	464.73	435.73	418.73	401.73	383.73	366.73
2,660	**2,680**	627.26	596.26	564.26	533.26	502.26	471.26	440.26	423.26	406.26	388.26	371.26

$2,680 and over Do not use this table. See page 34 for instructions.

SINGLE Persons—SEMIMONTHLY Payroll Period

(For Wages Paid in 2002)

And the wages are–		And the number of withholding allowances claimed is—										
At least	But less than	0	1	2	3	4	5	6	7	8	9	10
		The amount of income, social security, and Medicare taxes to be withheld is—										
$0	**$115**	7.65%	7.65%	7.65%	7.65%	7.65%	7.65%	7.65%	7.65%	7.65%	7.65%	7.65%
115	**120**	$9.99	$8.99	$8.99	$8.99	$8.99	$8.99	$8.99	$8.99	$8.99	$8.99	$8.99
120	**125**	10.37	9.37	9.37	9.37	9.37	9.37	9.37	9.37	9.37	9.37	9.37
125	**130**	11.75	9.75	9.75	9.75	9.75	9.75	9.75	9.75	9.75	9.75	9.75
130	**135**	12.14	10.14	10.14	10.14	10.14	10.14	10.14	10.14	10.14	10.14	10.14
135	**140**	13.52	10.52	10.52	10.52	10.52	10.52	10.52	10.52	10.52	10.52	10.52
140	**145**	13.90	10.90	10.90	10.90	10.90	10.90	10.90	10.90	10.90	10.90	10.90
145	**150**	15.28	11.28	11.28	11.28	11.28	11.28	11.28	11.28	11.28	11.28	11.28
150	**155**	15.67	11.67	11.67	11.67	11.67	11.67	11.67	11.67	11.67	11.67	11.67
155	**160**	17.05	12.05	12.05	12.05	12.05	12.05	12.05	12.05	12.05	12.05	12.05
160	**165**	17.43	12.43	12.43	12.43	12.43	12.43	12.43	12.43	12.43	12.43	12.43
165	**170**	18.81	12.81	12.81	12.81	12.81	12.81	12.81	12.81	12.81	12.81	12.81
170	**175**	19.20	13.20	13.20	13.20	13.20	13.20	13.20	13.20	13.20	13.20	13.20
175	**180**	20.58	13.58	13.58	13.58	13.58	13.58	13.58	13.58	13.58	13.58	13.58
180	**185**	20.96	13.96	13.96	13.96	13.96	13.96	13.96	13.96	13.96	13.96	13.96
185	**190**	22.34	14.34	14.34	14.34	14.34	14.34	14.34	14.34	14.34	14.34	14.34
190	**195**	22.73	14.73	14.73	14.73	14.73	14.73	14.73	14.73	14.73	14.73	14.73
195	**200**	24.11	15.11	15.11	15.11	15.11	15.11	15.11	15.11	15.11	15.11	15.11
200	**205**	24.49	15.49	15.49	15.49	15.49	15.49	15.49	15.49	15.49	15.49	15.49
205	**210**	25.87	15.87	15.87	15.87	15.87	15.87	15.87	15.87	15.87	15.87	15.87
210	**215**	26.26	16.26	16.26	16.26	16.26	16.26	16.26	16.26	16.26	16.26	16.26
215	**220**	27.64	16.64	16.64	16.64	16.64	16.64	16.64	16.64	16.64	16.64	16.64
220	**225**	28.02	17.02	17.02	17.02	17.02	17.02	17.02	17.02	17.02	17.02	17.02
225	**230**	29.40	17.40	17.40	17.40	17.40	17.40	17.40	17.40	17.40	17.40	17.40
230	**235**	29.79	17.79	17.79	17.79	17.79	17.79	17.79	17.79	17.79	17.79	17.79
235	**240**	31.17	18.17	18.17	18.17	18.17	18.17	18.17	18.17	18.17	18.17	18.17
240	**245**	31.55	19.55	18.55	18.55	18.55	18.55	18.55	18.55	18.55	18.55	18.55
245	**250**	32.93	19.93	18.93	18.93	18.93	18.93	18.93	18.93	18.93	18.93	18.93
250	**260**	33.51	21.51	19.51	19.51	19.51	19.51	19.51	19.51	19.51	19.51	19.51
260	**270**	35.27	23.27	20.27	20.27	20.27	20.27	20.27	20.27	20.27	20.27	20.27
270	**280**	37.04	25.04	21.04	21.04	21.04	21.04	21.04	21.04	21.04	21.04	21.04
280	**290**	38.80	26.80	21.80	21.80	21.80	21.80	21.80	21.80	21.80	21.80	21.80
290	**300**	40.57	28.57	22.57	22.57	22.57	22.57	22.57	22.57	22.57	22.57	22.57
300	**310**	42.33	30.33	23.33	23.33	23.33	23.33	23.33	23.33	23.33	23.33	23.33
310	**320**	44.10	32.10	24.10	24.10	24.10	24.10	24.10	24.10	24.10	24.10	24.10
320	**330**	45.86	33.86	24.86	24.86	24.86	24.86	24.86	24.86	24.86	24.86	24.86
330	**340**	47.63	35.63	25.63	25.63	25.63	25.63	25.63	25.63	25.63	25.63	25.63
340	**350**	49.39	37.39	26.39	26.39	26.39	26.39	26.39	26.39	26.39	26.39	26.39
350	**360**	51.16	39.16	27.16	27.16	27.16	27.16	27.16	27.16	27.16	27.16	27.16
360	**370**	53.92	40.92	27.92	27.92	27.92	27.92	27.92	27.92	27.92	27.92	27.92
370	**380**	55.69	42.69	29.69	28.69	28.69	28.69	28.69	28.69	28.69	28.69	28.69
380	**390**	58.45	44.45	31.45	29.45	29.45	29.45	29.45	29.45	29.45	29.45	29.45
390	**400**	60.22	46.22	33.22	30.22	30.22	30.22	30.22	30.22	30.22	30.22	30.22
400	**410**	62.98	47.98	34.98	30.98	30.98	30.98	30.98	30.98	30.98	30.98	30.98
410	**420**	64.75	49.75	36.75	31.75	31.75	31.75	31.75	31.75	31.75	31.75	31.75
420	**430**	67.51	51.51	38.51	32.51	32.51	32.51	32.51	32.51	32.51	32.51	32.51
430	**440**	69.28	53.28	40.28	33.28	33.28	33.28	33.28	33.28	33.28	33.28	33.28
440	**450**	72.04	55.04	42.04	34.04	34.04	34.04	34.04	34.04	34.04	34.04	34.04
450	**460**	73.81	56.81	43.81	34.81	34.81	34.81	34.81	34.81	34.81	34.81	34.81
460	**470**	76.57	58.57	45.57	35.57	35.57	35.57	35.57	35.57	35.57	35.57	35.57
470	**480**	78.34	60.34	47.34	36.34	36.34	36.34	36.34	36.34	36.34	36.34	36.34
480	**490**	81.10	62.10	49.10	37.10	37.10	37.10	37.10	37.10	37.10	37.10	37.10
490	**500**	82.87	64.87	50.87	38.87	37.87	37.87	37.87	37.87	37.87	37.87	37.87
500	**520**	87.02	68.02	54.02	41.02	39.02	39.02	39.02	39.02	39.02	39.02	39.02
520	**540**	91.55	72.55	57.55	44.55	40.55	40.55	40.55	40.55	40.55	40.55	40.55
540	**560**	96.08	77.08	61.08	48.08	42.08	42.08	42.08	42.08	42.08	42.08	42.08
560	**580**	100.61	81.61	64.61	51.61	43.61	43.61	43.61	43.61	43.61	43.61	43.61
580	**600**	105.14	86.14	68.14	55.14	45.14	45.14	45.14	45.14	45.14	45.14	45.14
600	**620**	109.67	90.67	71.67	58.67	46.67	46.67	46.67	46.67	46.67	46.67	46.67
620	**640**	114.20	95.20	76.20	62.20	50.20	48.20	48.20	48.20	48.20	48.20	48.20
640	**660**	118.73	99.73	80.73	65.73	53.73	49.73	49.73	49.73	49.73	49.73	49.73
660	**680**	123.26	104.26	85.26	69.26	57.26	51.26	51.26	51.26	51.26	51.26	51.26
680	**700**	127.79	108.79	89.79	72.79	60.79	52.79	52.79	52.79	52.79	52.79	52.79
700	**720**	132.32	113.32	94.32	76.32	64.32	54.32	54.32	54.32	54.32	54.32	54.32
720	**740**	136.85	117.85	98.85	79.85	67.85	55.85	55.85	55.85	55.85	55.85	55.85
740	**760**	141.38	122.38	103.38	84.38	71.38	58.38	57.38	57.38	57.38	57.38	57.38
760	**780**	145.91	126.91	107.91	88.91	74.91	61.91	58.91	58.91	58.91	58.91	58.91
780	**800**	150.44	131.44	112.44	93.44	78.44	65.44	60.44	60.44	60.44	60.44	60.44
800	**820**	154.97	135.97	116.97	97.97	81.97	68.97	61.97	61.97	61.97	61.97	61.97
820	**840**	159.50	140.50	121.50	102.50	85.50	72.50	63.50	63.50	63.50	63.50	63.50

(Continued on next page)

SINGLE Persons—**SEMIMONTHLY** Payroll Period

(For Wages Paid in 2002)

And the wages are–		And the number of withholding allowances claimed is—										
At least	But less than	0	1	2	3	4	5	6	7	8	9	10
		The amount of income, social security, and Medicare taxes to be withheld is—										
$840	**$860**	$164.03	$145.03	$126.03	$107.03	$89.03	$76.03	$65.03	$65.03	$65.03	$65.03	$65.03
860	**880**	168.56	149.56	130.56	111.56	93.56	79.56	67.56	66.56	66.56	66.56	66.56
880	**900**	173.09	154.09	135.09	116.09	98.09	83.09	71.09	68.09	68.09	68.09	68.09
900	**920**	177.62	158.62	139.62	120.62	102.62	86.62	74.62	69.62	69.62	69.62	69.62
920	**940**	182.15	163.15	144.15	125.15	107.15	90.15	78.15	71.15	71.15	71.15	71.15
940	**960**	186.68	167.68	148.68	129.68	111.68	93.68	81.68	72.68	72.68	72.68	72.68
960	**980**	191.21	172.21	153.21	134.21	116.21	97.21	85.21	74.21	74.21	74.21	74.21
980	**1,000**	195.74	176.74	157.74	138.74	120.74	101.74	88.74	75.74	75.74	75.74	75.74
1,000	**1,020**	200.27	181.27	162.27	143.27	125.27	106.27	92.27	79.27	77.27	77.27	77.27
1,020	**1,040**	204.80	185.80	166.80	147.80	129.80	110.80	95.80	82.80	78.80	78.80	78.80
1,040	**1,060**	209.33	190.33	171.33	152.33	134.33	115.33	99.33	86.33	80.33	80.33	80.33
1,060	**1,080**	213.86	194.86	175.86	156.86	138.86	119.86	102.86	89.86	81.86	81.86	81.86
1,080	**1,100**	218.39	199.39	180.39	161.39	143.39	124.39	106.39	93.39	83.39	83.39	83.39
1,100	**1,120**	222.92	203.92	184.92	165.92	147.92	128.92	109.92	96.92	84.92	84.92	84.92
1,120	**1,140**	227.45	208.45	189.45	170.45	152.45	133.45	114.45	100.45	88.45	86.45	86.45
1,140	**1,160**	231.98	212.98	193.98	174.98	156.98	137.98	118.98	103.98	91.98	87.98	87.98
1,160	**1,180**	236.51	217.51	198.51	179.51	161.51	142.51	123.51	107.51	95.51	89.51	89.51
1,180	**1,200**	241.04	222.04	203.04	184.04	166.04	147.04	128.04	111.04	99.04	91.04	91.04
1,200	**1,220**	245.57	226.57	207.57	188.57	170.57	151.57	132.57	114.57	102.57	92.57	92.57
1,220	**1,240**	250.10	231.10	212.10	193.10	175.10	156.10	137.10	118.10	106.10	94.10	94.10
1,240	**1,260**	255.63	235.63	216.63	197.63	179.63	160.63	141.63	122.63	109.63	96.63	95.63
1,260	**1,280**	263.16	240.16	221.16	202.16	184.16	165.16	146.16	127.16	113.16	100.16	97.16
1,280	**1,300**	269.69	244.69	225.69	206.69	188.69	169.69	150.69	131.69	116.69	103.69	98.69
1,300	**1,320**	277.22	249.22	230.22	211.22	193.22	174.22	155.22	136.22	120.22	107.22	100.22
1,320	**1,340**	283.75	253.75	234.75	215.75	197.75	178.75	159.75	140.75	123.75	110.75	101.75
1,340	**1,360**	290.28	258.28	239.28	220.28	202.28	183.28	164.28	145.28	127.28	114.28	103.28
1,360	**1,380**	297.81	263.81	243.81	224.81	206.81	187.81	168.81	149.81	131.81	117.81	105.81
1,380	**1,400**	304.34	270.34	248.34	229.34	211.34	192.34	173.34	154.34	136.34	121.34	109.34
1,400	**1,420**	311.87	277.87	252.87	233.87	215.87	196.87	177.87	158.87	140.87	124.87	112.87
1,420	**1,440**	318.40	284.40	257.40	238.40	220.40	201.40	182.40	163.40	145.40	128.40	116.40
1,440	**1,460**	324.93	291.93	261.93	242.93	224.93	205.93	186.93	167.93	149.93	131.93	119.93
1,460	**1,480**	332.46	298.46	266.46	247.46	229.46	210.46	191.46	172.46	154.46	135.46	123.46
1,480	**1,500**	338.99	304.99	271.99	251.99	233.99	214.99	195.99	176.99	158.99	139.99	126.99
1,500	**1,520**	346.52	312.52	278.52	256.52	238.52	219.52	200.52	181.52	163.52	144.52	130.52
1,520	**1,540**	353.05	319.05	285.05	261.05	243.05	224.05	205.05	186.05	168.05	149.05	134.05
1,540	**1,560**	359.58	326.58	292.58	265.58	247.58	228.58	209.58	190.58	172.58	153.58	137.58
1,560	**1,580**	367.11	333.11	299.11	270.11	252.11	233.11	214.11	195.11	177.11	158.11	141.11
1,580	**1,600**	373.64	339.64	306.64	274.64	256.64	237.64	218.64	199.64	181.64	162.64	144.64
1,600	**1,620**	381.17	347.17	313.17	279.17	261.17	242.17	223.17	204.17	186.17	167.17	148.17
1,620	**1,640**	387.70	353.70	319.70	286.70	265.70	246.70	227.70	208.70	190.70	171.70	152.70
1,640	**1,660**	394.23	361.23	327.23	293.23	270.23	251.23	232.23	213.23	195.23	176.23	157.23
1,660	**1,680**	401.76	367.76	333.76	300.76	274.76	255.76	236.76	217.76	199.76	180.76	161.76
1,680	**1,700**	408.29	374.29	341.29	307.29	279.29	260.29	241.29	222.29	204.29	185.29	166.29
1,700	**1,720**	415.82	381.82	347.82	313.82	283.82	264.82	245.82	226.82	208.82	189.82	170.82
1,720	**1,740**	422.35	388.35	354.35	321.35	288.35	269.35	250.35	231.35	213.35	194.35	175.35
1,740	**1,760**	428.88	395.88	361.88	327.88	293.88	273.88	254.88	235.88	217.88	198.88	179.88
1,760	**1,780**	436.41	402.41	368.41	335.41	301.41	278.41	259.41	240.41	222.41	203.41	184.41
1,780	**1,800**	442.94	408.94	375.94	341.94	307.94	282.94	263.94	244.94	226.94	207.94	188.94
1,800	**1,820**	450.47	416.47	382.47	348.47	315.47	287.47	268.47	249.47	231.47	212.47	193.47
1,820	**1,840**	457.00	423.00	389.00	356.00	322.00	292.00	273.00	254.00	236.00	217.00	198.00
1,840	**1,860**	463.53	430.53	396.53	362.53	328.53	296.53	277.53	258.53	240.53	221.53	202.53
1,860	**1,880**	471.06	437.06	403.06	370.06	336.06	302.06	282.06	263.06	245.06	226.06	207.06
1,880	**1,900**	477.59	443.59	410.59	376.59	342.59	308.59	286.59	267.59	249.59	230.59	211.59
1,900	**1,920**	485.12	451.12	417.12	383.12	350.12	316.12	291.12	272.12	254.12	235.12	216.12
1,920	**1,940**	491.65	457.65	423.65	390.65	356.65	322.65	295.65	276.65	258.65	239.65	220.65
1,940	**1,960**	498.18	465.18	431.18	397.18	363.18	330.18	300.18	281.18	263.18	244.18	225.18
1,960	**1,980**	505.71	471.71	437.71	404.71	370.71	336.71	304.71	285.71	267.71	248.71	229.71
1,980	**2,000**	512.24	478.24	445.24	411.24	377.24	343.24	310.24	290.24	272.24	253.24	234.24
2,000	**2,020**	519.77	485.77	451.77	417.77	384.77	350.77	316.77	294.77	276.77	257.77	238.77
2,020	**2,040**	526.30	492.30	458.30	425.30	391.30	357.30	323.30	299.30	281.30	262.30	243.30
2,040	**2,060**	532.83	499.83	465.83	431.83	397.83	364.83	330.83	303.83	285.83	266.83	247.83
2,060	**2,080**	540.36	506.36	472.36	439.36	405.36	371.36	337.36	308.36	290.36	271.36	252.36
2,080	**2,100**	546.89	512.89	479.89	445.89	411.89	377.89	344.89	312.89	294.89	275.89	256.89
2,100	**2,120**	554.42	520.42	486.42	452.42	419.42	385.42	351.42	317.42	299.42	280.42	261.42
2,120	**2,140**	560.95	526.95	492.95	459.95	425.95	391.95	357.95	324.95	303.95	284.95	265.95

$2,140 and over — Do not use this table. See page 34 for instructions.

Page 44

MARRIED Persons—**SEMIMONTHLY** Payroll Period

(For Wages Paid in 2002)

And the wages are–		And the number of withholding allowances claimed is—										
At least	But less than	0	1	2	3	4	5	6	7	8	9	10
		The amount of income, social security, and Medicare taxes to be withheld is—										
$0	**$270**	7.65%	7.65%	7.65%	7.65%	7.65%	7.65%	7.65%	7.65%	7.65%	7.65%	7.65%
270	**280**	$22.04	$21.04	$21.04	$21.04	$21.04	$21.04	$21.04	$21.04	$21.04	$21.04	$21.04
280	**290**	23.80	21.80	21.80	21.80	21.80	21.80	21.80	21.80	21.80	21.80	21.80
290	**300**	25.57	22.57	22.57	22.57	22.57	22.57	22.57	22.57	22.57	22.57	22.57
300	**310**	27.33	23.33	23.33	23.33	23.33	23.33	23.33	23.33	23.33	23.33	23.33
310	**320**	29.10	24.10	24.10	24.10	24.10	24.10	24.10	24.10	24.10	24.10	24.10
320	**330**	30.86	24.86	24.86	24.86	24.86	24.86	24.86	24.86	24.86	24.86	24.86
330	**340**	32.63	25.63	25.63	25.63	25.63	25.63	25.63	25.63	25.63	25.63	25.63
340	**350**	34.39	26.39	26.39	26.39	26.39	26.39	26.39	26.39	26.39	26.39	26.39
350	**360**	36.16	27.16	27.16	27.16	27.16	27.16	27.16	27.16	27.16	27.16	27.16
360	**370**	37.92	27.92	27.92	27.92	27.92	27.92	27.92	27.92	27.92	27.92	27.92
370	**380**	39.69	28.69	28.69	28.69	28.69	28.69	28.69	28.69	28.69	28.69	28.69
380	**390**	41.45	29.45	29.45	29.45	29.45	29.45	29.45	29.45	29.45	29.45	29.45
390	**400**	43.22	30.22	30.22	30.22	30.22	30.22	30.22	30.22	30.22	30.22	30.22
400	**410**	44.98	31.98	30.98	30.98	30.98	30.98	30.98	30.98	30.98	30.98	30.98
410	**420**	46.75	33.75	31.75	31.75	31.75	31.75	31.75	31.75	31.75	31.75	31.75
420	**430**	48.51	35.51	32.51	32.51	32.51	32.51	32.51	32.51	32.51	32.51	32.51
430	**440**	50.28	37.28	33.28	33.28	33.28	33.28	33.28	33.28	33.28	33.28	33.28
440	**450**	52.04	39.04	34.04	34.04	34.04	34.04	34.04	34.04	34.04	34.04	34.04
450	**460**	53.81	40.81	34.81	34.81	34.81	34.81	34.81	34.81	34.81	34.81	34.81
460	**470**	55.57	42.57	35.57	35.57	35.57	35.57	35.57	35.57	35.57	35.57	35.57
470	**480**	57.34	44.34	36.34	36.34	36.34	36.34	36.34	36.34	36.34	36.34	36.34
480	**490**	59.10	46.10	37.10	37.10	37.10	37.10	37.10	37.10	37.10	37.10	37.10
490	**500**	60.87	47.87	37.87	37.87	37.87	37.87	37.87	37.87	37.87	37.87	37.87
500	**520**	63.02	51.02	39.02	39.02	39.02	39.02	39.02	39.02	39.02	39.02	39.02
520	**540**	66.55	54.55	41.55	40.55	40.55	40.55	40.55	40.55	40.55	40.55	40.55
540	**560**	70.08	58.08	45.08	42.08	42.08	42.08	42.08	42.08	42.08	42.08	42.08
560	**580**	73.61	61.61	48.61	43.61	43.61	43.61	43.61	43.61	43.61	43.61	43.61
580	**600**	77.14	65.14	52.14	45.14	45.14	45.14	45.14	45.14	45.14	45.14	45.14
600	**620**	80.67	68.67	55.67	46.67	46.67	46.67	46.67	46.67	46.67	46.67	46.67
620	**640**	84.20	72.20	59.20	48.20	48.20	48.20	48.20	48.20	48.20	48.20	48.20
640	**660**	87.73	75.73	62.73	50.73	49.73	49.73	49.73	49.73	49.73	49.73	49.73
660	**680**	91.26	79.26	66.26	54.26	51.26	51.26	51.26	51.26	51.26	51.26	51.26
680	**700**	94.79	82.79	69.79	57.79	52.79	52.79	52.79	52.79	52.79	52.79	52.79
700	**720**	98.32	86.32	73.32	61.32	54.32	54.32	54.32	54.32	54.32	54.32	54.32
720	**740**	101.85	89.85	76.85	64.85	55.85	55.85	55.85	55.85	55.85	55.85	55.85
740	**760**	105.38	93.38	80.38	68.38	57.38	57.38	57.38	57.38	57.38	57.38	57.38
760	**780**	108.91	96.91	83.91	71.91	58.91	58.91	58.91	58.91	58.91	58.91	58.91
780	**800**	113.44	100.44	87.44	75.44	62.44	60.44	60.44	60.44	60.44	60.44	60.44
800	**820**	117.97	103.97	90.97	78.97	65.97	61.97	61.97	61.97	61.97	61.97	61.97
820	**840**	122.50	107.50	94.50	82.50	69.50	63.50	63.50	63.50	63.50	63.50	63.50
840	**860**	127.03	111.03	98.03	86.03	73.03	65.03	65.03	65.03	65.03	65.03	65.03
860	**880**	131.56	114.56	101.56	89.56	76.56	66.56	66.56	66.56	66.56	66.56	66.56
880	**900**	136.09	118.09	105.09	93.09	80.09	68.09	68.09	68.09	68.09	68.09	68.09
900	**920**	140.62	121.62	108.62	96.62	83.62	71.62	69.62	69.62	69.62	69.62	69.62
920	**940**	145.15	126.15	112.15	100.15	87.15	75.15	71.15	71.15	71.15	71.15	71.15
940	**960**	149.68	130.68	115.68	103.68	90.68	78.68	72.68	72.68	72.68	72.68	72.68
960	**980**	154.21	135.21	119.21	107.21	94.21	82.21	74.21	74.21	74.21	74.21	74.21
980	**1,000**	158.74	139.74	122.74	110.74	97.74	85.74	75.74	75.74	75.74	75.74	75.74
1,000	**1,020**	163.27	144.27	126.27	114.27	101.27	89.27	77.27	77.27	77.27	77.27	77.27
1,020	**1,040**	167.80	148.80	130.80	117.80	104.80	92.80	79.80	78.80	78.80	78.80	78.80
1,040	**1,060**	172.33	153.33	135.33	121.33	108.33	96.33	83.33	80.33	80.33	80.33	80.33
1,060	**1,080**	176.86	157.86	139.86	124.86	111.86	99.86	86.86	81.86	81.86	81.86	81.86
1,080	**1,100**	181.39	162.39	144.39	128.39	115.39	103.39	90.39	83.39	83.39	83.39	83.39
1,100	**1,120**	185.92	166.92	148.92	131.92	118.92	106.92	93.92	84.92	84.92	84.92	84.92
1,120	**1,140**	190.45	171.45	153.45	135.45	122.45	110.45	97.45	86.45	86.45	86.45	86.45
1,140	**1,160**	194.98	175.98	157.98	138.98	125.98	113.98	100.98	88.98	87.98	87.98	87.98
1,160	**1,180**	199.51	180.51	162.51	143.51	129.51	117.51	104.51	92.51	89.51	89.51	89.51
1,180	**1,200**	204.04	185.04	167.04	148.04	133.04	121.04	108.04	96.04	91.04	91.04	91.04
1,200	**1,220**	208.57	189.57	171.57	152.57	136.57	124.57	111.57	99.57	92.57	92.57	92.57
1,220	**1,240**	213.10	194.10	176.10	157.10	140.10	128.10	115.10	103.10	94.10	94.10	94.10
1,240	**1,260**	217.63	198.63	180.63	161.63	143.63	131.63	118.63	106.63	95.63	95.63	95.63
1,260	**1,280**	222.16	203.16	185.16	166.16	147.16	135.16	122.16	110.16	97.16	97.16	97.16
1,280	**1,300**	226.69	207.69	189.69	170.69	151.69	138.69	125.69	113.69	100.69	98.69	98.69
1,300	**1,320**	231.22	212.22	194.22	175.22	156.22	142.22	129.22	117.22	104.22	100.22	100.22
1,320	**1,340**	235.75	216.75	198.75	179.75	160.75	145.75	132.75	120.75	107.75	101.75	101.75
1,340	**1,360**	240.28	221.28	203.28	184.28	165.28	149.28	136.28	124.28	111.28	103.28	103.28
1,360	**1,380**	244.81	225.81	207.81	188.81	169.81	152.81	139.81	127.81	114.81	104.81	104.81
1,380	**1,400**	249.34	230.34	212.34	193.34	174.34	156.34	143.34	131.34	118.34	106.34	106.34
1,400	**1,420**	253.87	234.87	216.87	197.87	178.87	159.87	146.87	134.87	121.87	109.87	107.87

(Continued on next page)

MARRIED Persons—**SEMIMONTHLY** Payroll Period

(For Wages Paid in 2002)

And the wages are–		And the number of withholding allowances claimed is—										
At least	But less than	0	1	2	3	4	5	6	7	8	9	10
		The amount of income, social security, and Medicare taxes to be withheld is—										
$1,420	**$1,440**	$258.40	$239.40	$221.40	$202.40	$183.40	$164.40	$150.40	$138.40	$125.40	$113.40	$109.40
1,440	**1,460**	262.93	243.93	225.93	206.93	187.93	168.93	153.93	141.93	128.93	116.93	110.93
1,460	**1,480**	267.46	248.46	230.46	211.46	192.46	173.46	157.46	145.46	132.46	120.46	112.46
1,480	**1,500**	271.99	252.99	234.99	215.99	196.99	177.99	160.99	148.99	135.99	123.99	113.99
1,500	**1,520**	276.52	257.52	239.52	220.52	201.52	182.52	164.52	152.52	139.52	127.52	115.52
1,520	**1,540**	281.05	262.05	244.05	225.05	206.05	187.05	169.05	156.05	143.05	131.05	118.05
1,540	**1,560**	285.58	266.58	248.58	229.58	210.58	191.58	173.58	159.58	146.58	134.58	121.58
1,560	**1,580**	290.11	271.11	253.11	234.11	215.11	196.11	178.11	163.11	150.11	138.11	125.11
1,580	**1,600**	294.64	275.64	257.64	238.64	219.64	200.64	182.64	166.64	153.64	141.64	128.64
1,600	**1,620**	299.17	280.17	262.17	243.17	224.17	205.17	187.17	170.17	157.17	145.17	132.17
1,620	**1,640**	303.70	284.70	266.70	247.70	228.70	209.70	191.70	173.70	160.70	148.70	135.70
1,640	**1,660**	308.23	289.23	271.23	252.23	233.23	214.23	196.23	177.23	164.23	152.23	139.23
1,660	**1,680**	312.76	293.76	275.76	256.76	237.76	218.76	200.76	181.76	167.76	155.76	142.76
1,680	**1,700**	317.29	298.29	280.29	261.29	242.29	223.29	205.29	186.29	171.29	159.29	146.29
1,700	**1,720**	321.82	302.82	284.82	265.82	246.82	227.82	209.82	190.82	174.82	162.82	149.82
1,720	**1,740**	326.35	307.35	289.35	270.35	251.35	232.35	214.35	195.35	178.35	166.35	153.35
1,740	**1,760**	330.88	311.88	293.88	274.88	255.88	236.88	218.88	199.88	181.88	169.88	156.88
1,760	**1,780**	335.41	316.41	298.41	279.41	260.41	241.41	223.41	204.41	185.41	173.41	160.41
1,780	**1,800**	339.94	320.94	302.94	283.94	264.94	245.94	227.94	208.94	189.94	176.94	163.94
1,800	**1,820**	344.47	325.47	307.47	288.47	269.47	250.47	232.47	213.47	194.47	180.47	167.47
1,820	**1,840**	349.00	330.00	312.00	293.00	274.00	255.00	237.00	218.00	199.00	184.00	171.00
1,840	**1,860**	353.53	334.53	316.53	297.53	278.53	259.53	241.53	222.53	203.53	187.53	174.53
1,860	**1,880**	358.06	339.06	321.06	302.06	283.06	264.06	246.06	227.06	208.06	191.06	178.06
1,880	**1,900**	362.59	343.59	325.59	306.59	287.59	268.59	250.59	231.59	212.59	194.59	181.59
1,900	**1,920**	367.12	348.12	330.12	311.12	292.12	273.12	255.12	236.12	217.12	198.12	185.12
1,920	**1,940**	371.65	352.65	334.65	315.65	296.65	277.65	259.65	240.65	221.65	202.65	188.65
1,940	**1,960**	376.18	357.18	339.18	320.18	301.18	282.18	264.18	245.18	226.18	207.18	192.18
1,960	**1,980**	380.71	361.71	343.71	324.71	305.71	286.71	268.71	249.71	230.71	211.71	195.71
1,980	**2,000**	385.24	366.24	348.24	329.24	310.24	291.24	273.24	254.24	235.24	216.24	199.24
2,000	**2,020**	389.77	370.77	352.77	333.77	314.77	295.77	277.77	258.77	239.77	220.77	202.77
2,020	**2,040**	394.30	375.30	357.30	338.30	319.30	300.30	282.30	263.30	244.30	225.30	207.30
2,040	**2,060**	398.83	379.83	361.83	342.83	323.83	304.83	286.83	267.83	248.83	229.83	211.83
2,060	**2,080**	403.36	384.36	366.36	347.36	328.36	309.36	291.36	272.36	253.36	234.36	216.36
2,080	**2,100**	407.89	388.89	370.89	351.89	332.89	313.89	295.89	276.89	257.89	238.89	220.89
2,100	**2,120**	412.42	393.42	375.42	356.42	337.42	318.42	300.42	281.42	262.42	243.42	225.42
2,120	**2,140**	416.95	397.95	379.95	360.95	341.95	322.95	304.95	285.95	266.95	247.95	229.95
2,140	**2,160**	421.48	402.48	384.48	365.48	346.48	327.48	309.48	290.48	271.48	252.48	234.48
2,160	**2,180**	429.01	407.01	389.01	370.01	351.01	332.01	314.01	295.01	276.01	257.01	239.01
2,180	**2,200**	435.54	411.54	393.54	374.54	355.54	336.54	318.54	299.54	280.54	261.54	243.54
2,200	**2,220**	443.07	416.07	398.07	379.07	360.07	341.07	323.07	304.07	285.07	266.07	248.07
2,220	**2,240**	449.60	420.60	402.60	383.60	364.60	345.60	327.60	308.60	289.60	270.60	252.60
2,240	**2,260**	456.13	425.13	407.13	388.13	369.13	350.13	332.13	313.13	294.13	275.13	257.13
2,260	**2,280**	463.66	429.66	411.66	392.66	373.66	354.66	336.66	317.66	298.66	279.66	261.66
2,280	**2,300**	470.19	436.19	416.19	397.19	378.19	359.19	341.19	322.19	303.19	284.19	266.19
2,300	**2,320**	477.72	443.72	420.72	401.72	382.72	363.72	345.72	326.72	307.72	288.72	270.72
2,320	**2,340**	484.25	450.25	425.25	406.25	387.25	368.25	350.25	331.25	312.25	293.25	275.25
2,340	**2,360**	490.78	457.78	429.78	410.78	391.78	372.78	354.78	335.78	316.78	297.78	279.78
2,360	**2,380**	498.31	464.31	434.31	415.31	396.31	377.31	359.31	340.31	321.31	302.31	284.31
2,380	**2,400**	504.84	470.84	438.84	419.84	400.84	381.84	363.84	344.84	325.84	306.84	288.84
2,400	**2,420**	512.37	478.37	444.37	424.37	405.37	386.37	368.37	349.37	330.37	311.37	293.37
2,420	**2,440**	518.90	484.90	451.90	428.90	409.90	390.90	372.90	353.90	334.90	315.90	297.90
2,440	**2,460**	525.43	492.43	458.43	433.43	414.43	395.43	377.43	358.43	339.43	320.43	302.43
2,460	**2,480**	532.96	498.96	464.96	437.96	418.96	399.96	381.96	362.96	343.96	324.96	306.96
2,480	**2,500**	539.49	505.49	472.49	442.49	423.49	404.49	386.49	367.49	348.49	329.49	311.49
2,500	**2,520**	547.02	513.02	479.02	447.02	428.02	409.02	391.02	372.02	353.02	334.02	316.02
2,520	**2,540**	553.55	519.55	486.55	452.55	432.55	413.55	395.55	376.55	357.55	338.55	320.55
2,540	**2,560**	560.08	527.08	493.08	459.08	437.08	418.08	400.08	381.08	362.08	343.08	325.08
2,560	**2,580**	567.61	533.61	499.61	466.61	441.61	422.61	404.61	385.61	366.61	347.61	329.61
2,580	**2,600**	574.14	540.14	507.14	473.14	446.14	427.14	409.14	390.14	371.14	352.14	334.14
2,600	**2,620**	581.67	547.67	513.67	479.67	450.67	431.67	413.67	394.67	375.67	356.67	338.67
2,620	**2,640**	588.20	554.20	521.20	487.20	455.20	436.20	418.20	399.20	380.20	361.20	343.20
2,640	**2,660**	594.73	561.73	527.73	493.73	459.73	440.73	422.73	403.73	384.73	365.73	347.73
2,660	**2,680**	602.26	568.26	534.26	501.26	467.26	445.26	427.26	408.26	389.26	370.26	352.26
2,680	**2,700**	608.79	574.79	541.79	507.79	473.79	449.79	431.79	412.79	393.79	374.79	356.79
2,700	**2,720**	616.32	582.32	548.32	514.32	481.32	454.32	436.32	417.32	398.32	379.32	361.32

$2,720 and over — Do not use this table. See page 34 for instructions.

Page 46

SINGLE Persons—**MONTHLY** Payroll Period

(For Wages Paid in 2002)

And the wages are–		And the number of withholding allowances claimed is—										
At least	But less than	0	1	2	3	4	5	6	7	8	9	10
		The amount of income, social security, and Medicare taxes to be withheld is—										
$0	**$220**	7.65%	7.65%	7.65%	7.65%	7.65%	7.65%	7.65%	7.65%	7.65%	7.65%	7.65%
220	**230**	$17.21	$17.21	$17.21	$17.21	$17.21	$17.21	$17.21	$17.21	$17.21	$17.21	$17.21
230	**240**	18.98	17.98	17.98	17.98	17.98	17.98	17.98	17.98	17.98	17.98	17.98
240	**250**	20.74	18.74	18.74	18.74	18.74	18.74	18.74	18.74	18.74	18.74	18.74
250	**260**	22.51	19.51	19.51	19.51	19.51	19.51	19.51	19.51	19.51	19.51	19.51
260	**270**	24.27	20.27	20.27	20.27	20.27	20.27	20.27	20.27	20.27	20.27	20.27
270	**280**	26.04	21.04	21.04	21.04	21.04	21.04	21.04	21.04	21.04	21.04	21.04
280	**290**	27.80	21.80	21.80	21.80	21.80	21.80	21.80	21.80	21.80	21.80	21.80
290	**300**	29.57	22.57	22.57	22.57	22.57	22.57	22.57	22.57	22.57	22.57	22.57
300	**320**	32.72	23.72	23.72	23.72	23.72	23.72	23.72	23.72	23.72	23.72	23.72
320	**340**	36.25	25.25	25.25	25.25	25.25	25.25	25.25	25.25	25.25	25.25	25.25
340	**360**	39.78	26.78	26.78	26.78	26.78	26.78	26.78	26.78	26.78	26.78	26.78
360	**380**	43.31	28.31	28.31	28.31	28.31	28.31	28.31	28.31	28.31	28.31	28.31
380	**400**	46.84	29.84	29.84	29.84	29.84	29.84	29.84	29.84	29.84	29.84	29.84
400	**420**	50.37	31.37	31.37	31.37	31.37	31.37	31.37	31.37	31.37	31.37	31.37
420	**440**	53.90	32.90	32.90	32.90	32.90	32.90	32.90	32.90	32.90	32.90	32.90
440	**460**	57.43	34.43	34.43	34.43	34.43	34.43	34.43	34.43	34.43	34.43	34.43
460	**480**	60.96	35.96	35.96	35.96	35.96	35.96	35.96	35.96	35.96	35.96	35.96
480	**500**	64.49	39.49	37.49	37.49	37.49	37.49	37.49	37.49	37.49	37.49	37.49
500	**520**	68.02	43.02	39.02	39.02	39.02	39.02	39.02	39.02	39.02	39.02	39.02
520	**540**	71.55	46.55	40.55	40.55	40.55	40.55	40.55	40.55	40.55	40.55	40.55
540	**560**	75.08	50.08	42.08	42.08	42.08	42.08	42.08	42.08	42.08	42.08	42.08
560	**580**	78.61	53.61	43.61	43.61	43.61	43.61	43.61	43.61	43.61	43.61	43.61
580	**600**	82.14	57.14	45.14	45.14	45.14	45.14	45.14	45.14	45.14	45.14	45.14
600	**640**	87.43	62.43	47.43	47.43	47.43	47.43	47.43	47.43	47.43	47.43	47.43
640	**680**	94.49	69.49	50.49	50.49	50.49	50.49	50.49	50.49	50.49	50.49	50.49
680	**720**	101.55	76.55	53.55	53.55	53.55	53.55	53.55	53.55	53.55	53.55	53.55
720	**760**	109.61	83.61	58.61	56.61	56.61	56.61	56.61	56.61	56.61	56.61	56.61
760	**800**	118.67	90.67	65.67	59.67	59.67	59.67	59.67	59.67	59.67	59.67	59.67
800	**840**	127.73	97.73	72.73	62.73	62.73	62.73	62.73	62.73	62.73	62.73	62.73
840	**880**	136.79	104.79	79.79	65.79	65.79	65.79	65.79	65.79	65.79	65.79	65.79
880	**920**	145.85	111.85	86.85	68.85	68.85	68.85	68.85	68.85	68.85	68.85	68.85
920	**960**	154.91	118.91	93.91	71.91	71.91	71.91	71.91	71.91	71.91	71.91	71.91
960	**1,000**	163.97	126.97	100.97	75.97	74.97	74.97	74.97	74.97	74.97	74.97	74.97
1,000	**1,040**	173.03	136.03	108.03	83.03	78.03	78.03	78.03	78.03	78.03	78.03	78.03
1,040	**1,080**	182.09	145.09	115.09	90.09	81.09	81.09	81.09	81.09	81.09	81.09	81.09
1,080	**1,120**	191.15	154.15	122.15	97.15	84.15	84.15	84.15	84.15	84.15	84.15	84.15
1,120	**1,160**	200.21	163.21	129.21	104.21	87.21	87.21	87.21	87.21	87.21	87.21	87.21
1,160	**1,200**	209.27	172.27	136.27	111.27	90.27	90.27	90.27	90.27	90.27	90.27	90.27
1,200	**1,240**	218.33	181.33	143.33	118.33	93.33	93.33	93.33	93.33	93.33	93.33	93.33
1,240	**1,280**	227.39	190.39	152.39	125.39	100.39	96.39	96.39	96.39	96.39	96.39	96.39
1,280	**1,320**	236.45	199.45	161.45	132.45	107.45	99.45	99.45	99.45	99.45	99.45	99.45
1,320	**1,360**	245.51	208.51	170.51	139.51	114.51	102.51	102.51	102.51	102.51	102.51	102.51
1,360	**1,400**	254.57	217.57	179.57	146.57	121.57	105.57	105.57	105.57	105.57	105.57	105.57
1,400	**1,440**	263.63	226.63	188.63	153.63	128.63	108.63	108.63	108.63	108.63	108.63	108.63
1,440	**1,480**	272.69	235.69	197.69	160.69	135.69	111.69	111.69	111.69	111.69	111.69	111.69
1,480	**1,520**	281.75	244.75	206.75	169.75	142.75	117.75	114.75	114.75	114.75	114.75	114.75
1,520	**1,560**	290.81	253.81	215.81	178.81	149.81	124.81	117.81	117.81	117.81	117.81	117.81
1,560	**1,600**	299.87	262.87	224.87	187.87	156.87	131.87	120.87	120.87	120.87	120.87	120.87
1,600	**1,640**	308.93	271.93	233.93	196.93	163.93	138.93	123.93	123.93	123.93	123.93	123.93
1,640	**1,680**	317.99	280.99	242.99	205.99	170.99	145.99	126.99	126.99	126.99	126.99	126.99
1,680	**1,720**	327.05	290.05	252.05	215.05	178.05	153.05	130.05	130.05	130.05	130.05	130.05
1,720	**1,760**	336.11	299.11	261.11	224.11	186.11	160.11	135.11	133.11	133.11	133.11	133.11
1,760	**1,800**	345.17	308.17	270.17	233.17	195.17	167.17	142.17	136.17	136.17	136.17	136.17
1,800	**1,840**	354.23	317.23	279.23	242.23	204.23	174.23	149.23	139.23	139.23	139.23	139.23
1,840	**1,880**	363.29	326.29	288.29	251.29	213.29	181.29	156.29	142.29	142.29	142.29	142.29
1,880	**1,920**	372.35	335.35	297.35	260.35	222.35	188.35	163.35	145.35	145.35	145.35	145.35
1,920	**1,960**	381.41	344.41	306.41	269.41	231.41	195.41	170.41	148.41	148.41	148.41	148.41
1,960	**2,000**	390.47	353.47	315.47	278.47	240.47	203.47	177.47	152.47	151.47	151.47	151.47
2,000	**2,040**	399.53	362.53	324.53	287.53	249.53	212.53	184.53	159.53	154.53	154.53	154.53
2,040	**2,080**	408.59	371.59	333.59	296.59	258.59	221.59	191.59	166.59	157.59	157.59	157.59
2,080	**2,120**	417.65	380.65	342.65	305.65	267.65	230.65	198.65	173.65	160.65	160.65	160.65
2,120	**2,160**	426.71	389.71	351.71	314.71	276.71	239.71	205.71	180.71	163.71	163.71	163.71
2,160	**2,200**	435.77	398.77	360.77	323.77	285.77	248.77	212.77	187.77	166.77	166.77	166.77
2,200	**2,240**	444.83	407.83	369.83	332.83	294.83	257.83	219.83	194.83	169.83	169.83	169.83
2,240	**2,280**	453.89	416.89	378.89	341.89	303.89	266.89	228.89	201.89	176.89	172.89	172.89
2,280	**2,320**	462.95	425.95	387.95	350.95	312.95	275.95	237.95	208.95	183.95	175.95	175.95
2,320	**2,360**	472.01	435.01	397.01	360.01	322.01	285.01	247.01	216.01	191.01	179.01	179.01
2,360	**2,400**	481.07	444.07	406.07	369.07	331.07	294.07	256.07	223.07	198.07	182.07	182.07
2,400	**2,440**	490.13	453.13	415.13	378.13	340.13	303.13	265.13	230.13	205.13	185.13	185.13

(Continued on next page)

SINGLE Persons—MONTHLY Payroll Period

(For Wages Paid in 2002)

And the wages are–		And the number of withholding allowances claimed is—										
At least	But less than	0	1	2	3	4	5	6	7	8	9	10
		The amount of income, social security, and Medicare taxes to be withheld is—										
$2,440	**$2,480**	$499.19	$462.19	$424.19	$387.19	$349.19	$312.19	$274.19	$237.19	$212.19	$188.19	$188.19
2,480	**2,520**	512.25	471.25	433.25	396.25	358.25	321.25	283.25	246.25	219.25	194.25	191.25
2,520	**2,560**	526.31	480.31	442.31	405.31	367.31	330.31	292.31	255.31	226.31	201.31	194.31
2,560	**2,600**	539.37	489.37	451.37	414.37	376.37	339.37	301.37	264.37	233.37	208.37	197.37
2,600	**2,640**	553.43	498.43	460.43	423.43	385.43	348.43	310.43	273.43	240.43	215.43	200.43
2,640	**2,680**	567.49	507.49	469.49	432.49	394.49	357.49	319.49	282.49	247.49	222.49	203.49
2,680	**2,720**	581.55	516.55	478.55	441.55	403.55	366.55	328.55	291.55	254.55	229.55	206.55
2,720	**2,760**	595.61	527.61	487.61	450.61	412.61	375.61	337.61	300.61	262.61	236.61	211.61
2,760	**2,800**	608.67	541.67	496.67	459.67	421.67	384.67	346.67	309.67	271.67	243.67	218.67
2,800	**2,840**	622.73	555.73	505.73	468.73	430.73	393.73	355.73	318.73	280.73	250.73	225.73
2,840	**2,880**	636.79	568.79	514.79	477.79	439.79	402.79	364.79	327.79	289.79	257.79	232.79
2,880	**2,920**	650.85	582.85	523.85	486.85	448.85	411.85	373.85	336.85	298.85	264.85	239.85
2,920	**2,960**	664.91	596.91	532.91	495.91	457.91	420.91	382.91	345.91	307.91	271.91	246.91
2,960	**3,000**	677.97	610.97	542.97	504.97	466.97	429.97	391.97	354.97	316.97	279.97	253.97
3,000	**3,040**	692.03	625.03	557.03	514.03	476.03	439.03	401.03	364.03	326.03	289.03	261.03
3,040	**3,080**	706.09	638.09	571.09	523.09	485.09	448.09	410.09	373.09	335.09	298.09	268.09
3,080	**3,120**	720.15	652.15	585.15	532.15	494.15	457.15	419.15	382.15	344.15	307.15	275.15
3,120	**3,160**	734.21	666.21	599.21	541.21	503.21	466.21	428.21	391.21	353.21	316.21	282.21
3,160	**3,200**	747.27	680.27	612.27	550.27	512.27	475.27	437.27	400.27	362.27	325.27	289.27
3,200	**3,240**	761.33	694.33	626.33	559.33	521.33	484.33	446.33	409.33	371.33	334.33	296.33
3,240	**3,280**	775.39	707.39	640.39	572.39	530.39	493.39	455.39	418.39	380.39	343.39	305.39
3,280	**3,320**	789.45	721.45	654.45	586.45	539.45	502.45	464.45	427.45	389.45	352.45	314.45
3,320	**3,360**	803.51	735.51	668.51	600.51	548.51	511.51	473.51	436.51	398.51	361.51	323.51
3,360	**3,400**	816.57	749.57	681.57	614.57	557.57	520.57	482.57	445.57	407.57	370.57	332.57
3,400	**3,440**	830.63	763.63	695.63	628.63	566.63	529.63	491.63	454.63	416.63	379.63	341.63
3,440	**3,480**	844.69	776.69	709.69	641.69	575.69	538.69	500.69	463.69	425.69	388.69	350.69
3,480	**3,520**	858.75	790.75	723.75	655.75	588.75	547.75	509.75	472.75	434.75	397.75	359.75
3,520	**3,560**	872.81	804.81	737.81	669.81	602.81	556.81	518.81	481.81	443.81	406.81	368.81
3,560	**3,600**	885.87	818.87	750.87	683.87	615.87	565.87	527.87	490.87	452.87	415.87	377.87
3,600	**3,640**	899.93	832.93	764.93	697.93	629.93	574.93	536.93	499.93	461.93	424.93	386.93
3,640	**3,680**	913.99	845.99	778.99	710.99	643.99	583.99	545.99	508.99	470.99	433.99	395.99
3,680	**3,720**	928.05	860.05	793.05	725.05	658.05	593.05	555.05	518.05	480.05	443.05	405.05
3,720	**3,760**	942.11	874.11	807.11	739.11	672.11	604.11	564.11	527.11	489.11	452.11	414.11
3,760	**3,800**	955.17	888.17	820.17	753.17	685.17	618.17	573.17	536.17	498.17	461.17	423.17
3,800	**3,840**	969.23	902.23	834.23	767.23	699.23	632.23	582.23	545.23	507.23	470.23	432.23
3,840	**3,880**	983.29	915.29	848.29	780.29	713.29	645.29	591.29	554.29	516.29	479.29	441.29
3,880	**3,920**	997.35	929.35	862.35	794.35	727.35	659.35	600.35	563.35	525.35	488.35	450.35
3,920	**3,960**	1,011.41	943.41	876.41	808.41	741.41	673.41	609.41	572.41	534.41	497.41	459.41
3,960	**4,000**	1,024.47	957.47	889.47	822.47	754.47	687.47	619.47	581.47	543.47	506.47	468.47
4,000	**4,040**	1,038.53	971.53	903.53	836.53	768.53	701.53	633.53	590.53	552.53	515.53	477.53
4,040	**4,080**	1,052.59	984.59	917.59	849.59	782.59	714.59	647.59	599.59	561.59	524.59	486.59
4,080	**4,120**	1,066.65	998.65	931.65	863.65	796.65	728.65	661.65	608.65	570.65	533.65	495.65
4,120	**4,160**	1,080.71	1,012.71	945.71	877.71	810.71	742.71	675.71	617.71	579.71	542.71	504.71
4,160	**4,200**	1,093.77	1,026.77	958.77	891.77	823.77	756.77	688.77	626.77	588.77	551.77	513.77
4,200	**4,240**	1,107.83	1,040.83	972.83	905.83	837.83	770.83	702.83	635.83	597.83	560.83	522.83
4,240	**4,280**	1,121.89	1,053.89	986.89	918.89	851.89	783.89	716.89	648.89	606.89	569.89	531.89
4,280	**4,320**	1,135.95	1,067.95	1,000.95	932.95	865.95	797.95	730.95	662.95	615.95	578.95	540.95
4,320	**4,360**	1,150.01	1,082.01	1,015.01	947.01	880.01	812.01	745.01	677.01	625.01	588.01	550.01
4,360	**4,400**	1,163.07	1,096.07	1,028.07	961.07	893.07	826.07	758.07	691.07	634.07	597.07	559.07
4,400	**4,440**	1,177.13	1,110.13	1,042.13	975.13	907.13	840.13	772.13	705.13	643.13	606.13	568.13
4,440	**4,480**	1,191.19	1,123.19	1,056.19	988.19	921.19	853.19	786.19	718.19	652.19	615.19	577.19
4,480	**4,520**	1,205.25	1,137.25	1,070.25	1,002.25	935.25	867.25	800.25	732.25	665.25	624.25	586.25
4,520	**4,560**	1,219.31	1,151.31	1,084.31	1,016.31	949.31	881.31	814.31	746.31	679.31	633.31	595.31
4,560	**4,600**	1,232.37	1,165.37	1,097.37	1,030.37	962.37	895.37	827.37	760.37	692.37	642.37	604.37
4,600	**4,640**	1,246.43	1,179.43	1,111.43	1,044.43	976.43	909.43	841.43	774.43	706.43	651.43	613.43
4,640	**4,680**	1,260.49	1,192.49	1,125.49	1,057.49	990.49	922.49	855.49	787.49	720.49	660.49	622.49
4,680	**4,720**	1,274.55	1,206.55	1,139.55	1,071.55	1,004.55	936.55	869.55	801.55	734.55	669.55	631.55
4,720	**4,760**	1,288.61	1,220.61	1,153.61	1,085.61	1,018.61	950.61	883.61	815.61	748.61	680.61	640.61
4,760	**4,800**	1,301.67	1,234.67	1,166.67	1,099.67	1,031.67	964.67	896.67	829.67	761.67	694.67	649.67
4,800	**4,840**	1,315.73	1,248.73	1,180.73	1,113.73	1,045.73	978.73	910.73	843.73	775.73	708.73	658.73
4,840	**4,880**	1,329.79	1,261.79	1,194.79	1,126.79	1,059.79	991.79	924.79	856.79	789.79	721.79	667.79
4,880	**4,920**	1,343.85	1,275.85	1,208.85	1,140.85	1,073.85	1,005.85	938.85	870.85	803.85	735.85	676.85
4,920	**4,960**	1,357.91	1,289.91	1,222.91	1,154.91	1,087.91	1,019.91	952.91	884.91	817.91	749.91	685.91
4,960	**5,000**	1,370.97	1,303.97	1,235.97	1,168.97	1,100.97	1,033.97	965.97	898.97	830.97	763.97	695.97
5,000	**5,040**	1,385.03	1,318.03	1,250.03	1,183.03	1,115.03	1,048.03	980.03	913.03	845.03	778.03	710.03

$5,040 and over   Do not use this table. See page 34 for instructions.

MARRIED Persons—**MONTHLY** Payroll Period

(For Wages Paid in 2002)

And the wages are–		And the number of withholding allowances claimed is—										
At least	But less than	0	1	2	3	4	5	6	7	8	9	10
		The amount of income, social security, and Medicare taxes to be withheld is—										
$0	**$540**	7.65%	7.65%	7.65%	7.65%	7.65%	7.65%	7.65%	7.65%	7.65%	7.65%	7.65%
540	**560**	$43.08	$42.08	$42.08	$42.08	$42.08	$42.08	$42.08	$42.08	$42.08	$42.08	$42.08
560	**580**	46.61	43.61	43.61	43.61	43.61	43.61	43.61	43.61	43.61	43.61	43.61
580	**600**	50.14	45.14	45.14	45.14	45.14	45.14	45.14	45.14	45.14	45.14	45.14
600	**640**	55.43	47.43	47.43	47.43	47.43	47.43	47.43	47.43	47.43	47.43	47.43
640	**680**	62.49	50.49	50.49	50.49	50.49	50.49	50.49	50.49	50.49	50.49	50.49
680	**720**	69.55	53.55	53.55	53.55	53.55	53.55	53.55	53.55	53.55	53.55	53.55
720	**760**	76.61	56.61	56.61	56.61	56.61	56.61	56.61	56.61	56.61	56.61	56.61
760	**800**	83.67	59.67	59.67	59.67	59.67	59.67	59.67	59.67	59.67	59.67	59.67
800	**840**	90.73	65.73	62.73	62.73	62.73	62.73	62.73	62.73	62.73	62.73	62.73
840	**880**	97.79	72.79	65.79	65.79	65.79	65.79	65.79	65.79	65.79	65.79	65.79
880	**920**	104.85	79.85	68.85	68.85	68.85	68.85	68.85	68.85	68.85	68.85	68.85
920	**960**	111.91	86.91	71.91	71.91	71.91	71.91	71.91	71.91	71.91	71.91	71.91
960	**1,000**	118.97	93.97	74.97	74.97	74.97	74.97	74.97	74.97	74.97	74.97	74.97
1,000	**1,040**	126.03	101.03	78.03	78.03	78.03	78.03	78.03	78.03	78.03	78.03	78.03
1,040	**1,080**	133.09	108.09	83.09	81.09	81.09	81.09	81.09	81.09	81.09	81.09	81.09
1,080	**1,120**	140.15	115.15	90.15	84.15	84.15	84.15	84.15	84.15	84.15	84.15	84.15
1,120	**1,160**	147.21	122.21	97.21	87.21	87.21	87.21	87.21	87.21	87.21	87.21	87.21
1,160	**1,200**	154.27	129.27	104.27	90.27	90.27	90.27	90.27	90.27	90.27	90.27	90.27
1,200	**1,240**	161.33	136.33	111.33	93.33	93.33	93.33	93.33	93.33	93.33	93.33	93.33
1,240	**1,280**	168.39	143.39	118.39	96.39	96.39	96.39	96.39	96.39	96.39	96.39	96.39
1,280	**1,320**	175.45	150.45	125.45	100.45	99.45	99.45	99.45	99.45	99.45	99.45	99.45
1,320	**1,360**	182.51	157.51	132.51	107.51	102.51	102.51	102.51	102.51	102.51	102.51	102.51
1,360	**1,400**	189.57	164.57	139.57	114.57	105.57	105.57	105.57	105.57	105.57	105.57	105.57
1,400	**1,440**	196.63	171.63	146.63	121.63	108.63	108.63	108.63	108.63	108.63	108.63	108.63
1,440	**1,480**	203.69	178.69	153.69	128.69	111.69	111.69	111.69	111.69	111.69	111.69	111.69
1,480	**1,520**	210.75	185.75	160.75	135.75	114.75	114.75	114.75	114.75	114.75	114.75	114.75
1,520	**1,560**	217.81	192.81	167.81	142.81	117.81	117.81	117.81	117.81	117.81	117.81	117.81
1,560	**1,600**	226.87	199.87	174.87	149.87	124.87	120.87	120.87	120.87	120.87	120.87	120.87
1,600	**1,640**	235.93	206.93	181.93	156.93	131.93	123.93	123.93	123.93	123.93	123.93	123.93
1,640	**1,680**	244.99	213.99	188.99	163.99	138.99	126.99	126.99	126.99	126.99	126.99	126.99
1,680	**1,720**	254.05	221.05	196.05	171.05	146.05	130.05	130.05	130.05	130.05	130.05	130.05
1,720	**1,760**	263.11	228.11	203.11	178.11	153.11	133.11	133.11	133.11	133.11	133.11	133.11
1,760	**1,800**	272.17	235.17	210.17	185.17	160.17	136.17	136.17	136.17	136.17	136.17	136.17
1,800	**1,840**	281.23	244.23	217.23	192.23	167.23	142.23	139.23	139.23	139.23	139.23	139.23
1,840	**1,880**	290.29	253.29	224.29	199.29	174.29	149.29	142.29	142.29	142.29	142.29	142.29
1,880	**1,920**	299.35	262.35	231.35	206.35	181.35	156.35	145.35	145.35	145.35	145.35	145.35
1,920	**1,960**	308.41	271.41	238.41	213.41	188.41	163.41	148.41	148.41	148.41	148.41	148.41
1,960	**2,000**	317.47	280.47	245.47	220.47	195.47	170.47	151.47	151.47	151.47	151.47	151.47
2,000	**2,040**	326.53	289.53	252.53	227.53	202.53	177.53	154.53	154.53	154.53	154.53	154.53
2,040	**2,080**	335.59	298.59	260.59	234.59	209.59	184.59	159.59	157.59	157.59	157.59	157.59
2,080	**2,120**	344.65	307.65	269.65	241.65	216.65	191.65	166.65	160.65	160.65	160.65	160.65
2,120	**2,160**	353.71	316.71	278.71	248.71	223.71	198.71	173.71	163.71	163.71	163.71	163.71
2,160	**2,200**	362.77	325.77	287.77	255.77	230.77	205.77	180.77	166.77	166.77	166.77	166.77
2,200	**2,240**	371.83	334.83	296.83	262.83	237.83	212.83	187.83	169.83	169.83	169.83	169.83
2,240	**2,280**	380.89	343.89	305.89	269.89	244.89	219.89	194.89	172.89	172.89	172.89	172.89
2,280	**2,320**	389.95	352.95	314.95	277.95	251.95	226.95	201.95	176.95	175.95	175.95	175.95
2,320	**2,360**	399.01	362.01	324.01	287.01	259.01	234.01	209.01	184.01	179.01	179.01	179.01
2,360	**2,400**	408.07	371.07	333.07	296.07	266.07	241.07	216.07	191.07	182.07	182.07	182.07
2,400	**2,440**	417.13	380.13	342.13	305.13	273.13	248.13	223.13	198.13	185.13	185.13	185.13
2,440	**2,480**	426.19	389.19	351.19	314.19	280.19	255.19	230.19	205.19	188.19	188.19	188.19
2,480	**2,520**	435.25	398.25	360.25	323.25	287.25	262.25	237.25	212.25	191.25	191.25	191.25
2,520	**2,560**	444.31	407.31	369.31	332.31	294.31	269.31	244.31	219.31	194.31	194.31	194.31
2,560	**2,600**	453.37	416.37	378.37	341.37	303.37	276.37	251.37	226.37	201.37	197.37	197.37
2,600	**2,640**	462.43	425.43	387.43	350.43	312.43	283.43	258.43	233.43	208.43	200.43	200.43
2,640	**2,680**	471.49	434.49	396.49	359.49	321.49	290.49	265.49	240.49	215.49	203.49	203.49
2,680	**2,720**	480.55	443.55	405.55	368.55	330.55	297.55	272.55	247.55	222.55	206.55	206.55
2,720	**2,760**	489.61	452.61	414.61	377.61	339.61	304.61	279.61	254.61	229.61	209.61	209.61
2,760	**2,800**	498.67	461.67	423.67	386.67	348.67	311.67	286.67	261.67	236.67	212.67	212.67
2,800	**2,840**	507.73	470.73	432.73	395.73	357.73	320.73	293.73	268.73	243.73	218.73	215.73
2,840	**2,880**	516.79	479.79	441.79	404.79	366.79	329.79	300.79	275.79	250.79	225.79	218.79
2,880	**2,920**	525.85	488.85	450.85	413.85	375.85	338.85	307.85	282.85	257.85	232.85	221.85
2,920	**2,960**	534.91	497.91	459.91	422.91	384.91	347.91	314.91	289.91	264.91	239.91	224.91
2,960	**3,000**	543.97	506.97	468.97	431.97	393.97	356.97	321.97	296.97	271.97	246.97	227.97
3,000	**3,040**	553.03	516.03	478.03	441.03	403.03	366.03	329.03	304.03	279.03	254.03	231.03
3,040	**3,080**	562.09	525.09	487.09	450.09	412.09	375.09	337.09	311.09	286.09	261.09	236.09
3,080	**3,120**	571.15	534.15	496.15	459.15	421.15	384.15	346.15	318.15	293.15	268.15	243.15
3,120	**3,160**	580.21	543.21	505.21	468.21	430.21	393.21	355.21	325.21	300.21	275.21	250.21
3,160	**3,200**	589.27	552.27	514.27	477.27	439.27	402.27	364.27	332.27	307.27	282.27	257.27
3,200	**3,240**	598.33	561.33	523.33	486.33	448.33	411.33	373.33	339.33	314.33	289.33	264.33

(Continued on next page)

MARRIED Persons—MONTHLY Payroll Period

(For Wages Paid in 2002)

And the wages are–		And the number of withholding allowances claimed is—										
At least	But less than	0	1	2	3	4	5	6	7	8	9	10
		The amount of income, social security, and Medicare taxes to be withheld is—										
$3,240	**$3,280**	$607.39	$570.39	$532.39	$495.39	$457.39	$420.39	$382.39	$346.39	$321.39	$296.39	$271.39
3,280	**3,320**	616.45	579.45	541.45	504.45	466.45	429.45	391.45	354.45	328.45	303.45	278.45
3,320	**3,360**	625.51	588.51	550.51	513.51	475.51	438.51	400.51	363.51	335.51	310.51	285.51
3,360	**3,400**	634.57	597.57	559.57	522.57	484.57	447.57	409.57	372.57	342.57	317.57	292.57
3,400	**3,440**	643.63	606.63	568.63	531.63	493.63	456.63	418.63	381.63	349.63	324.63	299.63
3,440	**3,480**	652.69	615.69	577.69	540.69	502.69	465.69	427.69	390.69	356.69	331.69	306.69
3,480	**3,520**	661.75	624.75	586.75	549.75	511.75	474.75	436.75	399.75	363.75	338.75	313.75
3,520	**3,560**	670.81	633.81	595.81	558.81	520.81	483.81	445.81	408.81	370.81	345.81	320.81
3,560	**3,600**	679.87	642.87	604.87	567.87	529.87	492.87	454.87	417.87	379.87	352.87	327.87
3,600	**3,640**	688.93	651.93	613.93	576.93	538.93	501.93	463.93	426.93	388.93	359.93	334.93
3,640	**3,680**	697.99	660.99	622.99	585.99	547.99	510.99	472.99	435.99	397.99	366.99	341.99
3,680	**3,720**	707.05	670.05	632.05	595.05	557.05	520.05	482.05	445.05	407.05	374.05	349.05
3,720	**3,760**	716.11	679.11	641.11	604.11	566.11	529.11	491.11	454.11	416.11	381.11	356.11
3,760	**3,800**	725.17	688.17	650.17	613.17	575.17	538.17	500.17	463.17	425.17	388.17	363.17
3,800	**3,840**	734.23	697.23	659.23	622.23	584.23	547.23	509.23	472.23	434.23	397.23	370.23
3,840	**3,880**	743.29	706.29	668.29	631.29	593.29	556.29	518.29	481.29	443.29	406.29	377.29
3,880	**3,920**	752.35	715.35	677.35	640.35	602.35	565.35	527.35	490.35	452.35	415.35	384.35
3,920	**3,960**	761.41	724.41	686.41	649.41	611.41	574.41	536.41	499.41	461.41	424.41	391.41
3,960	**4,000**	770.47	733.47	695.47	658.47	620.47	583.47	545.47	508.47	470.47	433.47	398.47
4,000	**4,040**	779.53	742.53	704.53	667.53	629.53	592.53	554.53	517.53	479.53	442.53	405.53
4,040	**4,080**	788.59	751.59	713.59	676.59	638.59	601.59	563.59	526.59	488.59	451.59	413.59
4,080	**4,120**	797.65	760.65	722.65	685.65	647.65	610.65	572.65	535.65	497.65	460.65	422.65
4,120	**4,160**	806.71	769.71	731.71	694.71	656.71	619.71	581.71	544.71	506.71	469.71	431.71
4,160	**4,200**	815.77	778.77	740.77	703.77	665.77	628.77	590.77	553.77	515.77	478.77	440.77
4,200	**4,240**	824.83	787.83	749.83	712.83	674.83	637.83	599.83	562.83	524.83	487.83	449.83
4,240	**4,280**	833.89	796.89	758.89	721.89	683.89	646.89	608.89	571.89	533.89	496.89	458.89
4,280	**4,320**	843.95	805.95	767.95	730.95	692.95	655.95	617.95	580.95	542.95	505.95	467.95
4,320	**4,360**	858.01	815.01	777.01	740.01	702.01	665.01	627.01	590.01	552.01	515.01	477.01
4,360	**4,400**	871.07	824.07	786.07	749.07	711.07	674.07	636.07	599.07	561.07	524.07	486.07
4,400	**4,440**	885.13	833.13	795.13	758.13	720.13	683.13	645.13	608.13	570.13	533.13	495.13
4,440	**4,480**	899.19	842.19	804.19	767.19	729.19	692.19	654.19	617.19	579.19	542.19	504.19
4,480	**4,520**	913.25	851.25	813.25	776.25	738.25	701.25	663.25	626.25	588.25	551.25	513.25
4,520	**4,560**	927.31	860.31	822.31	785.31	747.31	710.31	672.31	635.31	597.31	560.31	522.31
4,560	**4,600**	940.37	873.37	831.37	794.37	756.37	719.37	681.37	644.37	606.37	569.37	531.37
4,600	**4,640**	954.43	887.43	840.43	803.43	765.43	728.43	690.43	653.43	615.43	578.43	540.43
4,640	**4,680**	968.49	901.49	849.49	812.49	774.49	737.49	699.49	662.49	624.49	587.49	549.49
4,680	**4,720**	982.55	914.55	858.55	821.55	783.55	746.55	708.55	671.55	633.55	596.55	558.55
4,720	**4,760**	996.61	928.61	867.61	830.61	792.61	755.61	717.61	680.61	642.61	605.61	567.61
4,760	**4,800**	1,009.67	942.67	876.67	839.67	801.67	764.67	726.67	689.67	651.67	614.67	576.67
4,800	**4,840**	1,023.73	956.73	888.73	848.73	810.73	773.73	735.73	698.73	660.73	623.73	585.73
4,840	**4,880**	1,037.79	970.79	902.79	857.79	819.79	782.79	744.79	707.79	669.79	632.79	594.79
4,880	**4,920**	1,051.85	983.85	916.85	866.85	828.85	791.85	753.85	716.85	678.85	641.85	603.85
4,920	**4,960**	1,065.91	997.91	930.91	875.91	837.91	800.91	762.91	725.91	687.91	650.91	612.91
4,960	**5,000**	1,078.97	1,011.97	943.97	884.97	846.97	809.97	771.97	734.97	696.97	659.97	621.97
5,000	**5,040**	1,093.03	1,026.03	958.03	894.03	856.03	819.03	781.03	744.03	706.03	669.03	631.03
5,040	**5,080**	1,107.09	1,040.09	972.09	905.09	865.09	828.09	790.09	753.09	715.09	678.09	640.09
5,080	**5,120**	1,121.15	1,053.15	986.15	918.15	874.15	837.15	799.15	762.15	724.15	687.15	649.15
5,120	**5,160**	1,135.21	1,067.21	1,000.21	932.21	883.21	846.21	808.21	771.21	733.21	696.21	658.21
5,160	**5,200**	1,148.27	1,081.27	1,013.27	946.27	892.27	855.27	817.27	780.27	742.27	705.27	667.27
5,200	**5,240**	1,162.33	1,095.33	1,027.33	960.33	901.33	864.33	826.33	789.33	751.33	714.33	676.33
5,240	**5,280**	1,176.39	1,109.39	1,041.39	974.39	910.39	873.39	835.39	798.39	760.39	723.39	685.39
5,280	**5,320**	1,190.45	1,122.45	1,055.45	987.45	920.45	882.45	844.45	807.45	769.45	732.45	694.45
5,320	**5,360**	1,204.51	1,136.51	1,069.51	1,001.51	934.51	891.51	853.51	816.51	778.51	741.51	703.51
5,360	**5,400**	1,217.57	1,150.57	1,082.57	1,015.57	947.57	900.57	862.57	825.57	787.57	750.57	712.57
5,400	**5,440**	1,231.63	1,164.63	1,096.63	1,029.63	961.63	909.63	871.63	834.63	796.63	759.63	721.63
5,440	**5,480**	1,245.69	1,178.69	1,110.69	1,043.69	975.69	918.69	880.69	843.69	805.69	768.69	730.69
5,480	**5,520**	1,259.75	1,191.75	1,124.75	1,056.75	989.75	927.75	889.75	852.75	814.75	777.75	739.75
5,520	**5,560**	1,273.81	1,205.81	1,138.81	1,070.81	1,003.81	936.81	898.81	861.81	823.81	786.81	748.81
5,560	**5,600**	1,286.87	1,219.87	1,151.87	1,084.87	1,016.87	949.87	907.87	870.87	832.87	795.87	757.87
5,600	**5,640**	1,300.93	1,233.93	1,165.93	1,098.93	1,030.93	963.93	916.93	879.93	841.93	804.93	766.93
5,640	**5,680**	1,314.99	1,247.99	1,179.99	1,112.99	1,044.99	977.99	925.99	888.99	850.99	813.99	775.99
5,680	**5,720**	1,329.05	1,261.05	1,194.05	1,126.05	1,059.05	991.05	935.05	898.05	860.05	823.05	785.05
5,720	**5,760**	1,343.11	1,275.11	1,208.11	1,140.11	1,073.11	1,005.11	944.11	907.11	869.11	832.11	794.11
5,760	**5,800**	1,356.17	1,289.17	1,221.17	1,154.17	1,086.17	1,019.17	953.17	916.17	878.17	841.17	803.17
5,800	**5,840**	1,370.23	1,303.23	1,235.23	1,168.23	1,100.23	1,033.23	965.23	925.23	887.23	850.23	812.23

$5,840 and over — Do not use this table. See page 34 for instructions.

Page 50

SINGLE Persons—DAILY Payroll Period

(For Wages Paid in 2002)

And the wages are–		And the number of withholding allowances claimed is—										
At least	But less than	0	1	2	3	4	5	6	7	8	9	10
		The amount of income, social security, and Medicare taxes to be withheld is—										
$0	**$12**	7.65%	7.65%	7.65%	7.65%	7.65%	7.65%	7.65%	7.65%	7.65%	7.65%	7.65%
12	**15**	$1.03	$1.03	$1.03	$1.03	$1.03	$1.03	$1.03	$1.03	$1.03	$1.03	$1.03
15	**18**	2.26	1.26	1.26	1.26	1.26	1.26	1.26	1.26	1.26	1.26	1.26
18	**21**	2.49	1.49	1.49	1.49	1.49	1.49	1.49	1.49	1.49	1.49	1.49
21	**24**	2.72	1.72	1.72	1.72	1.72	1.72	1.72	1.72	1.72	1.72	1.72
24	**27**	3.95	1.95	1.95	1.95	1.95	1.95	1.95	1.95	1.95	1.95	1.95
27	**30**	4.18	3.18	2.18	2.18	2.18	2.18	2.18	2.18	2.18	2.18	2.18
30	**33**	4.41	3.41	2.41	2.41	2.41	2.41	2.41	2.41	2.41	2.41	2.41
33	**36**	5.64	3.64	2.64	2.64	2.64	2.64	2.64	2.64	2.64	2.64	2.64
36	**39**	5.87	4.87	2.87	2.87	2.87	2.87	2.87	2.87	2.87	2.87	2.87
39	**42**	6.10	5.10	4.10	3.10	3.10	3.10	3.10	3.10	3.10	3.10	3.10
42	**45**	7.33	5.33	4.33	3.33	3.33	3.33	3.33	3.33	3.33	3.33	3.33
45	**48**	7.56	6.56	4.56	3.56	3.56	3.56	3.56	3.56	3.56	3.56	3.56
48	**51**	8.79	6.79	5.79	3.79	3.79	3.79	3.79	3.79	3.79	3.79	3.79
51	**54**	9.02	7.02	6.02	5.02	4.02	4.02	4.02	4.02	4.02	4.02	4.02
54	**57**	10.25	8.25	6.25	5.25	4.25	4.25	4.25	4.25	4.25	4.25	4.25
57	**60**	10.48	8.48	7.48	5.48	4.48	4.48	4.48	4.48	4.48	4.48	4.48
60	**63**	11.70	9.70	7.70	6.70	5.70	4.70	4.70	4.70	4.70	4.70	4.70
63	**66**	11.93	9.93	8.93	6.93	5.93	4.93	4.93	4.93	4.93	4.93	4.93
66	**69**	12.16	11.16	9.16	7.16	6.16	5.16	5.16	5.16	5.16	5.16	5.16
69	**72**	13.39	11.39	9.39	8.39	6.39	5.39	5.39	5.39	5.39	5.39	5.39
72	**75**	13.62	12.62	10.62	8.62	7.62	6.62	5.62	5.62	5.62	5.62	5.62
75	**78**	14.85	12.85	10.85	9.85	7.85	6.85	5.85	5.85	5.85	5.85	5.85
78	**81**	15.08	14.08	12.08	10.08	8.08	7.08	6.08	6.08	6.08	6.08	6.08
81	**84**	16.31	14.31	12.31	11.31	9.31	7.31	6.31	6.31	6.31	6.31	6.31
84	**87**	16.54	14.54	13.54	11.54	9.54	8.54	7.54	6.54	6.54	6.54	6.54
87	**90**	17.77	15.77	13.77	11.77	10.77	8.77	7.77	6.77	6.77	6.77	6.77
90	**93**	18.00	16.00	15.00	13.00	11.00	9.00	8.00	7.00	7.00	7.00	7.00
93	**96**	19.23	17.23	15.23	13.23	12.23	10.23	9.23	7.23	7.23	7.23	7.23
96	**99**	19.46	17.46	16.46	14.46	12.46	10.46	9.46	8.46	7.46	7.46	7.46
99	**102**	19.69	18.69	16.69	14.69	12.69	11.69	9.69	8.69	7.69	7.69	7.69
102	**105**	20.92	18.92	16.92	15.92	13.92	11.92	9.92	8.92	7.92	7.92	7.92
105	**108**	21.15	20.15	18.15	16.15	14.15	13.15	11.15	10.15	8.15	8.15	8.15
108	**111**	22.38	20.38	18.38	17.38	15.38	13.38	11.38	10.38	9.38	8.38	8.38
111	**114**	22.61	20.61	19.61	17.61	15.61	14.61	12.61	10.61	9.61	8.61	8.61
114	**117**	23.84	21.84	19.84	17.84	16.84	14.84	12.84	11.84	9.84	8.84	8.84
117	**120**	25.07	22.07	21.07	19.07	17.07	15.07	14.07	12.07	11.07	9.07	9.07
120	**123**	25.29	23.29	21.29	19.29	18.29	16.29	14.29	12.29	11.29	10.29	9.29
123	**126**	26.52	23.52	22.52	20.52	18.52	16.52	15.52	13.52	11.52	10.52	9.52
126	**129**	27.75	24.75	22.75	20.75	19.75	17.75	15.75	13.75	12.75	10.75	9.75
129	**132**	28.98	25.98	22.98	21.98	19.98	17.98	16.98	14.98	12.98	11.98	9.98
132	**135**	30.21	27.21	24.21	22.21	20.21	19.21	17.21	15.21	14.21	12.21	11.21
135	**138**	31.44	27.44	24.44	23.44	21.44	19.44	17.44	16.44	14.44	12.44	11.44
138	**141**	31.67	28.67	25.67	23.67	21.67	20.67	18.67	16.67	14.67	13.67	11.67
141	**144**	32.90	29.90	26.90	24.90	22.90	20.90	18.90	17.90	15.90	13.90	12.90
144	**147**	34.13	31.13	28.13	25.13	23.13	22.13	20.13	18.13	16.13	15.13	13.13
147	**150**	35.36	32.36	29.36	25.36	24.36	22.36	20.36	18.36	17.36	15.36	13.36
150	**153**	36.59	32.59	29.59	26.59	24.59	22.59	21.59	19.59	17.59	15.59	14.59
153	**156**	36.82	33.82	30.82	27.82	25.82	23.82	21.82	19.82	18.82	16.82	14.82
156	**159**	38.05	35.05	32.05	29.05	26.05	24.05	23.05	21.05	19.05	17.05	16.05
159	**162**	39.28	36.28	33.28	30.28	27.28	25.28	23.28	21.28	20.28	18.28	16.28
162	**165**	40.51	37.51	34.51	30.51	27.51	25.51	23.51	22.51	20.51	18.51	17.51
165	**168**	41.74	37.74	34.74	31.74	28.74	26.74	24.74	22.74	20.74	19.74	17.74
168	**171**	41.97	38.97	35.97	32.97	29.97	26.97	24.97	23.97	21.97	19.97	17.97
171	**174**	43.20	40.20	37.20	34.20	31.20	28.20	26.20	24.20	22.20	21.20	19.20
174	**177**	44.43	41.43	38.43	35.43	32.43	28.43	26.43	25.43	23.43	21.43	19.43
177	**180**	45.66	42.66	39.66	36.66	32.66	29.66	27.66	25.66	23.66	22.66	20.66
180	**183**	46.88	43.88	39.88	36.88	33.88	30.88	27.88	25.88	24.88	22.88	20.88
183	**186**	47.11	44.11	41.11	38.11	35.11	32.11	29.11	27.11	25.11	23.11	22.11
186	**189**	48.34	45.34	42.34	39.34	36.34	33.34	30.34	27.34	26.34	24.34	22.34
189	**192**	49.57	46.57	43.57	40.57	37.57	34.57	30.57	28.57	26.57	24.57	23.57
192	**195**	50.80	47.80	44.80	41.80	37.80	34.80	31.80	28.80	27.80	25.80	23.80
195	**198**	52.03	49.03	45.03	42.03	39.03	36.03	33.03	30.03	28.03	26.03	25.03
198	**201**	53.26	49.26	46.26	43.26	40.26	37.26	34.26	31.26	28.26	27.26	25.26
201	**204**	53.49	50.49	47.49	44.49	41.49	38.49	35.49	32.49	29.49	27.49	25.49
204	**207**	54.72	51.72	48.72	45.72	42.72	39.72	35.72	32.72	29.72	28.72	26.72
207	**210**	55.95	52.95	49.95	46.95	42.95	39.95	36.95	33.95	30.95	28.95	26.95
210	**213**	57.18	54.18	51.18	47.18	44.18	41.18	38.18	35.18	32.18	29.18	28.18
213	**216**	58.41	54.41	51.41	48.41	45.41	42.41	39.41	36.41	33.41	30.41	28.41
216	**219**	58.64	55.64	52.64	49.64	46.64	43.64	40.64	37.64	33.64	30.64	29.64
219	**222**	59.87	56.87	53.87	50.87	47.87	44.87	40.87	37.87	34.87	31.87	29.87

(Continued on next page)

SINGLE Persons—**DAILY** Payroll Period

(For Wages Paid in 2002)

And the wages are–		And the number of withholding allowances claimed is—										
At least	But less than	0	1	2	3	4	5	6	7	8	9	10
		The amount of income, social security, and Medicare taxes to be withheld is—										
$222	**$225**	$61.10	$58.10	$55.10	$52.10	$49.10	$45.10	$42.10	$39.10	$36.10	$33.10	$31.10
225	**228**	62.33	59.33	56.33	52.33	49.33	46.33	43.33	40.33	37.33	34.33	31.33
228	**231**	63.56	60.56	56.56	53.56	50.56	47.56	44.56	41.56	38.56	35.56	31.56
231	**234**	63.79	60.79	57.79	54.79	51.79	48.79	45.79	42.79	39.79	35.79	32.79
234	**237**	65.02	62.02	59.02	56.02	53.02	50.02	47.02	43.02	40.02	37.02	34.02
237	**240**	66.25	63.25	60.25	57.25	54.25	50.25	47.25	44.25	41.25	38.25	35.25
240	**243**	67.47	64.47	61.47	58.47	54.47	51.47	48.47	45.47	42.47	39.47	36.47
243	**246**	68.70	65.70	61.70	58.70	55.70	52.70	49.70	46.70	43.70	40.70	37.70
246	**249**	68.93	65.93	62.93	59.93	56.93	53.93	50.93	47.93	44.93	40.93	37.93
249	**252**	70.16	67.16	64.16	61.16	58.16	55.16	52.16	48.16	45.16	42.16	39.16
252	**255**	71.39	68.39	65.39	62.39	59.39	56.39	52.39	49.39	46.39	43.39	40.39
255	**258**	72.62	69.62	66.62	63.62	59.62	56.62	53.62	50.62	47.62	44.62	41.62
258	**261**	73.85	70.85	66.85	63.85	60.85	57.85	54.85	51.85	48.85	45.85	42.85
261	**264**	75.08	71.08	68.08	65.08	62.08	59.08	56.08	53.08	50.08	46.08	43.08
264	**267**	76.31	72.31	69.31	66.31	63.31	60.31	57.31	54.31	50.31	47.31	44.31
267	**270**	77.54	73.54	70.54	67.54	64.54	61.54	57.54	54.54	51.54	48.54	45.54
270	**273**	78.77	74.77	71.77	68.77	64.77	61.77	58.77	55.77	52.77	49.77	46.77
273	**276**	80.00	76.00	73.00	69.00	66.00	63.00	60.00	57.00	54.00	51.00	48.00
276	**279**	80.23	77.23	73.23	70.23	67.23	64.23	61.23	58.23	55.23	52.23	48.23
279	**282**	81.46	78.46	74.46	71.46	68.46	65.46	62.46	59.46	55.46	52.46	49.46
282	**285**	82.69	79.69	75.69	72.69	69.69	66.69	63.69	59.69	56.69	53.69	50.69
285	**288**	83.92	80.92	76.92	73.92	70.92	66.92	63.92	60.92	57.92	54.92	51.92
288	**291**	85.15	82.15	78.15	75.15	71.15	68.15	65.15	62.15	59.15	56.15	53.15
291	**294**	86.38	82.38	79.38	76.38	72.38	69.38	66.38	63.38	60.38	57.38	53.38
294	**297**	87.61	83.61	80.61	76.61	73.61	70.61	67.61	64.61	61.61	57.61	54.61
297	**300**	88.84	84.84	81.84	77.84	74.84	71.84	68.84	64.84	61.84	58.84	55.84
300	**303**	90.06	86.06	83.06	79.06	76.06	72.06	69.06	66.06	63.06	60.06	57.06
303	**306**	91.29	87.29	84.29	80.29	77.29	73.29	70.29	67.29	64.29	61.29	58.29
306	**309**	91.52	88.52	84.52	81.52	78.52	74.52	71.52	68.52	65.52	62.52	59.52
309	**312**	92.75	89.75	85.75	82.75	78.75	75.75	72.75	69.75	66.75	62.75	59.75
312	**315**	93.98	90.98	86.98	83.98	79.98	76.98	73.98	69.98	66.98	63.98	60.98
315	**318**	95.21	92.21	88.21	85.21	81.21	78.21	74.21	71.21	68.21	65.21	62.21
318	**321**	96.44	93.44	89.44	86.44	82.44	79.44	75.44	72.44	69.44	66.44	63.44
321	**324**	97.67	93.67	90.67	87.67	83.67	80.67	76.67	73.67	70.67	67.67	64.67
324	**327**	98.90	94.90	91.90	87.90	84.90	81.90	77.90	74.90	71.90	67.90	64.90
327	**330**	100.13	96.13	93.13	89.13	86.13	82.13	79.13	76.13	72.13	69.13	66.13
330	**333**	101.36	97.36	94.36	90.36	87.36	83.36	80.36	76.36	73.36	70.36	67.36
333	**336**	102.59	98.59	95.59	91.59	88.59	84.59	81.59	77.59	74.59	71.59	68.59
336	**339**	102.82	99.82	95.82	92.82	89.82	85.82	82.82	78.82	75.82	72.82	69.82
339	**341**	104.01	101.01	97.01	94.01	90.01	87.01	83.01	80.01	77.01	73.01	70.01
341	**343**	105.16	101.16	98.16	94.16	91.16	87.16	84.16	81.16	77.16	74.16	71.16
343	**345**	105.32	102.32	98.32	95.32	92.32	88.32	85.32	81.32	78.32	74.32	71.32
345	**347**	106.47	103.47	99.47	96.47	92.47	89.47	85.47	82.47	78.47	75.47	72.47
347	**349**	107.62	103.62	100.62	96.62	93.62	89.62	86.62	82.62	79.62	76.62	72.62
349	**351**	107.78	104.78	100.78	97.78	93.78	90.78	86.78	83.78	79.78	76.78	73.78
351	**353**	108.93	104.93	101.93	97.93	94.93	90.93	87.93	84.93	80.93	77.93	74.93
353	**355**	109.08	106.08	102.08	99.08	96.08	92.08	89.08	85.08	82.08	78.08	75.08
355	**357**	110.23	107.23	103.23	100.23	96.23	93.23	89.23	86.23	82.23	79.23	76.23
357	**359**	111.39	107.39	104.39	100.39	97.39	93.39	90.39	86.39	83.39	79.39	76.39
359	**361**	111.54	108.54	104.54	101.54	97.54	94.54	90.54	87.54	83.54	80.54	77.54
361	**363**	112.69	108.69	105.69	101.69	98.69	94.69	91.69	88.69	84.69	81.69	77.69
363	**365**	112.85	109.85	105.85	102.85	99.85	95.85	92.85	88.85	85.85	81.85	78.85
365	**367**	114.00	111.00	107.00	104.00	100.00	97.00	93.00	90.00	86.00	83.00	79.00
367	**369**	115.15	111.15	108.15	104.15	101.15	97.15	94.15	90.15	87.15	83.15	80.15
369	**371**	115.31	112.31	108.31	105.31	101.31	98.31	94.31	91.31	87.31	84.31	81.31
371	**373**	116.46	112.46	109.46	105.46	102.46	98.46	95.46	92.46	88.46	85.46	81.46
373	**375**	116.61	113.61	109.61	106.61	103.61	99.61	96.61	92.61	89.61	85.61	82.61
375	**377**	117.76	114.76	110.76	107.76	103.76	100.76	96.76	93.76	89.76	86.76	82.76
377	**379**	118.92	114.92	111.92	107.92	104.92	100.92	97.92	93.92	90.92	86.92	83.92
379	**381**	119.07	116.07	112.07	109.07	105.07	102.07	98.07	95.07	91.07	88.07	85.07
381	**383**	120.22	116.22	113.22	109.22	106.22	102.22	99.22	96.22	92.22	89.22	85.22
383	**385**	120.38	117.38	113.38	110.38	107.38	103.38	100.38	96.38	93.38	89.38	86.38
385	**387**	121.53	118.53	114.53	111.53	107.53	104.53	100.53	97.53	93.53	90.53	86.53
387	**389**	122.68	118.68	115.68	111.68	108.68	104.68	101.68	97.68	94.68	90.68	87.68
389	**391**	122.84	119.84	115.84	112.84	108.84	105.84	101.84	98.84	94.84	91.84	88.84

$391 and over — Do not use this table. See page 34 for instructions.

Page 52

APPENDIX A-1

MARRIED Persons—**DAILY** Payroll Period

(For Wages Paid in 2002)

And the wages are—		And the number of withholding allowances claimed is—										
At least	But less than	0	1	2	3	4	5	6	7	8	9	10
		The amount of income, social security, and Medicare taxes to be withheld is—										
$0	$27	7.65%	7.65%	7.65%	7.65%	7.65%	7.65%	7.65%	7.65%	7.65%	7.65%	7.65%
27	30	$2.18	$2.18	$2.18	$2.18	$2.18	$2.18	$2.18	$2.18	$2.18	$2.18	$2.18
30	33	3.41	2.41	2.41	2.41	2.41	2.41	2.41	2.41	2.41	2.41	2.41
33	36	3.64	2.64	2.64	2.64	2.64	2.64	2.64	2.64	2.64	2.64	2.64
36	39	3.87	2.87	2.87	2.87	2.87	2.87	2.87	2.87	2.87	2.87	2.87
39	42	5.10	3.10	3.10	3.10	3.10	3.10	3.10	3.10	3.10	3.10	3.10
42	45	5.33	4.33	3.33	3.33	3.33	3.33	3.33	3.33	3.33	3.33	3.33
45	48	5.56	4.56	3.56	3.56	3.56	3.56	3.56	3.56	3.56	3.56	3.56
48	51	5.79	4.79	3.79	3.79	3.79	3.79	3.79	3.79	3.79	3.79	3.79
51	54	7.02	6.02	4.02	4.02	4.02	4.02	4.02	4.02	4.02	4.02	4.02
54	57	7.25	6.25	5.25	4.25	4.25	4.25	4.25	4.25	4.25	4.25	4.25
57	60	7.48	6.48	5.48	4.48	4.48	4.48	4.48	4.48	4.48	4.48	4.48
60	63	8.70	7.70	5.70	4.70	4.70	4.70	4.70	4.70	4.70	4.70	4.70
63	66	8.93	7.93	6.93	5.93	4.93	4.93	4.93	4.93	4.93	4.93	4.93
66	69	9.16	8.16	7.16	6.16	5.16	5.16	5.16	5.16	5.16	5.16	5.16
69	72	10.39	8.39	7.39	6.39	5.39	5.39	5.39	5.39	5.39	5.39	5.39
72	75	10.62	9.62	8.62	6.62	5.62	5.62	5.62	5.62	5.62	5.62	5.62
75	78	10.85	9.85	8.85	7.85	6.85	5.85	5.85	5.85	5.85	5.85	5.85
78	81	12.08	10.08	9.08	8.08	7.08	6.08	6.08	6.08	6.08	6.08	6.08
81	84	12.31	11.31	9.31	8.31	7.31	6.31	6.31	6.31	6.31	6.31	6.31
84	87	13.54	11.54	10.54	9.54	7.54	6.54	6.54	6.54	6.54	6.54	6.54
87	90	13.77	12.77	10.77	9.77	8.77	7.77	6.77	6.77	6.77	6.77	6.77
90	93	15.00	13.00	11.00	10.00	9.00	8.00	7.00	7.00	7.00	7.00	7.00
93	96	15.23	13.23	12.23	11.23	9.23	8.23	7.23	7.23	7.23	7.23	7.23
96	99	16.46	14.46	12.46	11.46	10.46	9.46	7.46	7.46	7.46	7.46	7.46
99	102	16.69	14.69	13.69	11.69	10.69	9.69	8.69	7.69	7.69	7.69	7.69
102	105	16.92	15.92	13.92	11.92	10.92	9.92	8.92	7.92	7.92	7.92	7.92
105	108	18.15	16.15	14.15	13.15	12.15	10.15	9.15	8.15	8.15	8.15	8.15
108	111	18.38	17.38	15.38	13.38	12.38	11.38	10.38	8.38	8.38	8.38	8.38
111	114	19.61	17.61	15.61	14.61	12.61	11.61	10.61	9.61	8.61	8.61	8.61
114	117	19.84	18.84	16.84	14.84	12.84	11.84	10.84	9.84	8.84	8.84	8.84
117	120	21.07	19.07	17.07	16.07	14.07	13.07	11.07	10.07	9.07	9.07	9.07
120	123	21.29	19.29	18.29	16.29	14.29	13.29	12.29	11.29	9.29	9.29	9.29
123	126	22.52	20.52	18.52	16.52	15.52	13.52	12.52	11.52	10.52	9.52	9.52
126	129	22.75	20.75	19.75	17.75	15.75	14.75	12.75	11.75	10.75	9.75	9.75
129	132	23.98	21.98	19.98	17.98	16.98	14.98	13.98	11.98	10.98	9.98	9.98
132	135	24.21	22.21	21.21	19.21	17.21	15.21	14.21	13.21	12.21	10.21	10.21
135	138	24.44	23.44	21.44	19.44	18.44	16.44	14.44	13.44	12.44	11.44	10.44
138	141	25.67	23.67	21.67	20.67	18.67	16.67	15.67	13.67	12.67	11.67	10.67
141	144	25.90	24.90	22.90	20.90	18.90	17.90	15.90	14.90	13.90	11.90	10.90
144	147	27.13	25.13	23.13	22.13	20.13	18.13	16.13	15.13	14.13	13.13	12.13
147	150	27.36	26.36	24.36	22.36	20.36	19.36	17.36	15.36	14.36	13.36	12.36
150	153	28.59	26.59	24.59	23.59	21.59	19.59	17.59	16.59	14.59	13.59	12.59
153	156	28.82	26.82	25.82	23.82	21.82	19.82	18.82	16.82	15.82	14.82	12.82
156	159	30.05	28.05	26.05	24.05	23.05	21.05	19.05	17.05	16.05	15.05	14.05
159	162	30.28	28.28	27.28	25.28	23.28	21.28	20.28	18.28	16.28	15.28	14.28
162	165	30.51	29.51	27.51	25.51	24.51	22.51	20.51	18.51	17.51	15.51	14.51
165	168	31.74	29.74	27.74	26.74	24.74	22.74	21.74	19.74	17.74	16.74	15.74
168	171	31.97	30.97	28.97	26.97	24.97	23.97	21.97	19.97	18.97	16.97	15.97
171	174	33.20	31.20	29.20	28.20	26.20	24.20	22.20	21.20	19.20	17.20	16.20
174	177	33.43	32.43	30.43	28.43	26.43	25.43	23.43	21.43	19.43	18.43	17.43
177	180	34.66	32.66	30.66	29.66	27.66	25.66	23.66	22.66	20.66	18.66	17.66
180	183	34.88	32.88	31.88	29.88	27.88	26.88	24.88	22.88	20.88	19.88	17.88
183	186	36.11	34.11	32.11	30.11	29.11	27.11	25.11	24.11	22.11	20.11	18.11
186	189	36.34	34.34	33.34	31.34	29.34	27.34	26.34	24.34	22.34	21.34	19.34
189	192	37.57	35.57	33.57	31.57	30.57	28.57	26.57	24.57	23.57	21.57	19.57
192	195	37.80	35.80	34.80	32.80	30.80	28.80	27.80	25.80	23.80	21.80	20.80
195	198	38.03	37.03	35.03	33.03	32.03	30.03	28.03	26.03	25.03	23.03	21.03
198	201	39.26	37.26	35.26	34.26	32.26	30.26	29.26	27.26	25.26	23.26	22.26
201	204	40.49	38.49	36.49	34.49	32.49	31.49	29.49	27.49	26.49	24.49	22.49
204	207	41.72	38.72	36.72	35.72	33.72	31.72	29.72	28.72	26.72	24.72	22.72
207	210	41.95	39.95	37.95	35.95	33.95	32.95	30.95	28.95	26.95	25.95	23.95
210	213	43.18	40.18	38.18	37.18	35.18	33.18	31.18	30.18	28.18	26.18	24.18
213	216	44.41	41.41	39.41	37.41	35.41	33.41	32.41	30.41	28.41	27.41	25.41
216	219	45.64	42.64	39.64	37.64	36.64	34.64	32.64	30.64	29.64	27.64	25.64
219	222	46.87	43.87	40.87	38.87	36.87	34.87	33.87	31.87	29.87	27.87	26.87
222	225	48.10	44.10	41.10	39.10	38.10	36.10	34.10	32.10	31.10	29.10	27.10
225	228	48.33	45.33	42.33	40.33	38.33	36.33	35.33	33.33	31.33	29.33	28.33
228	231	49.56	46.56	43.56	40.56	38.56	37.56	35.56	33.56	32.56	30.56	28.56
231	234	50.79	47.79	44.79	41.79	39.79	37.79	35.79	34.79	32.79	30.79	29.79

(Continued on next page)

MARRIED Persons—**DAILY** Payroll Period

(For Wages Paid in 2002)

And the wages are–		And the number of withholding allowances claimed is—										
At least	But less than	0	1	2	3	4	5	6	7	8	9	10
		The amount of income, social security, and Medicare taxes to be withheld is—										
$234	**$237**	$52.02	$49.02	$46.02	$42.02	$40.02	$39.02	$37.02	$35.02	$33.02	$32.02	$30.02
237	**240**	53.25	49.25	46.25	43.25	41.25	39.25	37.25	36.25	34.25	32.25	30.25
240	**243**	53.47	50.47	47.47	44.47	41.47	40.47	38.47	36.47	34.47	33.47	31.47
243	**246**	54.70	51.70	48.70	45.70	42.70	40.70	38.70	37.70	35.70	33.70	31.70
246	**249**	55.93	52.93	49.93	46.93	43.93	40.93	39.93	37.93	35.93	34.93	32.93
249	**252**	57.16	54.16	51.16	47.16	44.16	42.16	40.16	38.16	37.16	35.16	33.16
252	**255**	58.39	55.39	51.39	48.39	45.39	42.39	41.39	39.39	37.39	35.39	34.39
255	**258**	58.62	55.62	52.62	49.62	46.62	43.62	41.62	39.62	38.62	36.62	34.62
258	**261**	59.85	56.85	53.85	50.85	47.85	44.85	42.85	40.85	38.85	36.85	35.85
261	**264**	61.08	58.08	55.08	52.08	49.08	45.08	43.08	41.08	40.08	38.08	36.08
264	**267**	62.31	59.31	56.31	53.31	49.31	46.31	43.31	42.31	40.31	38.31	36.31
267	**270**	63.54	60.54	56.54	53.54	50.54	47.54	44.54	42.54	40.54	39.54	37.54
270	**273**	63.77	60.77	57.77	54.77	51.77	48.77	45.77	43.77	41.77	39.77	37.77
273	**276**	65.00	62.00	59.00	56.00	53.00	50.00	47.00	44.00	42.00	41.00	39.00
276	**279**	66.23	63.23	60.23	57.23	54.23	51.23	47.23	44.23	43.23	41.23	39.23
279	**282**	67.46	64.46	61.46	58.46	54.46	51.46	48.46	45.46	43.46	41.46	40.46
282	**285**	68.69	65.69	61.69	58.69	55.69	52.69	49.69	46.69	44.69	42.69	40.69
285	**288**	69.92	65.92	62.92	59.92	56.92	53.92	50.92	47.92	44.92	42.92	41.92
288	**291**	70.15	67.15	64.15	61.15	58.15	55.15	52.15	49.15	46.15	44.15	42.15
291	**294**	71.38	68.38	65.38	62.38	59.38	56.38	52.38	49.38	46.38	44.38	43.38
294	**297**	72.61	69.61	66.61	63.61	60.61	56.61	53.61	50.61	47.61	45.61	43.61
297	**300**	73.84	70.84	67.84	63.84	60.84	57.84	54.84	51.84	48.84	45.84	43.84
300	**303**	75.06	71.06	68.06	65.06	62.06	59.06	56.06	53.06	50.06	47.06	45.06
303	**306**	75.29	72.29	69.29	66.29	63.29	60.29	57.29	54.29	50.29	47.29	45.29
306	**309**	76.52	73.52	70.52	67.52	64.52	61.52	58.52	54.52	51.52	48.52	46.52
309	**312**	77.75	74.75	71.75	68.75	65.75	61.75	58.75	55.75	52.75	49.75	46.75
312	**315**	78.98	75.98	72.98	68.98	65.98	62.98	59.98	56.98	53.98	50.98	47.98
315	**318**	80.21	77.21	73.21	70.21	67.21	64.21	61.21	58.21	55.21	52.21	48.21
318	**321**	80.44	77.44	74.44	71.44	68.44	65.44	62.44	59.44	56.44	52.44	49.44
321	**324**	81.67	78.67	75.67	72.67	69.67	66.67	63.67	59.67	56.67	53.67	50.67
324	**327**	82.90	79.90	76.90	73.90	70.90	66.90	63.90	60.90	57.90	54.90	51.90
327	**330**	84.13	81.13	78.13	75.13	71.13	68.13	65.13	62.13	59.13	56.13	53.13
330	**333**	85.36	82.36	78.36	75.36	72.36	69.36	66.36	63.36	60.36	57.36	54.36
333	**336**	85.59	82.59	79.59	76.59	73.59	70.59	67.59	64.59	61.59	57.59	54.59
336	**339**	86.82	83.82	80.82	77.82	74.82	71.82	68.82	64.82	61.82	58.82	55.82
339	**341**	88.01	85.01	82.01	79.01	76.01	72.01	69.01	66.01	63.01	60.01	57.01
341	**343**	89.16	85.16	82.16	79.16	76.16	73.16	70.16	67.16	64.16	60.16	57.16
343	**345**	89.32	86.32	83.32	80.32	77.32	73.32	70.32	67.32	64.32	61.32	58.32
345	**347**	90.47	86.47	83.47	80.47	77.47	74.47	71.47	68.47	65.47	62.47	58.47
347	**349**	90.62	87.62	84.62	81.62	78.62	75.62	71.62	68.62	65.62	62.62	59.62
349	**351**	91.78	88.78	84.78	81.78	78.78	75.78	72.78	69.78	66.78	63.78	60.78
351	**353**	91.93	88.93	85.93	82.93	79.93	76.93	73.93	69.93	66.93	63.93	60.93
353	**355**	93.08	90.08	87.08	83.08	80.08	77.08	74.08	71.08	68.08	65.08	62.08
355	**357**	93.23	90.23	87.23	84.23	81.23	78.23	75.23	71.23	68.23	65.23	62.23
357	**359**	94.39	91.39	88.39	84.39	81.39	78.39	75.39	72.39	69.39	66.39	63.39
359	**361**	94.54	91.54	88.54	85.54	82.54	79.54	76.54	73.54	69.54	66.54	63.54
361	**363**	95.69	92.69	89.69	86.69	82.69	79.69	76.69	73.69	70.69	67.69	64.69
363	**365**	95.85	92.85	89.85	86.85	83.85	80.85	77.85	74.85	71.85	67.85	64.85
365	**367**	97.00	94.00	91.00	88.00	85.00	81.00	78.00	75.00	72.00	69.00	66.00
367	**369**	98.15	94.15	91.15	88.15	85.15	82.15	79.15	76.15	73.15	70.15	66.15
369	**371**	98.31	95.31	92.31	89.31	86.31	83.31	79.31	76.31	73.31	70.31	67.31
371	**373**	99.46	96.46	92.46	89.46	86.46	83.46	80.46	77.46	74.46	71.46	67.46
373	**375**	99.61	96.61	93.61	90.61	87.61	84.61	80.61	77.61	74.61	71.61	68.61
375	**377**	100.76	97.76	93.76	90.76	87.76	84.76	81.76	78.76	75.76	72.76	69.76
377	**379**	100.92	97.92	94.92	91.92	88.92	85.92	82.92	78.92	75.92	72.92	69.92
379	**381**	102.07	99.07	96.07	92.07	89.07	86.07	83.07	80.07	77.07	74.07	71.07
381	**383**	102.22	99.22	96.22	93.22	90.22	87.22	84.22	81.22	77.22	74.22	71.22
383	**385**	103.38	100.38	97.38	94.38	90.38	87.38	84.38	81.38	78.38	75.38	72.38
385	**387**	103.53	100.53	97.53	94.53	91.53	88.53	85.53	82.53	78.53	75.53	72.53
387	**389**	104.68	101.68	98.68	95.68	91.68	88.68	85.68	82.68	79.68	76.68	73.68
389	**391**	104.84	101.84	98.84	95.84	92.84	89.84	86.84	83.84	80.84	76.84	73.84
391	**393**	105.99	102.99	99.99	96.99	93.99	89.99	86.99	83.99	80.99	77.99	74.99
393	**395**	107.14	103.14	100.14	97.14	94.14	91.14	88.14	85.14	82.14	79.14	75.14
395	**397**	107.29	104.29	101.29	98.29	95.29	92.29	88.29	85.29	82.29	79.29	76.29
397	**399**	108.45	105.45	101.45	98.45	95.45	92.45	89.45	86.45	83.45	80.45	76.45

$399 and over — Do not use this table. See page 34 for instructions.

11. Tables for Withholding on Distributions of Indian Gaming Profits to Tribal Members

If you make certain payments to members of Indian tribes from gaming profits, you must withhold Federal income tax. You must withhold if (1) the total payment to a member for the year is over $7,700 and (2) the payment is from the net revenues of class II or class III gaming activities (classified by the Indian Gaming Regulatory Act) conducted or licensed by the tribes.

A class I gaming activity is ***not subject to this withholding requirement.*** Class I activities are social games solely for prizes of minimal value or traditional forms of Indian gaming engaged in as part of tribal ceremonies or celebrations.

Class II. Class II includes (1) bingo and similar games, such as pull tabs, punch boards, tip jars, lotto, and instant bingo, and (2) card games that are authorized by the state or that are not explicitly prohibited by the state and played at a location within the state.

Class III. A class III gaming activity is any gaming that is not class I or class II. Class III includes horse racing, dog racing, jai alai, casino gaming, and slot machines.

Withholding Tables

To figure the amount of tax to withhold each time you make a payment, use the table on page 56 for the period for which you make payments. For example, if you make payments weekly, use table 1; if you make payments monthly, use table 4. If the total payments to an individual for the year are $7,700 or less, no withholding is required.

Example: A tribal member is paid monthly. The monthly payment is $5,000. Using Table 4, Monthly Distribution Period, figure the withholding as follows:

Subtract $2,971 from the $5,000 payment for a remainder of $2,029. Multiply this amount by 27%, for a total of $547.83. Add $324.35, for total withholding of $872.18.

Depositing and reporting withholding. Combine the Indian gaming withholding with all other nonpayroll withholding (e.g., backup withholding and withholding on gambling winnings). Generally, you must deposit the amounts withheld by electronic funds transfer (see page 2) or at an authorized financial institution using **Form 8109,** Federal Tax Deposit Coupon. See **Circular E,** Employer's Tax Guide, for a detailed discussion of the deposit requirements.

Report Indian gaming withholding on **Form 945,** Annual Return of Withheld Federal Income Tax. For more information, see Form 945 and its instructions. Also, report the payments and withholding to tribal members and to the IRS on **Form 1099-MISC,** Miscellaneous Income (see the **Instructions for Forms 1098–MISC.**).

Tables for Withholding on Distributions of Indian Gaming Profits to Tribal Members

Tables for All Individuals

(For Payments Made in 2002)

Table 1—WEEKLY DISTRIBUTION PERIOD

If the amount of the payment is: Not over $148 — The amount of income tax to withhold is: $0

Over—	But not over—		of excess over—
$148	$263	10%	$148
$263	$686	$11.50 plus 15%	$263
$686	$1,450	$74.95 plus 27%	$686
$1,450	....	$281.23 plus 30%	$1,450

Table 2—BIWEEKLY DISTRIBUTION PERIOD

If the amount of the payment is: Not over $296 — The amount of income tax to withhold is: $0

Over—	But not over—		of excess over—
$296	$527	10%	$296
$527	$1,371	$23.10 plus 15%	$527
$1,371	$2,900	$149.70 plus 27%	$1,371
$2,900	....	$562.53 plus 30%	$2,900

Table 3—SEMIMONTHLY DISTRIBUTION PERIOD

If the amount of the payment is: Not over $321 — The amount of income tax to withhold is: $0

Over—	But not over—		of excess over—
$321	$571	10%	$321
$571	$1,485	$25.00 plus 15%	$571
$1,485	$3,142	$162.10 plus 27%	$1,485
$3,142	....	$609.49 plus 30%	$3,142

Table 4—MONTHLY DISTRIBUTION PERIOD

If the amount of the payment is: Not over $642 — The amount of income tax to withhold is: $0

Over—	But not over—		of excess over—
$642	$1,142	10%	$642
$1,142	$2,971	$50.00 plus 15%	$1,142
$2,971	$6,283	$324.35 plus 27%	$2,971
$6,283	....	$1,218.59 plus 30%	$6,283

Table 5—QUARTERLY DISTRIBUTION PERIOD

If the amount of the payment is: Not over $1,925 — The amount of income tax to withhold is: $0

Over—	But not over—		of excess over—
$1,925	$3,425	10%	$1,925
$3,425	$8,913	$150,00 plus 15%	$3,425
$8,913	$18,850	$973.20 plus 27%	$8,913
$18,850	....	$3,656.19 plus 30%	$18,850

Table 6—SEMIANNUAL DISTRIBUTION PERIOD

If the amount of the payment is: Not over $3,850 — The amount of income tax to withhold is: $0

Over—	But not over—		of excess over—
$3,850	$6,850	10%	$3,850
$6,850	$17,825	$300.00 plus 15%	$6,850
$17,825	$37,700	$1,946.25 plus 27%	$17,825
$37,700	....	$7,312.50 plus 30%	$37,700

Table 7—ANNUAL DISTRIBUTION PERIOD

If the amount of the payment is: Not over $7,700 — The amount of income tax to withhold is: $0

Over—	But not over—		of excess over—
$7,700	$13,700	10%	$7,700
$13,700	$35,650	$600.00 plus 15%	$13,700
$35,650	$75,400	$3,892.50 plus 27%	$35,650
$75,400	....	$14,625.00 plus 30%	$75,400

Table 8—DAILY or MISCELLANEOUS DISTRIBUTION PERIOD

If the amount of the payment is: Not over $29.60 — The amount of income tax to withhold is: $0

Over—	But not over—		of excess over—
$29.60	$52.70	10%	$29.60
$52.70	$137.10	$2.31 plus 15%	$52.70
$137.10	$290.00	$14.97 plus 27%	$137.10
$290.00	....	$56.25 plus 30%	$290.00

Index

■

Quick and Easy Access to Tax Help and Forms

PERSONAL COMPUTER

Access the IRS's Web Site 24 hours a day, 7 days a week at **www.irs.gov** *to do the following:*

- Download forms, instructions, and publications
- See answers to frequently asked tax questions
- Search publications on-line by topic or keyword
- Figure your withholding allowances using our W-4 calculator
- Send us comments or request help via e-mail
- Sign up to receive local and national tax news by e-mail

You can also reach us using File Transfer Protocol at ftp.irs.gov

FAX

You can get over 100 of the most requested forms and instructions 24 hours a day, 7 days a week, by fax. Just call ***703-368-9694*** *from the telephone connected to the fax machine.*

MAIL

You can order forms, instructions, and publications by completing the order blank in **Circular E,** Employer's Tax Guide. You should receive your order within 10 days after we receive your request.

PHONE

You can get forms, publications, and automated information 24 hours a day, 7 days a week, by phone.

Forms and Publications

Call **1-800-TAX-FORM** (1-800-829-3676) to order current and prior year forms, instructions, and publications. You should receive your order within 10 days.

WALK-IN

You can pick up some of the most requested forms, instructions, and publications at many post offices, libraries, and IRS offices. Some IRS offices, libraries, grocery stores, office supply stores, and copy centers have an extensive collection of products available to photocopy or print from a CD-ROM.

CD-ROM

Order ***Pub. 1796,*** *Federal Tax Products on CD-ROM, and get:*

- Current year forms, instructions, and publications
- Prior year forms, instructions, and publications
- Popular tax forms that may be filled in electronically, printed out for submission, and saved for recordkeeping
- Internal Revenue Bulletins

Buy the CD-ROM on the Internet at **www.irs.gov/cdorders** from the National Technical Information Service (NTIS) for $21 (no handling fee), or call **1-877-CDFORMS** (1-877-233-6767) toll-free to buy the CD-ROM for $21 (plus a $5 handling fee).

TABLE OF CASES

References are to sections and to forms.

TABLE OF STATUTES AND ADMINISTRATIVE ANNOUNCEMENTS

References are to sections and to forms.

INDEX

References are to sections and to forms.

L

N

P

R

S

T

W